DIFFERENTIAL GAMES AND CONTROL THEORY II

PURE AND APPLIED MATHEMATICS

A Program of Monographs, Textbooks, and Lecture Notes

Executive Editors — Monographs, Textbooks, and Lecture Notes

Earl J. Taft
Rutgers University
New Brunswick, New Jersey

Edwin Hewitt
University of Washington
Seattle, Washington

Chairman of the Editorial Board

S. Kobayashi
University of California, Berkeley
Berkeley, California

Editorial Board

Masanao Aoki
University of California, Los Angeles

Glen E. Bredon
Rutgers University

Sigurdur Helgason
Massachusetts Institute of Technology

G. Leitman
University of California, Berkeley

W.S. Massey
Yale University

Irving Reiner
University of Illinois at Urbana-Champaign

Paul J. Sally, Jr.
University of Chicago

Jane Cronin Scanlon
Rutgers University

Martin Schechter
Yeshiva University

Julius L. Shaneson
Rutgers University

LECTURE NOTES
IN PURE AND APPLIED MATHEMATICS

1. *N. Jacobson,* Exceptional Lie Algebras
2. *L.-Å. Lindahl and F. Poulsen,* Thin Sets in Harmonic Analysis
3. *I. Satake,* Classification Theory of Semi-Simple Algebraic Groups
4. *F. Hirzebruch, W. D. Newmann, and S. S. Koh,* Differentiable Manifolds and Quadratic Forms
5. *I. Chavel,* Riemannian Symmetric Spaces of Rank One
6. *R. B. Burckel,* Characterization of C(X) Among Its Subalgebras
7. *B. R. McDonald, A. R. Magid, and K. C. Smith,* Ring Theory: Proceedings of the Oklahoma Conference
8. *Y.-T. Siu,* Techniques of Extension of Analytic Objects
9. *S. R. Caradus, W. E. Pfaffenberger, and B. Yood,* Calkin Algebras and Algebras of Operators on Banach Spaces
10. *E. O. Roxin, P.-T. Liu, and R. L. Sternberg,* Differential Games and Control Theory
11. *M. Orzech and C. Small,* The Brauer Group of Commutative Rings
12. *S. Thomeier,* Topology and Its Applications
13. *J. M. López and K. A. Ross,* Sidon Sets
14. *W. W. Comfort and S. Negrepontis,* Continuous Pseudometrics
15. *K. McKennon and J. M. Robertson,* Locally Convex Spaces
16. *M. Carmeli and S. Malin,* Representations of the Rotation and Lorentz Groups: An Introduction
17. *G. B. Seligman,* Rational Methods in Lie Algebras
18. *D. G. de Figueiredo,* Functional Analysis: Proceedings of the Brazilian Mathematical Society Symposium
19. *L. Cesari, R. Kannan, and J. D. Schuur,* Nonlinear Functional Analysis of Differential Equations: Proceedings of the Michigan State University Conference
20. *J. J. Schäffer,* Geometry of Spheres in Normed Spaces
21. *K. Yano and M. Kon,* Anti-Invariant Submanifolds
22. *W. V. Vasconcelos,* The Rings of Dimension Two
23. *R. E. Chandler,* Hausdorff Compactifications
24. *S. P. Franklin and B. V. S. Thomas,* Topology: Proceedings of the Memphis State University Conference
25. *S. K. Jain,* Ring Theory: Proceedings of the Ohio University Conference
26. *B. R. McDonald and R. A. Morris,* Ring Theory II: Proceedings of the Second Oklahoma Conference
27. *R. B. Mura and A. Rhemtulla,* Orderable Groups
28. *J. R. Graef,* Stability of Dynamical Systems: Theory and Applications
29. *H.-C. Wang,* Homogeneous Banach Algebras
30. *E. O. Roxin, P. -T. Liu, and R. L. Sternberg,* Differential Games and Control Theory II

Other volumes in preparation

DIFFERENTIAL GAMES AND CONTROL THEORY II

Proceedings of the Second Kingston Conference

Held at University of Rhode Island
Kingston, Rhode Island
June 7 to 10, 1976

Theme: Stochastic Problems and Applications

Edited by
Emilio O. Roxin, Pan-Tai Liu
University of Rhode Island
Kingston, Rhode Island
and
Robert L. Sternberg
Office of Naval Research
Boston, Massachusetts

MARCEL DEKKER, INC. New York and Basel

Library of Congress Cataloging in Publication Data

Kingston Conference on Differential Games and Control Theory, 2d, University of Rhode Island, 1976.
Differential games and control theory II.

(Lecture notes in pure and applied mathematics ; v. 30)
1. Differential games--Congresses. 2. Control theory--Congresses. I. Roxin, Emilio O. II. Liu, Pan-Tai. III. Sternberg, Robert L. IV. Title.
QA272.K56 1976 519.3 77-4689
ISBN 0-8247-6549-4

MARCEL DEKKER, INC.

270 Madison Avenue, New York, New York 10016

Current printing (last digit):
10 9 8 7 6 5 4 3 2 1

PRINTED IN THE UNITED STATES OF AMERICA

To all those who
encouraged us to
proceed with the
Second Kingston Conference
on
Differential Games and Control Theory

LIST OF CONTRIBUTORS

N. U. AHMED, University of Ottawa, Ottawa, Canada

A. V. BALAKRISHNAN, University of California at Los Angeles, Los Angeles, California

TAMER BAŞAR, Marmara Research Institute, Gebye, Kocaeli, Turkey

A. BENSOUSSAN, University of Paris IX and LABORIA, Paris, France

R. S. BUCY, University of Southern California, Los Angeles, California

MOU-HSIUNG CHANG, University of Alabama at Huntsville, Huntsville, Alabama

ETHELBERT N. CHUKWU, Cleveland State University, Cleveland, Ohio

ROBERT J. ELLIOTT, University of Hull, Hull, England

A. EPHREMIDES, University of Maryland, College Park, Maryland

WENDELL H. FLEMING, Brown University, Providence, Rhode Island

W. M. GETZ, National Research Institute of Mathematical Sciences, Pretoria, South Africa

EDMOND GHANDOUR, University of Tel-Aviv, Ramat-Aviv, Israel

JAN M. GRONSKI, Cleveland State University, Cleveland, Ohio

M. HEYMANN, Technion - Israel Institute of Technology, Haifa, Israel

H. J. KELLEY, Analytical Mechanics Associates, Inc., Jericho, New York

G. LEDWICH, University of Newcastle, Newcastle, New South Wales, Australia

G. LEITMANN, University of California at Berkeley, Berkeley, California

N. LEVITT, Rutgers - State University of New Jersey, New Brunswick, New Jersey

J. L. LIONS, Collège de France and LABORIA, Paris, France

PAN-TAI LIU, University of Rhode Island, Kingston, Rhode Island

J. B. MOORE, University of Newcastle, Newcastle, New South Wales, Australia

GEERT JAN OLSDER, Twente University of Technology, Enschede, The Netherlands

M. PACHTER, Council for Scientific and Industrial Research, Pretoria, South Africa

WILLIAM J. PALM, University of Rhode Island, Kingston, Rhode Island

T. PARTHASARATHY, University of Illinois at Chicago Circle, Chicago, Illinois

EMILIO O. ROXIN, University of Rhode Island, Kingston, Rhode Island

RICHARD C. SCALZO, University of Illinois at Chicago Circle, Chicago, Illinois and Colby College, Waterville, Maine

K. D. SENNE, MIT-Lincoln Laboratory, Lexington, Massachusetts

M. STERN, Analytic Services, Inc., Falls Church, Virginia

R. J. STERN, Concordia University, Montreal, Canada

H. SUSSMANN, Rutgers - State University of New Jersey, New Brunswick, New Jersey

JON G. SUTINEN, University of Rhode Island, Kingston, Rhode Island

PAWEL J. SZABLOWSKI, Institute of Mathematics, Warsaw Technical University, Warsaw, Poland

H. W. WONG, University of Ottawa, Ottawa, Canada

H. YOUSSEF, Lockheed Aircraft Company, Burbank, California

FOREWORD

In this volume appear twenty-four of the thirty-six papers presented in person or by title at the Second Kingston Conference on Differential Games and Control Theory held at the University of Rhode Island in Kingston, June 7 to 10, 1976 under sponsorship of the University of Rhode Island with the participation of the International Federation of Automatic Control. Included are Invited Lectures, Contributed Papers, and four papers from the Adjunct Program which were read by title. The selection includes papers by widely known experts and also contributions from beginning scholars just starting out in this field.

While we should have liked to publish all of the papers from the Conference, we were unable to do so for a variety of reasons mostly connected with limitations of time and space, but we are grateful nevertheless for the enthusiastic formal participation in the Conference of Rufus Isaacs of Johns Hopkins University; Mary Ellen Bock of Purdue University; Max Mintz of the University of Pennsylvania; Hubert Hai-Ao Chin of York College of the City University of New York; Musa Yildiz of the University of New Hampshire; John Danskin of Universität Bonn and the École National Supérieure des Télécommunications; N. M. Olgac, R. W. Longman, and C. A. Cooper of Columbia University and the Bell Telephone Laboratories; Wolfgang Carmele of the Technische Universität Darmstadt; D. R. K. Rao of Jundi

Shapur University; Donald W. Tufts and J. T. Francis of University of Rhode Island and the Naval Underwater Systems Center; D. G. Lainiotis of the State University of New York at Buffalo; Howard Blum of Rutgers, State University of New Jersey; and A. G. Lindgren of the University of Rhode Island.

A major purpose of the Conference was to bring together mathematicians, scientists, and engineers from a variety of disciplines having a common interest in the Conference Topic and perhaps special interests in the Theme: Stochastic Problems and Applications. To what extent this effort met with success may perhaps be judged by a perusal of the varied topics of the papers in this book which range from almost purely mathematical considerations to applications in systems analysis, electrical engineering, resource economics, public policy, fisheries management, and harvesting strategies.

The Conference was organized by the three editors of this book with the assistance of Helen M. Sternberg of the University of Connecticut who served as Conference Secretary and Geert Jan Olsder of the Twente University of Technology who served as the IFAC Liaison Representative. Henry J. Kelley of Analytical Mechanics Associates, Inc., while not officially a member of the Organizing Committee, gave invaluable assistance and advice during the several months preparation for the Conference.

Marguerite Ellis prepared the final typescript in her customary exquisite fashion and also prepared most of the illustrations.

In closing, the writer wishes to express the appreciation of the Organizing Committee for the financial support for the Conference kindly provided by the office of the Academic Vice-President, the College of Engineering, the College of Arts and Sciences, the Division of University Extension, the Graduate School, the Development Council through a gift from the Eastman Kodak Company, and the Visiting Scholars Fund of the University of Rhode Island, and wishes to express his own indebtedness to W. R. Ferrante, Douglas Rosie, George J. Dillavou, L. D. Conta, A. A. Michael, C. J. Wilson, Virginia O'Brien, Norman J. Finizio, Rosalind Shumate, June Chandronet, Fred Jackson, Harold Fisher, Frank Dietz, Ghasi R. Verma, James T. Lewis, Charles D. Nash, Jr., Nathaniel McL. Sage, Jr., and Herman E. Sheets, also of the University of Rhode Island, and to Derrill J. Bordelon of the Naval Underwater Systems Center in Newport and A. L. Powell and Ruth Berrett of the Office of Naval Research in Boston for their encouragement, assistance, advice and support during the writer's private labors on the Organizing Committee and in the minutiae of editing these Proceedings.

R. L. S.

CONTENTS

DIFFERENTIAL GAMES AND CONTROL THEORY II

MARKOV GAMES - A SURVEY

T. Parthasarathy and M. Stern

University of Illinois at Chicago Circle
Chicago, Illinois

and

Analytic Services Inc.
Falls Church, Virginia

ABSTRACT

Markov games or stochastic games were first introduced by Shapley in an historically important paper that appeared in 1953. Since then many authors have extended their results in various directions. These extensions are discussed in some detail. Results relating to limiting average pay-offs (first considered by Gillette in his Ph.D. thesis around 1953) are discussed. Next, the algorithmic aspects of the problem under consideration are also discussed. In conclusion, some problems which are still open will be mentioned.

§0. INTRODUCTION

The relevant literature on sequential compounding of two-person games dates back to the early 1950's and since that time, independently, a number of workers have attacked variations on the theme of compounding. This of course has led to a good deal of redundancy, both conceptual and technical in nature. In one class of games (recursive and Markov or stochastic) a normalized game is played at each stage, and the player's strategies control not only the (monetary) pay-off but also the transition probabilities which govern the game to be played at the next stage. In another class (survival and attrition games) there is but one component game and it is repeated. The players have limited resources, and these fluctuate in time according to the outcomes of repeated plays of the given game. The overall game is concluded when one of the players is bankrupt. In still another class (compound decision problems) a given game is repeated, and each player attempts to control the average pay-off by exploiting the statistical records of his adversary's previous choices. The final class (economic ruin games) is characterized by the problem of corporate dividend policy: the more generous the dividend policy of the corporation, the less secure it is against future exigencies. An excellent introduction to these topics is given in Luce and Raiffa [72] who indicate some of the interrelations, namely, how the theory of Markov or stochastic games suggested that of recursive games which in turn, is related to the theory of survival and attrition games; how Blackwell's approachability theory, which was motivated by attrition games, can be used to analyze compound decision problems; and how approachability theory is technically similar to a generalization of the theory of survival games.

In this article we will discuss in some depth the theory of stochastic games or Markov games. The term Markov game is due to Zachrisson [138]. (Many authors following Shapley [111] use the term stochastic games.) The theory of Markov games was first introduced by Shapley in an historically important paper [111] during 1953. Around the same time Gillette [144] in his

Ph.D. thesis entitled "Representable infinite games" considered Markov games of perfect information in extensive form. A Markov game is an infinite game in which it is assumed that a pay-off to the players of the game is made at each move. Two types of pay-off are considered in the literature: in one the sum of all pay-offs at different moves--the total expected pay-off--is examined; in the other, a limit of the average expected pay-off over the number of moves made as the number of moves approaches infinity is examined. Each player tries to maximize his expected pay-off by playing optimally. Since the appearance of the Shapley-Gillette results, many authors have extended them in various directions. We will discuss these in some detail. We will also discuss the algorithmic aspects of the problem under consideration.

§1. ZERO-SUM (STOCHASTIC) MARKOV GAMES - STATE SPACE FINITE OR COUNTABLE

A Markov game is determined by five objects, S, A, B, q, and r. Here S denotes the state space of the system. The states will be denoted by s or s'. Once a day players I and II (for simplicity we consider two-person games; theory for n-person games is similar) observe the current state s of the system, and then player I chooses an action a from a finite set A of actions, and player II chooses an action b from a finite set B of actions. As a result of this, two things happen: (i) player I receives an immediate income r(s,a,b), depending on the current state s of the system and the actions a and b chosen, and (ii) the system moves to a new state s' with probability q(s'/s,a,b) which also depends on s,a,b. We assume that $|r(s,a,b)| \le M$ for all s,a,b. Payments accumulate throughout the course of the play.

Player I wants to maximize his accumulated income while player II wants to minimize the same. The problem is to choose a strategy for player I that will maximize his total expected income and to choose another strategy for player II that will minimize the income of player I.

In order that the total accumulated income be a well-defined number, we introduce a discount factor β, $0 \le \beta < 1$, so that the value of the unit income n days in the future is β^n. In other words, the total income to player I is equal to $\sum_{n-1}^{\infty} \beta^{n-1} r_n$ where r_n is the income to I on the nth day. Shapley assumes $\inf_{s,i,j} q_{ij}^s > 0$ where q_{ij}^s is the probability that the game stops if (i,j) are the actions chosen by the two players at state s. This means the game ends with probability one after a finite number of steps and hence the total accumulated income is well-defined. For simplicity we will use the discount factor to make the total income well-defined.

It follows from the Kuhn-Aumann [66,4] theorem that in a game of perfect recall (and consequently in Markov games) players can restrict themselves to playing only behavior strategies. A behavior strategy π for player I is a sequence $(\pi_1, \pi_2, \cdots, \pi_n)$ where π_n is a conditional probability distribution on A given the past history $h_n = (s_0, a_0, b_0, s_1, a_1, b_1, \cdots, s_{n-1}, a_{n-1}, b_{n-1}, s_n)$. A behavior strategy π is called stationary if $\pi_n = f$ for all $n \ge 1$. Similarly strategies are defined for player II.

The total expected pay-off for player I from (π, Γ) is denoted by $I(\pi, \Gamma)$; the sth coordinate of $I(\pi, \Gamma)$ is the income to player I if the initial state is s.

<u>Definition</u> Call a strategy π^* optimal for player I if $I(\pi^*, \Gamma)(s) \geq \inf \sup I(\pi, \Gamma)(s)$ for all Γ and s. A strategy Γ^* is optimal for II if $I(\pi, \Gamma^*)(s) \leq \sup \inf I(\pi, \Gamma)(s)$ for all π and s.

<u>Definition</u> A Markov game has a value if $\inf_{\Gamma} \sup_{\pi} I(\pi,\Gamma)(s) = \sup_{\pi} \inf_{\Gamma} I(\pi, \Gamma)(s)$ for all s. In that case we call inf sup $I(\pi, \Gamma)(s)$ as the value of the Markov game.

Let P_A be the space of probability measures on A. Similarly P_B is defined. We assume A, B to be finite. As such P_A and P_B are closed and bounded. With each $s \in S$ we associate the dummy game whose pay-off is given by

$$r(s,i,j) + \beta \sum_{s'} w^*(s')q(s'/s,i,j)$$

where w^* satisfies the functional equation

$$w^*(s) = \min_{\lambda} \max_{\mu}[r(s,\mu,\lambda) + \beta \sum_{s'} w^*(s')q(s'/s,\mu,\lambda)]$$

That w^* exists is a consequence of the fact that the operator T, defined below, which maps the bounded measurable functions on S into itself is a contraction.

$$(Tw)(s) = \min \max[r(s,\mu,\lambda) + \beta\sum_{s'} w(s')q(s'/s,\mu,\lambda)]$$

where $r(s,\mu,\lambda) = \sum_{i,j} r(s,i,j)\mu(i)\lambda(j)$ and $q(\cdot/s,\mu,\lambda) = \sum_{i,j} q(\cdot/s,i,j)\mu(i)\lambda(j)$ with $\mu \in P_A$ and $\lambda \in P_B$. Now Shapley's fundamental theorem on Markov games can be stated as follows.

<u>Theorem 1.1</u> Let S, A and B be finite. Then the discounted Markov game has a value and the two players have optimal stationary strategies. In fact the fixed point w^* associated with the operator T is the value of the Markov game.

Proof of this may be found in [111] or [92]; in the latter, proof is given using results from dynamic programming. In fact Zachrisson in his paper deals with the T-stage Markov game, with finite state space and compact action spaces. He shows under certain assumptions that the T-stage Markov game has a value, that optimal strategies exist and that they may be determined by a dynamic programming approach. Zachrisson obtains Shapley's theorem by letting T tend to infinity.

Remark 1.1 Iwamoto and Kai consider two-person Markov games where the discount factor is not always constant but depends on the state and the actions chosen by the two players. They prove Theorem 1.1 if each discount factor is less than one [57]. We have the following theorem when S is countable and A, B are arbitrary and r bounded.

Theorem 1.2 Suppose for each $s \in S$ and $w \in M(S)$ (= space of real-valued bounded measurable functions) the dummy game with the following pay-off has a value

$$r(s,\mu,\lambda) + \beta \sum_{s'} w(s')q(s'/s,\mu,\lambda)$$

Then the Markov game has a value and the two players have ε-optimal stationary strategies.

Remark 1.2 Theorem 1.2 includes Theorem 1 of Takahashi [122].

Remark 1.3 In Theorem 1.1 as well as Theorem 1.2 it is enough if the players use only stationary strategies. This is due to the fact that if there is an optimal π there is one which is stationary.

The following theorem which generalizes Theorem 1.1 is a sample of more general results for the total expected reward case obtained by J. Wessels when r is not necessarily bounded [152].

Theorem 1.3 Let S be countable and A, B be finite. Let μ be a given vector with positive components. Write b for the vector with entries $(1/\mu(s))$. Suppose there exists a constant $M > 0$ with $|r(s,i,j)| \leq M\, b(s)$ for all $i \in A$ and $j \in B$. Further suppose there is a $\beta \in (0.1)$ such that $\sum_{s'} q(s'/s,i,j)b(s') \leq \beta\, b(s)$ for all i, j, and s. Then the Markov game with total expected income ($I(\pi, \Gamma) = \sum_{n-1}^{\infty} r_n(\pi, \Gamma)$ where r_n is the income on the nth day to player I) has a value and the two players have optimal stationary strategies.

Remark 1.4 For characterizations in which the above assumption on transition probabilities is satisfied we refer to [153].

We shall now briefly discuss positive Markov games; that is, we shall assume $r(s,i,j)$ is nonnegative and $\beta = 1$. In other words $I(\pi, \Gamma) = \Sigma r_n(\pi, \Gamma)$. In the positive case $I(\pi, \Gamma)(s) \geq 0$ and it can take the value ∞. We have the following result due to D. Kamerud [59].

Theorem 1.5 Let S be countable and A, B be finite with $r(s,i,j) \geq 0$ for all s,i,j. If $V(s) = \lim_{\beta \to 1} V_\beta(s)$ where V_β = value of the discounted game, then the positive Markov game has value V and player II has a stationary optimal strategy. For each $\varepsilon > 0$ player I has an ε-optimal strategy which is semi-stationary (dependent only on initial state and the current state).

Remark 1.5 Proof of this theorem depends on a sequence of estimation lemmas involving truncation and discounting, together with Theorem 1.1. If S, A, B are finite and if $I(\pi, \Gamma)(s) \leq K$ for all π, Γ and s, then both players have optimal stationary strategies in the positive case. When S is countable, player I need not have optimal stationary strategy [94].

These theorems are true if instead of assuming that there is a uniform upper bound on the number of actions available in each state, we stipulate only that the number of actions available in each state be finite. Markov games in general do not have perfect information in the sense of Kuhn [66]. However, perfect information can be simulated within our framework by demanding that in each state one of the players has only one action. If the Markov game has perfect information, the two players have pure optimal stationary strategies when S, A, B are finite. Kushner and Chamberlain give a set of sufficient conditions for the Markov game--with finite state and compact action spaces under the criterion of total expected rewards--to have pure optimal strategies [69]. For related results on the existence of pure stationary optimal strategies see Frid [33] and Kifer [61].

When S is uncountable, measure structure is necessary and problems concerning the measurability of the value as a function of the state arise. We will tackle this problem in another section.

§2. MARKOV GAMES WITH LIMITING AVERAGE PAY-OFF

In this section we consider the limiting average pay-off to player I; this type of pay-off was considered first by Gillette in his thesis for perfect information games and later for cyclic games [38]. The limiting average pay-off is defined as follows:

$$L(\pi,\Gamma)(s) = \lim\inf \frac{1}{N} \sum_{i=1}^{N} r_i(\pi,\Gamma)(s)$$

where r_i denotes the pay-off to I at the ith stage. The lim inf is taken to insure existence. However, any convex combination of lim inf and lim sup may be taken without destroying the character of proofs of results that will be presented in this section.

Gillette was kind enough to inform one of the authors that the real life motivation for his work was the game of mumblety-peg. Webster's New Collegiate Dictionary defines mumblety-peg as a game in which players throw or flip a knife to stick in the ground. What Webster's does not say is that the sequence of knife throws is one of increasing difficulty. After throwing the knife in one way, a player continues to the next method. If he misses, he has two choices; (a) he may pass the knife to his opponent, and when the opponent passes the knife back to him after playing, he again attempts the difficult trick that he tried before; (b) he may make a second attempt. If he does stick the knife in the ground, he goes on to the next harder throw. However, if he misses in the second attempt, he must pass the knife to his opponent and when he gets it back, he must start again from the beginning. As one can see, this can easily be a non-terminating game. Webster's Dictionary also says that the origin of the name is in the penalty for the loser: he must pull a peg out of the ground with his teeth.

Throughout this section we will assume without loss of generality $r(s,i,j) > 0$ for all s,i,j. The results obtained so far in the theory of Markov games with limiting average pay-off are far from complete even when S, A, B are finite. If the Markov game has perfect information, Gillette has shown that the game has a value in the limiting average sense and that the two players have pure optimal stationary strategies, though the proof he supplied uses an incorrect extension of the Hardy-Littlewood theorem. Liggett and Lippman [71] prove this result using results from Blackwell [11]. See also Federgruen, Vrieze, and Wanrooij [145].

<u>Definition</u> A Markov game is cyclic if there exists a positive integer n such that for any state $s \in S$, there is a positive probability of being in state s at step n irrespective of the starting state or the strategies followed by players I and II.

<u>Definition</u> A Markov game is irreducible if when the players adopt stationary strategies, then the resulting transition probabilities on states determine an irreducible Markov process, no matter what the stationary strategies may be.

Now we have the following theorem due to Gillette, Hoffman, and Karp [46].

<u>Theorem 2.1</u> Let S, A, B be finite. Suppose the Markov game is cyclic or irreducible. Then the Markov game with limiting average pay-off has a value which is independent of

the initial state and the two players have optimal stationary strategies.

Stern [118] in his thesis proves the following, which is a slight generalization of Theorem 2.1.

Theorem 2.2 Let S, A, B be finite. Suppose there is a state s_o such that if the game is in any state, then with positive probability it will eventually reach state s_o, no matter what the strategies of the players might be. Then the Markov game with limiting average pay-off has a value which is independent of the initial state and the two players have optimal stationary strategies. Proof of this theorem depends on the following proposition due to Derman-Ross, and on Shapley's theorem on the discounted case. Assume S countable, A, B finite.

Proposition 2.1 Suppose there exist a constant v and a bounded function h(s) such that

$$\begin{aligned} v + h(s) &= \min_{\lambda} \max_{\mu} [r(s,\mu,\lambda) + \sum_{s'} h(s')q(s'/s,\mu,\lambda)] \\ &= \max_{\mu} [r(s,\mu,g^*)(s) + \sum_{s'} h(s')q(s'/s,\mu,g^*)(s)] \\ &= \min_{\lambda} [r(s,f^*(s),\lambda) + \sum_{s'} h(s')q(s'/s,f^*(s),\lambda)] \end{aligned}$$

Then

$$\begin{aligned} v &= \inf_{g} \sup_{f} \phi(f,g)(s) = \sup_{f} \inf_{g} \phi(f,g)(s) \\ &= \sup_{f} \phi(f,g^*)(s) = \inf_{g} \phi(f^*,g)(s). \end{aligned}$$

Here f, g denote stationary strategies for the two players.

<u>Remark 2.1</u> The following example due to Gillette shows that a Markov game may <u>not</u> have a value if the players play only stationary strategies. This is a clear cut departure from the discounted case.

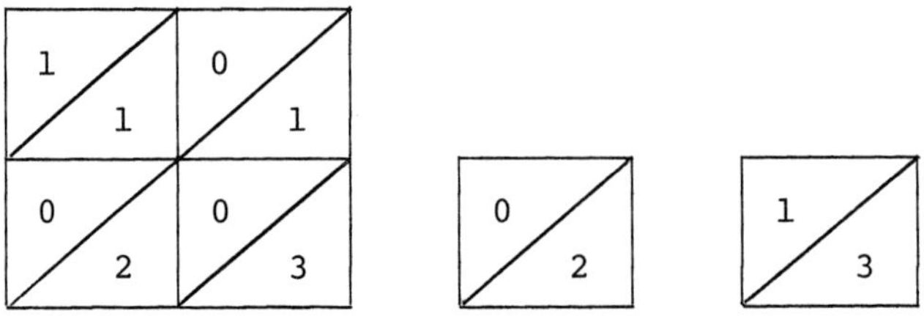

State 1 State 2 State 3

The notation 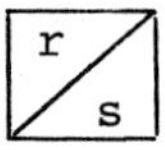

means that if the players choose the row and column corresponding to this box, then player II pays player I the amount r and the next state is s.

Blackwell and Ferguson [14] have shown that the game has a value in the limiting average pay-off sense in the bigger class, namely the class of behavior strategies. They also show that player I has no optimal strategy and player II has an optimal stationary strategy. However, if a Markov game has a value in the class of stationary strategies and if the two players have optimal stationary strategies, then, as in the discounted case, it is irrelevant to consider behavior strategies; for a proof we refer to [92].

In the above example, Q(f,g) [whose (s,s')th element is given by q(s'/s,f(s),g(s))] is multichained (closed irreducible sets of states) for some stationary pair (f,g). However, if Q(f,g) is unichained for every pair of pure stationary (f,g) then we know that the Markov game has a value and the players have optimal stationary strategies [115].

Let $Q^*(f,g)$ denote the Cesaro limit of the sequence $\{Q^n(f,g)\}$. In Gillette's example, $Q^*(f,g)$ is discontinuous in (f,g). Federgruen, et al. give two sets of sufficient conditions which guarantee the continuity of $Q^*(f,g)$ for every stationary pair (f,g) and consequently it follows that the Markov game has a value and the two players have optimal stationary strategies [145]. In fact, they prove results for noncooperative stochastic games which will be discussed in the next section.

Now we shall discuss the recent work by Bewley and Kohlberg [147]. Their work is important on two counts: (i) the proof they supply is algebraic in nature which relies heavily on Tarski's decision method for elementary algebra of real closed fields, and (ii) most of the known results in the theory of finite stochastic games (zero-sum) follow from their results. Their main contribution is a complete description of the asymptotic behavior of the value and optimal strategies, both in the n-step game as n goes to infinity, and in the discounted pay-off game as β goes to one. This description is valid for all stochastic games with zero-stop probabilities.

Let S, A, B be finite. If the initial state is s and play stops after n stages, one obtains a well-defined finite game which has a minimax value which we shall denote by v_{ns} and the vector $(v_{n1}, v_{n2}, \cdots, v_{ns})$ by v_n.

<u>Definition</u> If v_n/n converges to a vector v_0, then v_0 is called the asymptotic average value of the game; $\lim v_{ns}/n$ is called the asymptotic average value of the game with initial state s.

Theorem 2.3 Let S, A, B be finite. Then the Markov game has an asymptotic average value. Moreover, if v_β is the value of the Markov game with discount factor β, and if v_n is (as defined above) the minimax value after n steps, then

$$\lim_{\beta \downarrow 0} (1 - \beta) v_\beta = \lim_{n\to\infty} (v_n/n)$$

For a proof of this theorem we refer to Bewley and Kohlberg [147].

Another result which they deduce is the following due to Stern.

Theorem 2.4 Let S, A, B be finite. Further, assume that player I controls the transition probabilities; that is, $q(s'/s,i,j)$ does not depend on j for all i, s and s'. Then the Markov game has a value in the limiting average sense and the two players have optimal stationary strategies.

Another interesting class of Markov games for which a value exists in the limiting average sense is the so-called repeated games with absorbing states, i.e., a Markov game in which every state except one is absorbing. Kamerud has considered the following pay-off. With each history h are associated the rewards $r[s_n(h), a_n(h), b_n(h)]$. The accumulated reward to player I is given by

$$\limsup_{N\to\infty} \sum_{n=1}^{N} r[s_n(h), a_n(h), b_n(h)] = Z(h)$$

If π and Γ are the strategies used by the two players, we can define the expected pay-off $E(\pi,\Gamma)(Z) = \int Z(h)\, dP_{\pi,\Gamma}(h)$ where P is the measure induced by (π, Γ) on the history space given the initial state s. Call a strategy π' playable if $E_{\pi',\Gamma}(Z)$ exists (finite or infinite) for all Γ and s. Similarly, a playable strategy is defined for II. We define the lower value as $\sup_{\pi'} \inf_{\Gamma} E(\pi', \Gamma)(Z)$ where sup is taken over all playable strategies π' and the upper value as inf sup $E(\pi, \Gamma')(Z)$ where inf is taken over all playable strategies Γ'. Kamerud has given the following example to show that a value (that is, upper value equal to lower value) may not exist in general in the above sense even if the game is a repeated game with absorbing states. This is a departure from the result mentioned above for the limiting average pay-off.

<u>Example 2.2</u>

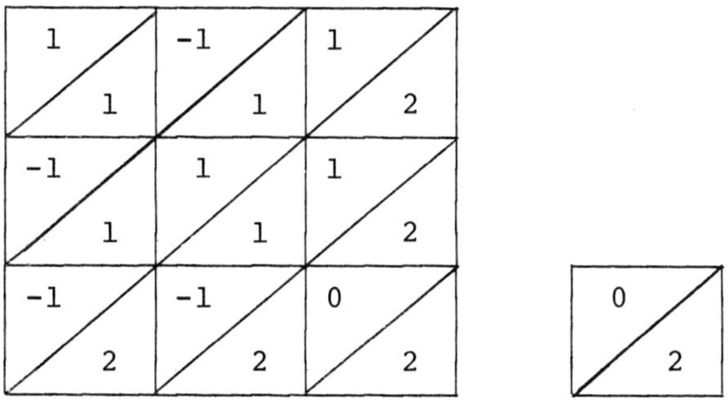

State 1 State 2

In this example, it turns out that the upper value is at least 1 and the lower value is at most -1.

<u>Definition</u> Call a vector v the average value in the strong sense if there exist strategies π^*, Γ^* for players I and II, respectively, such that

$$\liminf_{N\to\infty} \inf_{\Gamma} E_{\pi^*,\Gamma}\left(\frac{1}{N}\sum_{n=1}^{N} r_n\right) \geq v$$

and

$$\limsup_{N\to\infty} \sup_{\pi} E_{\pi,\Gamma^*}\left(\frac{1}{N}\sum_{n=1}^{N} r_n\right) \le v$$

The strategies π^* and Γ^* are optimals for I and II.

Remark 2.2 Whenever the average value in the strong sense exists, it must be equal to the asymptotic average value (as well as the value in the limiting average sense). Bewley and Kohlberg give a sufficient condition (see proposition 10.1 in [147]) for the existence of the average value in the strong sense.

We will close this section with the statement of a version of the Hardy-Littlewood theorem and a consequence of this in the study of Markov games with limiting average pay-off.

Hardy-Littlewood Theorem Let a_n be a bounded sequence of real numbers. Then

$$\liminf_{\beta\to 1} (1-\beta)\sum_{n=0}^{\infty} \beta^n a_n \ge \liminf_{N\to\infty} \frac{1}{N}\sum_{j=0}^{N} a_j$$

$$\limsup_{\beta\to 1} (1-\beta)\sum_{n=0} \beta^n a_n \le \limsup_{N\to\infty} \frac{1}{N}\sum_{j=0}^{N} a_j$$

Furthermore, if the limit on the left of the inequalities exists, then the limit on the right of the inequalities exists and the limits are equal; that is, if $\lim_{\beta\to 1}(1-\beta)\sum_{n=0}^{\infty} \beta^n a_n$ exists, then $\lim_{N\to\infty} \frac{1}{N}\sum_{j=0}^{N} a_j$ exists and

$$\lim (1-\beta)\sum_{n=0}^{\infty} \beta^n a_n = \lim \frac{1}{N}\sum_{j=0}^{N} a_j$$

For a proof see Hobson [45].

A proof of Gillette's theorem on cyclic games can be given

based on the Hardy-Littlewood theorem. The following result is also a consequence of this theorem.

Theorem 2.5 Let S, A, B be finite. Suppose there exists a pair of optimal stationary strategies (f^o, g^o) for the discounted Markov games for all β sufficiently near one. Then (f^o, g^o) is also optimal in the limiting average sense.

Remark 2.3 Conditions of Theorem 2.5 are met when the Markov game has perfect information.

§3. NONZERO SUM MARKOV GAMES

A nonzero sum Markov game is determined by S, A, B, q, r_1, and r_2. If the state is s and if I chooses $a \in A$ and II chooses $b \in B$, then I gets an income $r_1(s,a,b)$ and II gets $r_2(s,a,b)$ and the system moves to a new state s' according to the law of motion $q(s'/s,a,b)$. Payments accumulate throughout the course of the play. Here the problem is that both players want to maximize their accumulated income. It is not hard to see that if $r_2(s,a,b) = -r_1(s,a,b)$ we have the zero-sum case. As before we will make use of the discount factor β with $0 \leq \beta < 1$ in order for the total income for I and II to be well-defined.

Definition (following Nash) Call a pair of strategies (π^*, Γ^*) an equilibrium pair if $I_1(\pi^*, \Gamma^*) \geq I_1(\pi, \Gamma^*)$ for all π and s and $I_2(\pi^*, \Gamma^*) \geq I_2(\pi^*, \Gamma)$ for all Γ and s. Here $I_j(\pi, \Gamma)$ refers to the total income to player j for j = 1, 2.

Fink [1963] and Takahashi [1964] have studied the non-cooperative nonzero-sum Markov games. In fact, they have proved the following theorem.

Theorem 3.1 (Fink-Takahashi) Let S, A, B be finite. Then the discounted Markov game has at least one pair of stationary equilibrium strategies. Proof of this result follows readily from Kakutani's fixed point theorem (for details see []9]).

Remark 3.1 Theorem 3.1 has also been proved independently by Rogers [99] and Sobel [115]. Takahashi's result is more general; he proves the result when S is finite, A and B compact under certain topological assumptions on r and q.

Rogers [1969] and Sobel [1969] were the first ones to study noncooperative games in the limiting average sense of Gillette. We have the following theorem.

Theorem 3.2 Suppose for every pure stationary (f,g), Q(f,g) is unichained. Then there exists a pair of stationary equilibrium strategies in the limiting average sense.

Remark 3.2 Federgruen, et al. point out a minor error in Sobel when he attempts to prove the existence of a (g,w) or bias equilibrium. When S is countable, Federgruen, et al. give a number of recurrency conditions with respect to the transition probability matrices associated with the stationary policies that guarantee the existence of an equilibrium policy under the criterion of the average return per unit time. In their approach the average equilibrium strategies can be obtained as limit strategies from a sequence of β-discounted equilibrium strategies with β tending to one.

Remark 3.3 Stern proves Theorem 3.2 for noncooperative Markov games.

<u>Example 3-1</u> The following example given in [145] shows that in the case of nonzero sum Markov games with perfect information, there may not exist pure equilibrium strategies.

State 1

$(0,1)$ / $(\frac{2}{3},\frac{1}{3})$
$(0,-1)$ / $(\frac{1}{3},\frac{2}{3})$

State 2

$(1,0)$ / $(\frac{1}{3},\frac{2}{3})$	$(-1,0)$ / $(\frac{2}{3},\frac{1}{3})$

The notation (r_1,r_2) / $(\alpha,1-\alpha)$ means that if the players choose the row and column corresponding to this box, then the players I and II receive r_1, r_2, respectively, and with probabilities α and $(1 - \alpha)$ the game will move to states 1 and 2, respectively.

§4. ALGORITHMS FOR MARKOV GAMES

Throughout this section we will assume S, A, B to be finite. The following example given in Parthasarathy and Raghaven [95] shows that in Markov games the value function may not lie in the same field as that of the entries in the payoff matrices, transition probabilities and the discount factor.

State 1

1 / 1	0 / 2
0 / 2	3 / 1

State 2

0 / 2

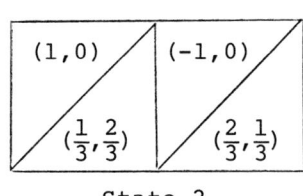

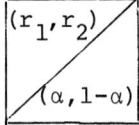

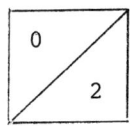

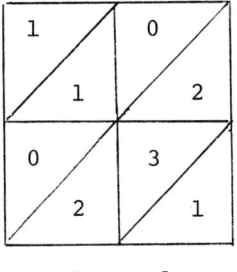

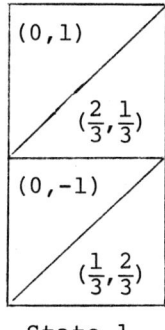

For $\beta = 1/2$, one easily checks that the value at state 1 is equal to $(-4 + 2\sqrt{13})/3$. Thus, it is clear that the theory of Markov games differs from that of the usual matrix games where it is known from a theorem of Weyl [154] that the minimax value lies in the same ordered field as that of the matrix entries. In light of this simple example, it follows that one can not hope to find an algorithm which will give the value function in finitely many rational steps, in general, even if the entries are rational numbers. However, for the special case when only one player, say the maximiser, controls the transition we have the following theorem [95].

<u>Theorem 4.1</u> Let S, A, B be finite. Suppose $q(s'/s,i,j)$ does not depend on j for all i, s, and s'. Then the components of the value function and a pair of optimal stationary strategies lie in the same ordered field as that of the entries in the pay-off matrices, transition probabilities and the discount factor.

<u>Remark 4.1</u> A similar result is true for irreducible Markov games in the limiting average sense (for details see [95]).

Shapley was the first to propose a (numerical) algorithm for finding the fixed point w* of the operator T defined in Section 1, based on successive approximations given by the following recurrence relation:

$$v^{(n+1)} = Tv^{(n)}$$

At each step of this procedure, N (where N is the cardinality of S) matrix games--in other words, N linear programs must be solved. This method does not utilize the information contained in the optimal strategies at each step, and therefore belongs to the class of iterations in the function space. Shapley provides a proof for the convergence of this algorithm.

Another method was suggested by Hoffman and Karp for Markov games with limiting average pay-off and should apply for the discounted case as well. We will describe the procedure for the discounted case.

(a) Choose a stationary strategy $g_0(s)$ for the second player.

(b) Given $g_r(s)$, compute $v_r(s)$ the unique solution of $v_r(s) = \max_i [r(s,i,g_r(s) + \sum_{s'} q(s'/s,i,g_r(s)v_r(s')]$ for all s.

This could be done using the algorithm of Howard [50] for Markovian decision processes with discounted rewards.

(c) Given $v_r(s)$ find an optimal strategy $g_{r+1}(s)$ for player II for the matrix game with pay-off given by $r(s,i,j) + \beta \sum_{s'} q(s'/s,i,j)v_r(s')$ for each s.

Remark 4.2 Rao, Chandrasekaran and Nair provide a proof for the convergence of this algorithm for the discounted case [97] while Hoffman and Karp provide an algorithm with proof for the limiting average pay-off case [46]. It is clear that the Hoffman-Karp method utilizes some of the information contained in the optimal strategies at each iteration and is a combination of iterations in the strategy and function spaces.

Pollatschek and Avi-Itzhak present a new approach, based on analysis and geometrical interpretations, to the solution of Markov (stochastic) games. Their algorithm, using iterations in the strategy space, turns out to be a Newton-Raphson type procedure. We will now briefly describe the procedure. Let us write $\phi_s(w)$ for $(Tw)(s)$. Let $\Phi_s(w) = \phi_s(w) - w(s)$. What we want is a w^* with $\Phi_s(w^*) = 0$ for all s. Using the Newton Raphson method for solving the above equation, we have

$$-\Phi_s\left(v^{(n)}\right) = \nabla\Phi_s\left(v^{(n)}\right)\left(v^{(n+1)} - v^{(n)}\right)$$

We also have

$$\partial\Phi_s(w)/\partial w(s') = c(s,s') - \delta_{s,s'}$$

where $c(x,s') = \beta\sum_i\sum_j q(s/s,i,j)f_s^0(i)g_s^0(j)$. (Here, f^0, g^0 stand for optimal strategies for the dummy game with pay off $r(s,i,j) + \beta\sum_{s'}w(s')q(s'/s,i,j)$ and $\delta_{s,s'}$ is Kronecker's delta.) Note that $c(s,s')$ depends on w. It follows that in matrix form,

$$\phi\left(v^{(n)}\right) - \left(v^{(n)}\right) = I - c(s,s')\left(v^{(n)}\right)\left(v^{(n+1)} - v^{(n)}\right)$$

Observe that

$$\phi_s(w) = \sum_{i}\sum_{j} r(s,i,j)f_s^0(i)g_s^0(j) + \sum_{s'} c(s,s')w(s')$$

$$= d_s(w) + \sum_{s'} c(s,s')w(s')$$

That is, we write $d_s(w)$ for $\sum_i\sum_j r(s,i,j)f_s^0(i)g_s^0(j)$. Substituting this in the above equation and solving for $v^{(n+1)}$, we obtain the following equation:

$$v^{(n+1)} = \left(I - c(s,s')\left(v^{(n)}\right)\right)^{-1}d\left(v^{(n)}\right)$$

This equation can also be written in the following form:

$$v^{(n+1)} = v^{(n)} + \left(I - c(s,s')\left(v^{(n)}\right)\right)^{-1}\Phi\left(v^{(n)}\right)$$

The following is a set of sufficient conditions for the convergence of the Newton-Raphson type process from any starting point in the discounted case.

Theorem 4.2 Let S, A, B be finite. Suppose the discount factor satisfies the following inequality:

$$0 \le \beta < 1/[1 + \max_{s} \sum_{s'} \{\max q(s'/s,i,j) - \min q(s'/s,i,j)\}]$$

Then the Newton-Raphson method will converge from any starting point.

Remarks 4.3 This theorem is proved in Pollatschek and Avi-Itzhak [96]. They also work out several examples covering the discounted case as well as limiting average pay-off case and point out through those examples that the Newton-Raphson method requires fewer iterations and less running time compared with Shapley's method and the Hoffman-Karp algorithm.

Condition of Theorem 4.2 is satisfied if $q(s'/s,i,j)$ is independent of i, j for all s, s'; in other words, the transition probabilities should not depend on the actions chosen by the two players. Very recently Van der Wal in his master's thesis has worked out a set of policy improvement-value approximation algorithms providing ε-optimal strategies and an approximation of the value of the (infinite horizon) Markov game with discounting. If the maximizing player has in each state only one admissible action, the set reduces to the set of successive approximation algorithms described by Van Nunen [126]. A special element of this set is Howard's policy iteration algorithm. The set of algorithms given in Van der Wal [127] is closely related to the algorithm suggested by Hoffman and Karp.

Moreover, for two classes of games (at the criterion of total expected rewards) a successive approximation algorithm is given yielding upper and lower bounds on the value of the game.

§5. MARKOV GAMES WITH GENERAL STATE SPACE

Throughout this section we shall assume S to be a nonempty Borel subset of a Polish space. Before we proceed to formulate the further conditions on S, A, B, q, r under which we shall tackle this problem, let us consider an example.

Example 5.1 Let A, B be nonempty Borel subsets of Polish spaces and let $k_s(a,b)$ be a bounded Borel function on A x B for each s. These objects define two-person zero-sum games G_s with A, B as the spaces of pure strategies for players I and II, respectively, and $k_s(a,b)$ as the pay-off function. Consider the following Markov game with state space S = [0, 1] with action spaces A, B for I and II, respectively, and $q(s/s,a,b) = 1$ for all s, a, b, and $r(s,a,b) = k_s(a,b)$. Now it is easy to see that the Markov game has a value if and only if for each s,

$$\sup_\mu \inf_\lambda \iint k_s(a,b)\,d\mu(a)\,d\lambda(b) = \inf_\lambda \sup_\mu \iint k_s(a,b)\,d\mu(a)\,d\lambda(b)$$

that is, if and only if each G_s has a value. Moreover, there are optimal strategies for players I and II in the Markov game if and only if players I and II have optimal strategies in the game G_s.

It is clear, then, from the above example that the Markov problem, as we have formulated it, is far too general to admit a value. Also, it is not clear that even if the value exists for each s, whether the value function (as well as the optimal strategies for the Markov game) is measurable. We shall therefore have to impose suitable additional restrictions on S, A, B, q, r to ensure that the Markov game has a value. Maitra and Parthasarathy [78] were the first to consider Markov games with general state space. We have the following theorem.

Theorem 5.1 Let S, A, B be compact metric spaces, let r be a continuous, real valued function on S x A x B, and assume, moreover, that whenever $(s_n,a_n,b_n) \to (s_o,a_o,b_o)$ in S x A x B, $q(\cdot/s_n,a_n,b_n)$ converges weakly to $q(\cdot/s_o,a_o,b_o)$. Then the discounted Markov game has a value, the value function is continuous, and players I and II have measurable optimal stationary strategies.

Remark 5.1 Proof makes use of results on dynamic programming from Blackwell [13] and the theory of selection theorems [93]. In view of a recent result obtained by Whitt [148], the optimal strategies of players I and II can be chosen to be functions such that the inverse images of open sets are F_σ sets. For related results under weaker hypothesis see [94].

Remark 5.2 Sengupta [110] has considered Shapley's Markov game in a more general setting, with the accumulated pay-off being regarded as a function on the space of infinite trajectories and the set of states of the system taken as a compact metric space. It has been shown that any game with a lower semi-continuous pay-off has a value and the minimizer has an optimal strategy.

Remark 5.3 Prabhakar in his Ph.D. thesis submitted to the Indian Statistical Institute [1972-73] has given an example to show that $\inf_{\Gamma} \sup_{\pi} I(\pi,\Gamma)(s)$ need not be universally measurable in general; this contrasts with a result in dynamic programming due to Strauch (see Theorem 7.1 in [119]).

Sobel using ideas from Denardo [20] has proved the following theorem for noncooperative discounted Markov games.

Theorem 5.2 Let S, A, B be compact metric spaces, let r_1, r_2 be two continuous, real valued functions on S x A x B and assume, moreover, that whenever $(s_n,a_n,b_n) \to (s_o,a_o,b_o)$ in S x A x B, then $q(\cdot/s_n,a_n,b_n)$ converges weakly to $q(\cdot/s_o,a_o,b_o)$. Then the discounted Markov game has a pair of equilibrium stationary strategies for the two players.

Remark 5.4 Sobel has considered all stationary policies, though not necessarily measurable ones. It appears that the proof of this theorem is incomplete--the trouble arises when he applies Schauder--Tychonoff fixed point theorem and in showing that a fixed point corresponds with an equilibrium point. For details see Vrieze [155]. In view of this, Theorem 5.2 is proved with an additional assumption in [145].

Theorem 5.3 Let S, A, B be compact metric spaces; let r_1, r_2 be two continuous, real valued functions on S x A x B. Suppose $q(\cdot/s_n,a_n,b_n)$ converges weakly to $q(\cdot/s_o,a_o,b_o)$ whenever $(s_n,a_n,b_n) \to (s_o,a_o,b_o)$. Further suppose $q(\cdot/s,a_n,b_n)$ converges setwise to $q(\cdot/s,a_o,b_o)$ whenever $(a_n,b_n) \to (a_o,b_o)$. Then the discounted Markov game has a pair of equilibrium stationary strategies.

Remark 5.5 Glicksberg's fixed point theorem is used to prove the existence of equilibrium strategies. Federgruen, et al. show that a certain mapping is upper semicontinuous in the sense of Kuratowski. In order to do that they use sequential arguments though this has not been justified.

Himmelberg, et al. introduced the notion of p-equilibrium strategies for Markov games similar to the one in dynamic programming introduced by Strauch [42]. Let p be a fixed probability distribution on S.

Definition 5.1 Call (π^*, Γ^*) a p-equilibrium pair if $p\{s : I_1(\pi^*,\Gamma^*)(s) \geq I_1(\pi,\Gamma^*)(s)$ for all π and $I_2(\pi^*,\Gamma^*)(s) \geq I_2(\pi^*,\Gamma)(s)$ for all $\Gamma\} = 1$.

In [42] the following theorem is proved.

Theorem 5.4 Let $S = [0, 1]$ and A, B be finite. Let $r_j(s,a,b) = h_j(s,a) + k_j(s,b)$ for $j = 1, 2$, where h_j, k_j are bounded measurable functions in s for $a \in A$, $b \in B$.

Let $q(\cdot/s,a,b) = [q'(\cdot/s,a) + q''(\cdot/s,b)]/2$ where q', q'' are probability measures and further, are measurable in s for each $a \in A$, $b \in B$. Suppose $q(\cdot/s,a,b)$ is absolutely continuous with respect to p for every (s,a,b). Then the discounted Markov game has a pair of p-equilibrium stationary strategies.

Remark 5.6 The real problem in the proof of Theorem 5.4 is to topologize the space of strategies so that it becomes a compact metric space and consequently one can use sequential arguments to apply Glicksberg's fixed point theorem. For a proof see [42].

We will close this section after a brief discussion on two-stage noncooperative discounted Markov games considered by Adam Idzik [51]. A two-stage Markov game is determined by S, A, B, q, r_1, r_2, ϕ, ψ, β. Here $\phi : S \times P_A \times P_B \to 2^A$ (= space of closed subsets of A) and $\psi : S \times P_A \times P_B \to 2^B$ (= space of closed subsets of B) are Borel measurable functions which restrict actions of players in such a way that $\phi(s,f(s),g(s))$ and $\psi(s,f(s),g(s))$ are, respectively, the sets of actions available to players I and II (the constraints) at state s where f and g are the strategies chosen by the two players. At the first stage of the play both players choose (independently of each other) strategies f and g, respectively, where $f : S \to P_A$ and $g : S \to P_B$ are Borel measurable functions. As a consequence of the choice associated with (f, g), a noncooperative Markov game is defined (in the sense of Parthasarathy [1973]) by the objects: S, A, B, $\phi(f,g)$, $\psi(f,g)$, q, r_1, r_2, and β where $\phi(f,g)(s) = \phi(s,f(s),g(s))$ and $\psi(f,g)(s) = \psi(s,f(s),g(s))$. Now, it follows that in the second stage of the play, the players behave further as in the (usual) Markov game defined above.

Definition 5.2 In the two-stage noncooperative discounted Markov game call a pair of strategies (f*, g*) an equilibrium pair if (f*, g*) is an equilibrium pair of policies for the noncooperative discounted Markov game associated with (f*, g*).

We have the following theorem due to Adam Idzik.

Theorem 5.5 Let S be a countable set, A, B be compact metric, and let r_1, r_2 be continuous on A x B for each $s \in S$. Suppose $\max_{a,b} r_j(s,a,b) \in C_o(S)$ (= space of all real valued functions on S which vanish at ∞) and $\max_{a,b} q(s'/s,a,b) \in C_o(S)$ for every s'. Further, suppose $q(s'/s,a_n,b_n) \to q(s'/s,a_o,b_o)$ whenever $(a_n,b_n) \to (a_o,b_o)$ for every s', $s \in S$, and for fixed $s \in S$, $\phi(s,\mu,\lambda)$ and $\psi(s,\mu,\lambda)$ are continuous functions of (μ, λ). Then, in the two-stage noncooperative discounted Markov game there exists a pair of equilibrium stationary strategies (f*, g*) for the two players.

§6. MISCELLANEOUS REMARKS AND OPEN PROBLEMS

Stochastic differential or sequential games occur in many ways--for example, in (i) many competitive situations in economics, (ii) pursuit-evasion problems for aircrafts, and (iii) control engineering problems.

Consider the problem of the submarine that wishes to determine the route from point A to point B which will minimize some cost while taking into account the probability of detection. On the other hand, the searcher wishes to determine the search procedure which will maximize the probability of detection. The best search procedure depends on the random laws governing the path, and the best path depends on the search procedure. Here, since the object is to select two 'controls' or 'policies,' the optimum values of which depend on the other, a game--or minimax--formulation is sometimes appropriate. In fact, Charnes and Schroeder [19] treat a similar problem and formulate it as a Markov problem.

Kirman and Sobel in [62] develop a dynamic model of

oligopoly and discuss the existence and characteristics of optimal policies in such a model. The firms are assumed to face a random demand so they hold inventories which fluctuate from one period to the next. This necessitates a dynamic model rather than a static one. Their extension of the equilibrium concept is founded on the results on nonzero-sum Markov games. Other applications of the theory of Markov games are discussed in Filar [28], Shubik and Whitt [112], and Zachrisson [138]. We will now mention some open problems.

Problem 6.1 Let S, A, B be finite. Find a set of sufficient conditions, at which for an arbitrary policy there exists an identical Markov policy π if we make no assumptions on the form of the pay-off as well as on the law of motion.

Remark 6.1 Bartoszynski in [8] treats this problem with reference to problems in dynamic programming. Call a policy $\pi = (f_1, f_2, \cdots, f_n, \cdots)$ a Markov policy if each $f_n\colon S \to P_A$. Iwamoto treats the finite horizon zero-sum two-person 'recursive' Markov games with nonstationary laws [56].

Problem 6.2 Let S, A, B be finite. Prove or disprove that the value exists (in the class of behavior strategies) in the limiting average sense of Gillette.

Remark 6.2 It will be interesting to settle this problem for the case when the transition probabilities are deterministic; that is, $q(s'/s,i,j) = 0$ or 1 for all s', s, i, j.

Problem 6.3 Let S, A, B be finite. If there exists a $\beta_o \in (0, 1)$ such that player I has a stationary strategy f_o

optimal in the β-discounted game for all $\beta \in (\beta_o, 1)$, does it mean that the value exists in the limiting average sense and if so, is this f_o optimal?

Remark 6.3 If both players have (f_o, g_o) optimal strategies for all $\beta \in (\beta_o, 1)$ then it is known that the value exists in the limiting average sense and (f_o, g_o) are also optimal for the limiting average game.

*Problem 6.4 Let S, A, B be finite. Does the Newton-Raphson method suggested by Pollatschek and Avi-Itzhak converge from any starting point (in general) for β-discounted Markov games?

Remark 6.4 Rao, Chandrasekaran and Nair in [97] attempt to prove that Newton-Raphson procedure always converges from any starting point; however, the proof they supply is incorrect. The conditions given in Theorem 4.2 for convergence are very restrictive in nature--it would be nice to find other suitable conditions for convergence.

Problem 6.5 Is Theorem 2.4 valid for noncooperative Markov games? In other words, does there exist a pair of equilibrium stationary strategies for limiting average pay-off games if $q(s'/s,i,j)$ does not depend on j for every (s', s, i)?

Problem 6.6 Find suitable conditions under which a (g, w) equilibrium pair of policies exist.

Remark 6.5 The concept of (g, w) equilibrium pair is due to Sobel and Federgruen, et al.

---

*Very recently Van der Wal has given a counter example to problem 6.4 [159].

Another very interesting area is the following.

Problem 6.7 Is it possible to apply the methods employed by Bewley-Kohlberg to find expansions that would represent equilibrium expected pay-offs in the game with pay-offs discounted by β when β is near one? In other words, give a description of the asymptotic behavior of equilibrium strategies both in the n-stage game as n tends to infinity and in the β discounted pay-off game as β goes to one.

Remark 6.6 It will be interesting to give an algebraic proof of Theorem 3.1. It is not clear Theorem 4.1 is valid for nonzero-sum discounted Markov games. In other words, if $q(s'/s,i,j)$ is independent of j for all (s', s, i), can we assert the existence of a pair of equilibrium stationary strategies whose components lie in the same ordered field as that of the entries in the pay-off matrices, transition probabilities and the discount factor. When the state space is an arbitrary Borel set the following problem is worth mentioning.

Problem 6.8 Let S, A, B be Borel subsets of Polish spaces. Let $I(\pi,\Gamma)(s) = \sum_{n=1}^{\infty} \beta^{n-1} r_n$ denote the discounted total pay-off. Assume $r(s,a,b)$ is a bounded Borel measurable function on S x A x B and moreover, assume $q(\cdot/s,a,b)$ is Borel measurable on S x A x B for every Borel subset of S. If inf sup $I(\pi,\Gamma)(s)$ = sup inf $I(\pi,\Gamma)(s)$ for every s, can be assert that the value function is universally measurable?

Remark 6.7 If S is an arbitrary Borel set, A, B are finite sets, then one can show that the value function exists and that it is Borel measurable. For a proof see [94].

Problem 6.9 Let S be an arbitrary Borel set, A and B finite. Give a set of suitable conditions such that the value function exists in the limiting average sense and the two players have optimal strategies.

Remark 6.8 It appears that one simple condition is the following: There is a state s_o and $\alpha > 0$ such that $q(s_o/s_j, i, j) \geq \alpha$ for all $s \in S$, $i \in A$, $j \in B$. For related results in the dynamic programming case when the state space S is arbitrary, see Iwamoto [54].

Another area of research is to find for two-person, nonzero-sum Markov games equilibrium points by the use of methods like the Howson-Lemke algorithm for bimatrix games. This problem was suggested by Hoffman. It is also worthwhile to develop successive approximation algorithms for the Hoffman-Karp procedure given for Markov games with limiting average pay-off. It would be interesting to compare and work out (mathematically) the efficiency or the rate of convergence of different algorithms for zero-sum discounted Markov games.

Recursive games developed by Everett [25] and stochastic games are closely related, the only difference being the form of the pay-off functions in the component games. In a recursive game two small, but important, modifications are made: (i) a pay-off of actual units occurs only when the play terminates, (ii) the probability of stopping the game is not necessarily positive. Orkin in [88] gives an inductive proof (of Everett's result) that these games have a value, with ε-optimal stationary strategies available to each player. It is not known whether the recursive game has a value in the sense of

Everett-Orkin if the number of states is countable. Such a result would have immediate application to the theory of Markov games with imperfect information (see also [16,156,157]). Due to lack of space we have not discussed certain topics like survival games and attrition games or games with incomplete information. We hope that we have given the reader a fair account of the theory of stochastic games or Markov games. We are responsible for any errors or omissions.

ACKNOWLEDGMENT

Various people have helped us in preparing this manuscript. We want to thank in particular Dr. T. E. S. Raghavan for several useful discussions on this and related topics and Dr. A. V. Lakshminarayanan for some useful comments regarding the presentation of this article.

BIBLIOGRAPHY

1. Aggarwal, V., Bimatrix Markovian Decision Processes and Stochastic Ratio Games, Ph.D. dissertation submitted to Case Western Reserve University (1973).

2. Aggarwal, V., R. Chandrasekavan, and K. P. K. Nair, Markov Ratio Decision Processes, to appear in Jour. Opt. Theory and its Appl. (1976).

3. Aggarwal, V., R. Chandrasekavan, and K. P. K. Nair, Discounted Stochastic Ratio Games, preprint.

4. Aumann, R. J., Mixed and Behaviour Strategies in Infinite Extensive Games, Advances in Game Theory, Ann. Math. Studies No. 52, Princeton University Press, Princeton, New Jersey (1964), 627-650.

5. Aumann, R. J. and M. Maschler, Repeated Games with Incomplete Information: The Zero-sum Extensive Case, Report to the U. S. Arms Control and Disarmament Agency,

Washington, D. C., final report on contract ACDA/ST-143, prepared by Mathematica, Princeton (1968), 25-108.

6. Barbosa Dantas, C. A., On the Existence of Stationary Optimal Plans, Ph.D. dissertation submitted to the University of California, Berkeley (1966).

7. Baron, S., D. L. Kleinman, and S. Serbin, A Study of the Markov Game Approach to Tactical Maneuvering Problems, NASA, Langley Research Center, prepared by Bolt, Beranek and Newman, Inc., Cambridge, Mass., nr NASA CR-1979 (1972).

8. Bartoszynski, R., On the Existence of Markov Policies in Dynamic Programs, Bulletin of the Polish Academy of Sciences 19 (1971), 403-406.

9. Bellman, R., Dynamic Programming, Princeton University Press, Princeton, New Jersey (1957).

10. Beniest, W., Jeux Stochastiques Totalment Cooperatifs Arbitres, Cashiers du Centre d'Etudes de Recherche Operationnelle 5 (1963), 124-138.

11. Blackwell, D., Discrete Dynamic Programming, Ann. Math. Stat. 33 (1962), 719-726.

12. Blackwell, D. and C. Ryll-Nardzewski, Non-existence of Everywhere Proper Conditional Distributions, Ann. Math. Stat. 34 (1963), 223-225.

13. Blackwell, D., Discounted Dynamic Programming, Ann. Math. Stat. 36 (1965), 226-235.

14. Blackwell, D. and T. S. Ferguson, The Big Match, Ann. Math. Stat. 39 (1968), 159-163.

15. Blackwell, D., Positive Dynamic Programming, Proc. 5th Berkeley Symposium on Math. Stat. Prob. I (1965), 415-418.

16. Blackwell, D., Infinite G Games with Imperfect Information, Zastosowania Matematyki Applications Mathematicae, Hugo Steinhaus Jubilee Volume X (1969), 99-101.

17. Blackwell, D., D. Freedman, and M. Orkin, The Optimal Reward Operator in Dynamic Programming, Ann. Prob. 2

18. Chamberlain, S. G., Stochastic Games, Ph.D. dissertation submitted to Brown University, Providence, Rhode Island (1969).

19. Charnes, A. M. and R. G. Schroeder, On Some Stochastic Tactical Antisubmarine Games, Naval Res. Log. Quarterly 14 (1967), 291-312.

20. Denardo, E. V., Contraction Mapping in the Theory Underlying Dynamic Programming, SIAM Rev. 9 (1967), 165-177.

21. Derman, C., Markovian Sequential Control Processes - Denumerable State Space, Jour. Math. Analy. Appl. 10 (1965), 295-302.

22. Dubins, L. and D. Freedman, Measurable Sets of Measures, Pac. Jour. Math. 14 (1964), 1211-1222.

23. Dubins, L. and L. J. Savage, How to Gamble If You Must, McGraw Hill, New York (1965).

24. Dunford, N. and J. T. Schwartz, Linear Operators, Part I General Theory, Interscience Publishers, Inc., New York (1967).

25. Everett, H., Recursive Games, Ann. Math. Studies No. 39, Princeton University Press, Princeton, New Jersey (1957), 47-78.

26. Fan Ky, Fixed Point and Minimax Theorems in Locally Convex Topological Linear Spaces, Proc. Nat. Academy Sci., U. S. A. 38 (1952).

27. Feller, W., An Introduction to Probability Theory and its Applications, Vol. I., John Wiley and Sons, Inc. (1950), second edition.

28. Filar, J. A., Markov Games and the Theory of Duopoly, M.S. dissertation submitted to Monash University, Australia (1975).

29. Fink, A. M., Equilibrium in a Stochastic n-person Game, J. Sci. Hiroshima Univ. Ser. A - I 28 (1964), 89-93.

30. Flynn, J., Averaging vs. Discounting in Dynamic Programming: A Counter Example, Ann. Stat. 2 (1974), 411-413.

31. Flynn, J., Conditions for the Equivalence of Optimality Criteria in Dynamic Programming, preprint.

32. Fox, B. L., Finite State Approximations to Denumerable State Dynamic Programs, Rand Corporation, R. M., 6195, Riz (1970).

33. Frid, E. B., The Optimal Stopping Rule for a Two-person Markov Chain with Opposing Interests, Theory of Prob. and its Applications 14 (1969), 714-716.

34. Frid, E. B., On Stochastic Games, Theory of Prob. and its Applications 18 (1973), 389-393.

35. Friedman, A., Stochastic Games and Variational Inequalities, preprint.

36. Friedman, A., Differential Games, Pure and Applied Mathematics, Vol. 25, Wiley-Interscience, New York (1971).

37. Furukawa, N., Markovian Decision Processes with Compact Action Spaces, Ann. Math. Stat. 43 (1972), 1612-1622.

38. Gillette, D., Stochastic Games with Zero-stop Probabilities, Ann. Math. Studies No. 39, Princeton University Press, Princeton, New Jersey (1957), 179-187.

39. Glicksberg, I. L., A Further Generalization of the Kakutani Fixed Point Theorem with Application to Nash Equilibrium Points, Proc. Amer. Math. Soc. 3 (1952), 170-174.

40. Himmelberg, C. J. and F. S. Van Vleck, Multifunctions with Values in a Space of Probability Measures, Jour. Math. Analy. and Appl. 50 (1975), 108-112.

41. Himmelberg, C. J., Measurable Relations, Fund. Math. 87 (1975), 53-72.

42. Himmelberg, C. J., T. Parthasarathy, T. E. S. Raghavan, and F. S. Van Vleck, Existence of p-equilibrium and Optimal Stationary Strategies in Stochastic Games, to appear in Proc. Amer. Math. Soc. (1976).

43. Himmelberg, C. J., T. Parthasarathy, and F. S. Van Vleck, Optimal Plans for Dynamic Programming Problems, to appear

in Math. Operations Res.

44. Hinderer, K., Foundations of Non-stationary Dynamic Programming with Discrete Time-parameter, Lecture notes in Operations Research and Mathematical Systems, Vol. 33, Berlin-Heidelberg-New York, Springer-Verlag (1970)

45. Hobson, E., The Theory of Functions of a Real Variable and the Theory of Fourier's Series, Cambridge University Press (1926).

46. Hoffman, A. D. and R. M. Karp, On Non-terminating Stochastic Games, Management Science 12 (1966), 359-370.

47. Hordijk, A., Dynamic Programming and Markov Potential Theory, Mathematical Centre Tracts, No. 51, Mathematisch Centrum, Amsterdam (1974).

48. Hordijk, A., P. J. Schweitzer, and H. C. Tijms, The Asymptotic Behaviour of the Minimal Total Expected Cost for the Denumerable State Markov Decision Model, J. Appl Prob. 12 (1975), 298-305.

49. Hordijk, A., Convergent Dynamic Programming, to appear in Mathematics of Operations Research.

50. Howard, R., Dynamic Programming and Markov Processes, Technology Press and Wiley, New York (1960).

51. Idzik, A., Two-stage Non-cooperative Discounted Stochastic Games, preprint.

52. Idzik, A., Personal Communication.

53. Isaacs, R., Differential Games, third edition and reprinted by Robert E. Krieger Publishing Company, Huntington, New York (1975).

54. Iwamoto, S., Average Reward Markovian Decision Processes in the Completely Ergodic Case, Bull. Math. Stat. 15 (1973), 55-68.

55. Iwamoto, S., and Yu Kai, On Deterministic Stationary Strategies for Markov Games, Bull. Math. Stat. 16 (1974), 71-82.

56. Iwamoto, S., Finite Horizon Markov Games with Recursive Pay-off Systems, Memoirs of the Faculty of Science, Kyushu University, Series A, Vol 29 (1975), 123-147.

57. Iwamoto, S. and Yu Kai, Discrete Markov Games with Recursive Additive Pay-off Systems, preprint.

58. Kai Yu, On Optimal Non-random Stationary Policies in Finite State Stochastic Games, Bull. Math. Stat. 15 (1973), 93-99.

59. Kamerud, D. B., Repeated Games and Positive Stochastic Games, Ph.D. dissertation submitted to the University of Minnesota (1975).

60. Karlin, S., Mathematical Methods and the Theory of Games, Vols. I and II, Addison Wesley, London (1959).

61. Kifer, T. I., Optimal Strategy in Games with an Unbounded Sequence of Moves, Theory of Prob. and its Appl. 14 (1969), 279-286.

62. Kirman, A. P. and M. J. Sobel, Dynamic Oligopoly with Inventories, Econometrica 42 (1974), 279-287.

63. Kohnberg, E., Repeated Games with Absorbing States, Ann. Stat. 2 (1974), 724-738.

64. Kohlberg, E. and S. Zamir, Repeated Games of Incomplete Information: The Symmetric Case, Ann. Stat. 2 (1974), 724-738.

65. Krylov, N. V., On the Existence of ε-optimal Homogeneous Markov Strategies for Controlled Chains, Doklady Academy Nauk 155 (1964), 747-750 (Russian).

66. Kuhn, H., Extensive Games and the Problem of Information, Ann. Math. Studies No. 28, Princeton University Press, Princeton, New Jersey (1953), 193-216.

67. Kuratowski, K. and C. Ryll-Nardzewski, A General Theorem on Selectors, Bull. Acad. Polon. Sci. Ser. Math Astronom. Phys. 13 (1965), 397-403.

68. Kuratowski, K., Topology, Vol. I, II, Academic Press (1966).

69. Kushner, H. J. and S. G. Chamberlain, Finite State Stochastic Games: Existence Theorems and Computational Procedures, IEEE, Trans. Automatic Control AC-14 (1969), 248-255.

70. Lemke, C. E. and J. T. Howson, Equilibrium Points of Bimatrix Games, SIAM Jour. on Appl. Math. 12 (1964), 413-423.

71. Liggett, T. M. and S. A. Lippman, Stochastic Games with Perfect Information and Time Average Pay-off, SIAM Review 11 (1969), 604-607.

72. Luce, R. D. and H. Raiffa, Games and Decisions, Wiley and Sons, New York (1957).

73. Mackey, G. W., Borel Structure in Groups and Their Duals, Trans. Amer. Math. Soc. 85 (1957), 134-165.

74. MacQueen, J., A Modified Dynamic Programming Method for Markovian Decision Problems, J. Math. Analy. Appl. 14 (1966), 38-43.

75. Maitra, A., A Note on Undiscounted Dynamic Programming, Ann. Math. Stat. 37 (1966), 1042-1044.

76. Maitra, A. and T. Parthasarathy, Dynamic Programming Approach to Stochastic Games with Countable State Space, Technical Report No. 1/67, Indian Statistical Institute, Calcutta (1967).

77. Maitra, A., Discounted Dynamic Programming on Compact Metric Spaces, SANKHYA Series A, 30 (1968), 211-216.

78. Maitra, A. and T. Parthasarathy, On Stochastic Games, Jour. Opt. Theory and its Appl. 5 (1970), 289-300.

79. Maitra, A. and T. Parthasarathy, On Stochastic Games II, Jour. Opt. Theory and its Appl. 8 (1971), 154-160.

80. Manne, A. S., Linear Programming and Sequential Decisions, Management Science 6 (1960), 259-267.

81. Mertens, J. F. and S. Zamir, The Value of Two-person Zero-sum Repeated Games with Lack of Information on Both Sides, Int. Jour. Game Theory 1 (1971), 39-64.

82. Michael, E., Topologies on Spaces of Subsets, Trans. Amer. Math. Soc. 71 (1951), 152-182.

83. Mine, H., K. Yamada, and S. Osaki, On Terminating Stochastic Games, Management Science 16 (1970), 560-571.

84. Mon, R. G., A Dynamic Theory of Zero-sum Two-person Games, Jour. Math. Analy. Appl. 29 (1970), 392-411.

85. Mon, R. G., Diffusions and Stochastic Games, Int. Jour. Control 13 (1971), 853-863.

86. Moulin, H., Iterated Games, submitted for publication in Int. Jour. Game Theory.

87. Nair, K. P. K., P. K. Banerjee, and S. Srinivasan, Symmetric Stochastic Games, unpublished manuscript.

88. Orkin, M., Recursive Matrix Games, Jour. Appl. Prob. 9 (1972), 813-820.

89. Owen, G., Game Theory, W. B. Saunders Company, Philadelphia (1968).

90. Parthasarathy, K. R., Probability Measures on Metric Spaces, Academic Press, New York (1967).

91. Parthasarathy, T., Discounted and Positive Stochastic Games, Bull. Amer. Math. Soc. 77 (1971), 134-136.

92. Parthasarathy, T. and T. E. S. Raghavan, Some Topics in Two-person Games, American Elsevier Publishing Company, New York (1971).

93. Parthasarathy, T., Selection Theorems and Their Applications, Lecture notes in Mathematics, No. 263, Springer-Verlag, Berlin-New York (1972).

94. Parthasarathy, T., Discounted, Positive and Non-cooperative Stochastic Games, Int. Jour. Game Theory 2 (1973), 25-37.

95. Parthasarathy, T. and T. E. S. Raghavan, Finite Algorithms for Stochastic Games, submitted for publication.

96. Pollatschek, M. A. and B. Avi-Itzhak, Algorithms for Stochastic Games, Management Science 15 (1969), 399-415.

97. Rao, S. S., R. Chandrasekaran, and K. P. K. Nair, Algorithms for Discounted Stochastic Games, Jour. Opt. Theory and its Appl. 11 (1973), 627-637.

98. Rios, S. and I. Yanez, Programmation Sequentielle en Concurrence, Research Papers in Statistics, Edited by F. N. David, John Wiley and Sons, London-New York-Sydney (1966), 289-299.

99. Rogers, P. D., Non-zero-sum Stochastic Games, Report ORC 69-8, Operations Research Centre, University of California, Berkeley, Ph.D. dissertation (1969).

100. Rosenfeld, J., Adaptive Competitive Decision, Ann. Math. Studies No. 52, Princeton University Press, Princeton, New Jersey (1964), 69-83.

101. Ross, S. M., Non-discounted Denumerable Markovian Decision Models, Ann. Math. Stat. 39 (1968), 412-423.

102. Ross, S. M., Arbitrary State Markovian Decision Processes, Ann. Math. Stat. 39 (1968), 2118-2122.

103. Ross, S. M., On the Non-existence of ε-optimal Randomized Stationary Policies in Average Cost Markov Decision Models, Ann. Math. Stat. 42 (1971), 1767-1768.

104. Satia, J. K. and R. E. Lave, Markovian Decision Processes with Uncertain Transition Probabilities, O. R. 21 (1973), 728-740.

105. Schal, M., On Continuous Dynamic Programming with Discrete Time Parameter, Z. Wahrscheinlichkeitstheorie Verw. Geb. 21 (1972), 279-288.

106. Schal, M., A Selection Theorem for Optimization Problems, Archiv Der Mathematik 25 (1974), 219-224.

107. Schal, M., Conditions for Optimality in Dynamic Programming and for the Limit of n-stage Optimal Policies to be Optimal, Z. Wahrscheinlichkeitstheorie Verw. Gebiete 32 (1975), 179-196.

108. Schroeder, R. G., Linear Programming Solutions to Ratio Games, Operations Research 18 (1970), 300-305.

109. Schweitzer, P. J., Annotated Bibliography on Markov Decision Processes, I.B.M. Watson Research Center, New York, working copy, September (1973).

110. Sengupta, S. K., Lower Semicontinuous Stochastic Games with Imperfect Information, Ann. Stat. 3 (1975), 554-558.

111. Shapley, L. S., Stochastic Games, Proc. Nat. Acad. Sci. 39 (1953), 1095-1100.

112. Shubik, M. and W. Whitt, Fiat Money in an Economy with One Nondurable Good and No Credit, Topics in Differential Games, Edited by A. Blaquiere, North Holland Publishing Company (1973), 401-448.

113. Shubik, M. and M. J. Sobel, Bibliography on Sequential Games (1974), Yale University.

114. Sion, M., On General Minimax Theorems, Pac. Jour. Math 8 (1958), 171-176.

115. Sobel, M. J., Non-cooperative Stochastic Games, Ann. Math. Stat. 42 (1971), 1930-1935.

116. Sobel, M. J., Continuous Stochastic Games, Jour. Appl. Prob. 10 (1973), 597-604.

117. Stearns, R. E., A Formal Information Concept for Games with Incomplete Information, Report to the U. S. Arms Control and Disarmament Agency, Washington, D. C., final report on contract ACDA/AT-116, prepared by Mathematica, Princeton, New Jersey (1967).

118. Stern, M. A., On Stochastic Games with Limiting Average Pay-off, Ph.D. dissertation submitted to the University of Illinois, Circle Campus, Chicago (1975).

119. Strauch, R. E., Negative Dynamic Programming, Ann. Math. Stat. 37 (1966), 871-890.

120. Suddereth, W. D., A Strong Law for Uniformly Bounded Functions of the Past, unpublished manuscript.

121. Sweat, C. W., Adaptive Competitive Decision in Repeated Play of a Matrix Game with Uncertain Entries, Naval Res. Logistics Quarterly 15 (1968), 425-448; Ph.D. dissertation submitted to the University of California, Berkeley.

122. Takahashi, M., Stochastic Games with Infinitely Many Strategies, Jour. Sci. Hiroshima Univ. Series A-I Mathematics 26 (1962), 123-134.

123. Takahashi, M., Recursive Games with Infinitely Many Strategies, Jour. Sci. Hiroshima Univ. Series A-I 27 (1963), 51-59.

124. Takahashi, M., Equilibrium Points of Stochastic Non-cooperative N-person Games, J. Sci. Hiroshima Univ. Ser. A-I 28 (1964), 95-99.

125. Taylor, H., Markovian Replacement Processes, Ann. Math. Stat. 36 (1965), 1677-1694.

126. Van Nunen, J. A. E., A Set of Successive Approximation Methods for Discounted Markovian Decision Problems, Zeitschrift Fur O. R. 19 (1975).

127. Van der Wal, J., The Solution of Markov Games by Successive Approximation, M.S. thesis submitted to Technological University, Eindhoven (1975).

128. Van der Wal, J. and J. Wessels, On Markov Games, Memorandum COSOR 75-12, Technological University, Eindhoven (1975), to appear in Statistica Neerlandica (1976).

129. Van der Wal, J., The Method of Successive Approximations for the Discounted Markov Games, Memorandum COSOR 75-02, Technological University, Eindhoven (1975), submitted for publication to the Int. Jour. Game Theory.

130. Van der Wal, J., Markov Games, An Annotated Bibliography, Memorandum COSOR 75-09, Technological University, Eindhoven (1975).

131. Veinott, Jr., A., Discrete Dynamic Programming with Sensitive Discount Optimality Criteria, Ann. Math. Stat. 40 (1969), 1635-1660.

132. Von Neumann, J. and O. Morgenstern, Theory of Games and Economic Behaviour, Princeton University Press, Princeton, New Jersey (1944).

133. Von Neumann, J., A Model of General Economic Equilibrium, Rev. Econ. Studies 13 (1945), 1-9.

134. Warga, J., Functions of Relaxed Controls, SIAM J. Control 5 (1967), 628-641.

135. Wessels, J. and J. A. E. E. Van Nunen, Discounted Semi-Markov Decision Processes: Linear Programming and Policy Iteration, Statistica Neerlandica 29 (1975), 1-7.

136. Wessels, J. and J. A. E. E. Van Nunen, A Principle for Generating Optimization Procedures for Discounted Markov Decision Processes, to appear in Colloquia Mathematica Societatis Janos Bolyai 12 (A. Prekopa, ed.), North Holland Publishing Company, Amsterdam (1975).

137. Wessels, J., Stopping Times and Markov Programming, Memorandum COSOR 74-09, Technological University, Eindhoven, to appear in Proceedings of the 1974 European Meeting of Statisticians and 7th Prague Conference (1975).

138. Zachrisson, L. E., Markov Games, Advances in Game Theory, Ann. Math. Studies No. 52, Princeton University Press, Princeton, New Jersey (1964), 211-253.

139. Zamir, S., On the Notion of Value for Games with Infinitely Many Stages, Ann. Stat., 1 (1973), 791-796.

140. Derman, C., Denumerable state Markovian decision processes-average cost criterion, Ann. Math. Stat. 37 (1966), 1545-1554.

141. Derman, C. and A. Veinott, A Solution to a Countable System of Equations arising in Markovian Decision Processes, Ann. Math. Stat. 38 (1967), 582-585.

142. Cook, W. D. and M. J. L. Kirby, Stochastic Games With Random Pay-offs and Multiple Goals, preprint.

143. Brown, L. D. and R. Purves, Measurable Selections of Extrema, Ann. Stat. 1 (1973), 902-912.

144. Gillette, D., Representable Infinite Games, Ph.D. thesis submitted to the University of California at Berkeley (May, 1953).

145. Federgruen, A., Vrieze, O. J., and G. L. Wanrooij, On the Existence of Discounted and Average Return Equilibrium Policies in N-person Stochastic Games, preprint.

146. Federgruen, A., Personal Communication.

147. Bewley, T. and E. Kohlberg, The Theory of Stochastic Games with Zero Stop Probabilities, Discussion Paper No. 456, Harvard Institute of Economic Research, Harvard University, January (1976).

148. Whitt, W., Baire Classification of Measurable Selections of Extrema, preprint (1976).

149. Whitt, W., Approximations of Dynamic Programs, to appear in Mathematics of Operations Research.

150. Wagner, D. H. and L. D. Stone, Necessity and Existence Results on Constrained Optimization of Separable Functionals by a Multiplier Rule, SIAM J. Control 12 (1974), 356-372.

151. Wagner, D. H., Survey of Measurable Selections Theory, Daniel H. Wagner Associates, Station Square One, Paoli, Pennsylvania (1976).

152. Wessels, J., Markov Games with Unbounded Rewards, Memorandum COSOR 76-05, Department of Mathematics, Eindhoven University of Technology (1976).

153. Van Hee, K. M., and J. Wessels, Markov Decision Processes and Strongly Excessive Functions, Memorandum COSOR 75-22, Department of Mathematics, Eindhoven University of Technology (1975).

154. Weyl, H., Elementary Proof of a Minimax Theorem Due to Von Neumann, Ann. Math. Studies No. 24 (1950), 19-25.

155. Vrieze, O. J., to appear.

156. Orkin, M., An Approximation Theorem for Infinite Games, Proc. Amer. Math. Soc. 36 (1972), 212-216.

157. Orkin, M., Infinite Games with Imperfect Information, Trans. Amer. Math. Soc. 171 (1972), 501-507.

158. Filar, J. A., Estimation of Strategies in a Markov Game, to appear in Naval Res. Logistics Quarterly.

159. Van der Wal, J., Successive Approximation and Discounted Markov Games, Memorandum No. 119, Twente University of Technology, Enschede, The Netherlands (1976).

STABILIZATION OF DYNAMICAL SYSTEMS UNDER BOUNDED INPUT DISTURBANCE AND PARAMETER UNCERTAINTY

G. Leitmann

University of California at Berkeley
Berkeley, California

§1. INTRODUCTION

The problem of designing feedback controllers for uncertain or disturbed dynamical systems has been addressed in many investigations, e.g., [1-7]. Here, as in [8], we are concerned with situations in which (i) no statistical information is available but an upper bound (either of the norm or of the components) of the disturbance or uncertainty is known (or can be assumed), and (ii) stable operation under all possible circumstances must be assured.

Using the concept of "variable structure systems," researchers, especially in the USSR [9-11], have endeavored to develop methods for improving the performance of linear systems. The resultant nonlinear controllers are sometimes stabilizing against bounded uncertainties or disturbances [12]. This latter technique involves the determination of an attractive hyperplane in state space, on which motion tends toward the origin. Since "attractivity" is a local property, additional conditions (known as "fall conditions") must be imposed to assure that trajectories tend toward the hyperplane. This latter fact, together with the difficulty of finding attractive hyperplanes in multivariable systems of high order, prompts us to seek alternative methods.

In view of our requirement to guarantee stability under all circumstances, including the worst possible ones, one is led naturally to the notion of "worst case" design, and it is this philosophy which we adopt here as we did in [8]. That is, we argue that we must design a controller that stabilizes the system in the presence of the "worst possible" uncertainty or disturbance, and that such a controller ought to be a stabilizing one against all other possible uncertainties of disturbances. It is important, perhaps, to stress that (i) we do not suppose that "Nature" is a true adversary and hence will do her worst, and (ii) "worst case" design is not unduly pessimistic, for if the worst circumstances do not arise, the consequent controller will do "better" under more favorable circumstances.

Here we deal with two cases, input disturbance and system uncertainty. In particular, we treat the following problems.

1.1 Input Disturbance Consider the dynamical system

$$\dot{x}(t) = A(t)x(t) + B(x(t),t)u(t) + C(x(t),t)v(t) \quad (1)$$

$$x(t_o) = x_o$$

where state is $x(t) \in R^n$, control $u(t) \in R^p$, disturbance $v(t) \in R^q$, and matrices A, B, C are of appropriate dimensions.

Given the norm-bound of the disturbance, that is,

$$\|v(t)\| \leq \rho(x,t) \qquad \forall \ (x,t) \in R^n \times R^1_+ \quad (2)$$

where $\rho(x,t)$ is prescribed for all (x,t), we seek a feedback controller

$$\tilde{p}(\cdot) : R^n \times R^1_+ \to R^p \tag{3}$$

where

$$u(t) = \tilde{p}(x(t),t)$$

such that the origin {0} is uniformly asymptotically stable in the large (Lyapunov), no matter what $v(t)$, $t \in R^1_+$.

<u>1.2 System Uncertainty</u> Consider the dynamical system

$$\dot{x}(t) = A(t)x(t) + \Delta A(t,w(t))x(t) + B(x(t),t)u(t) \tag{4}$$

$$x(t_o) = x_o$$

where state is $x(t) \in R^n$, control $u(t) \in R^p$, uncertainty $w(t) \in R^q$, $q = n^2$, and matrices A, ΔA, B are of appropriate dimensions.

The uncertainty matrix

$$\Delta A(t,w(t)) \triangleq [a_{ij}(t)w_{ij}(t)] \tag{5}$$

where the $a_{ij}(\cdot) : R^1_+ \to R^1_+$ are prescribed.

Given the bounds on parameter uncertainties, that is

$$w(t) \in W \triangleq \{w \in R^{n^2} \mid |w_{ij}| \le 1,\ i,j = 1, 2, \cdots, n\} \tag{6}$$

we seek a feedback control (3) such that {0} is uniformly asymptotically stable in the large no matter what $w(t)$, $t \in R^1_+$.

<u>1.3 Associated Differential Game</u> Since we adopt the point of view of worst case design, we pose a differential game whose solution provides candidates for the desired feedback controls. In particular, consider

$$\dot{x}(t) = A(t)x(t) + B(x(t),t)[u(t) + v(t)] \tag{7}$$

$$x(t_o) = x_o,\ t \in [t_o, T]$$

where

$$x(t) \in R^n$$

$$u(t),\ v(t) \in R^m$$

$$A(\cdot) \text{ is } n \times n \text{ and } C^1 \text{ on } (-\infty, T]$$

$$B(\cdot) \text{ is } n \times m \text{ and } C^1 \text{ on } R^n \times (-\infty, T]$$

Since we desire stability of the origin $\{0\}$, we consider a measure of deviation as cost; that is,

$$J = \int_{t_o}^{T} x'(t)Q(t)x(t)dt \tag{8}$$

where

$$Q(\cdot) \text{ is } n \times n, \text{ symmetric and } C^1 \text{ on } (-\infty, T].$$

The strategies of the controller and of "Nature," respectively, are

$$p(\cdot) : R^n \times (-\infty, T] \to U$$

$$e(\cdot) : R^n \times (-\infty, T] \to V \tag{9}$$

with

$$U = V \triangleq \{z \in R^m \mid \|z\| \le \rho(x,t), \text{ given } C^1 \text{ function}$$

$$\rho(\cdot) : R^n \times (-\infty, T] \to (0, \infty)\} \tag{10}$$

Then, with

$$u(t) = p(x(t),t), \qquad v(t) = e(x(t),t) \tag{11}$$

we have

$$J = J(x_o, t_o, p(\cdot), e(\cdot), x(\cdot)) \tag{12}$$

for which we seek a *saddlepoint* strategy pair $\{p^*(\cdot), e^*(\cdot)\}$, that is, such that for all $(x_o, t_o) \in R^n \times (-\infty, T]$

$$J(x_o, t_o, p^*(\cdot), e(\cdot), x^e(\cdot))$$
$$\leq J(x_o, t_o, p^*(\cdot), e^*(\cdot), x^*(\cdot))$$
$$\leq J(x_o, t_o, p(\cdot), e^*(\cdot), x^p(\cdot)) \tag{13}$$

for all terminating plays [13], where $x^e(\cdot)$, $x^*(\cdot)$, $x^p(\cdot)$ are solutions of (7) generated by $\{p^*(\cdot), e(\cdot)\}$, $\{p^*(\cdot), e^*(\cdot)\}$, $\{p(\cdot), e^*(\cdot)\}$, respectively.

Under applying necessary conditions [13-14] and sufficiency conditions [15], one establishes the saddlepoint

$$p^*(x,t) = -e^*(x,t) = -\frac{B'(x,t)P(t)x}{\|B'(x,t)P(t)x\|}\rho(x,t)$$
$$\text{for } (x,t) \notin N \tag{14}$$

and

$$p^*(x,t) \in U, \; e^*(x,t) \in V \quad \text{for } (x,t) \in N$$

where

$$N \triangleq \{(x,t) \mid B'(x,t)P(t)x = 0\} \tag{15}$$

and $P(\cdot)$ is the solution of

$$\dot{P}(t) + P(t)A(t) + A'(t)P(t) + Q(t) = 0, \; P(T) = 0 \tag{16}$$

§2. STABILITY

Having found a saddlepoint of the associated differential game, we utilize the controller strategy $p^*(\cdot)$ as a candidate for a stabilizing control in problems 1.1 and 1.2.

<u>2.1 Input Disturbance</u> Recall dynamical systems (1) and (2). We now state

Assumptions 2.1

i) $A(\cdot)$ is continuous on R^1_+ and uniformly asymptotically stable[1].

ii) $B(\cdot)$ and $C(\cdot)$ are continuous on $R^n \times R^1_+$.

iii) $Q(\cdot)$ is continuous on R^1_+, and $Q(t)$ is positive definite for all $t \in R^1_+$.

iv) For all $(x,t) \in R^n \times R^1_+$ there is a matrix $D(x,t)$ such that $C = BD$ and $D(\cdot)$ is continuous on $R^n \times R^1_+$.

v) $\rho(\cdot) : R^n \times R^1_+ \to (0, \infty)$ is continuous.

vi) $e(\cdot) : R^n \times R^1_+ \to R^q$ is continuous[2], where $v(t) = e(x(t),t)$.

Then we have

Theorem 2.1 Given dynamical system (1-2), if Assumptions 2.1 are met, then feedback control

$$\tilde{p}(\cdot) : R^n \times R^1_+ \to R^p$$

given by

$$\tilde{p}(x,t) = - \frac{B'(x,t)P(t)x}{\|B'(x,t)P(t)x\|} \|D(x,t)\,\rho(x,t)\| \qquad \forall\ (x,t) \notin N$$

$$\tilde{p}(x,t) = u \text{ with } \|u\| \leq \|D(x,t)\rho(x,t)\| \qquad \forall\ (x,t) \in N \tag{17}$$

---

[1]That is, for system $\dot{x}(t) = A(t)x(t)$ the origin is uniformly asymptotically stable in the large.

[2]This condition may be relaxed to piecewise continuity in t; that is, $e(x,\cdot)$ need be only piecewise continuous.

where

$$N \triangleq \{(x,t) \mid B'(x,t)P(t)x = 0\}$$

$$\dot{P}(t) + P(t)A(t) + A'(t)P(t) + Q(t) = 0, \quad \lim_{t\to\infty} P(t) = 0,$$

[that is [16], $P(t) = \int_t^\infty \Phi'(\tau,t)Q(\tau)\Phi(\tau,t)d\tau$, where $\Phi(\cdot)$ is the transition matrix of $\dot{x}(t) = A(t)x(t)$] guarantees uniform asymptotic stability in the large of $\{0\}$ for all admissible disturbances $e(\cdot)$.

Proof Since $\tilde{p}(\cdot)$ is discontinuous and hence not unique for $(x,t) \in N$, system (1-2) is a generalized dynamical system [17-22]

$$\dot{x}(t) \in K(x(t),t) \tag{18}$$

where the set-valued function $K(\cdot)$ is given by

$$K(x,t) \triangleq \{z \in R^n \mid z = A(t)x + B(x,t)\tilde{p}(x,t) + C(x,t)e(x,t)\}$$

Now it is readily shown [23-24] that

i) $K(x,t)$ is convex for all $(x,t) \in R^n \times R^1_+$.

ii) $K(x,t)$ is compact for all $(x,t) \in R^n \times R^1_+$.

iii) $K(\cdot)$ is upper semicontinuous on $R^n \times R^1_+$.

Thus, given any $(x_o,t_o) \in R^n \times R^1_+$, there exists at least one solution of (1-2). In view of the Lyapunov function exhibited below, no such solution has a finite escape time.

Now consider the function $V(\cdot) : R^n \times R^1_+ \to R^1$ given by

$$V(x,t) = x'P(t)x$$

Since $A(\cdot)$ is uniformly asymptotically stable and $Q(t)$ is positive definite, matrix $P(t)$ is positive definite [16]; thus,

$V(\cdot)$ is a Lyapunov function candidate. To show that it is indeed a Lyapunov function, consider a solution

$$x(\cdot) : [t_o, t_1] \to R^n$$

and the function

$$W(\cdot) : [t_o, t_1] \to R^1$$

defined by

$$W(t) = \frac{d}{dt} V \circ x(t) = \text{grad}_x V(x(t),t)\dot{x}(t) + \frac{\partial V(x(t),t)}{\partial t} \quad (20)$$

Now[3], for $(x(t),t) \notin N$,

$$\begin{aligned} W(t) &= 2x'P\left[Ax - B\frac{B'Px}{\|B'Px\|}\|D\rho\| + BDe\right] + x'\dot{P}x \\ &= x'[\dot{P} + PA + A'P]x - 2\|B'Px\|\|D\rho\| + 2x'PBDe \\ &\le -x'Qx - 2(\|D\rho\| - \|D\rho\|)\|B'Px\| \\ &= -x'Qx < 0 \end{aligned}$$

For $(x(t),t) \in N$ but $x(t) \neq 0$, we have at once

$$W(t) = -x'Qx < 0$$

This concludes the proof.

2.2 System Uncertainty Now recall the dynamical system (4)-(6). We state

Assumptions 2.2

i) $A(\cdot)$ is continuous on R^1_+ and uniformly asymptotically stable.

ii) For $(x,t,w) \in R^n \times R^1_+ \times W$ there is a matrix $E(x,t,w)$ such that $\Delta A = BE$ and $E(\cdot)$ is continuous on $R^n \times R^1_+ \times W$.

---

[3]For the sake of brevity, the arguments of functions are omitted.

iii) The $a_{ij}(\cdot)$ are continuous on R^1_+.

iv) $B(\cdot)$ is continuous on $R^n \times R^1_+$.

v) $Q(\cdot)$ is continuous on R^1_+, and $Q(t)$ is positive definite for all $t \in R^1_+$.

vi) $e(\cdot) : R^n \times R^1_+ \to R^{n^2}$ is continuous on $R^n \times R^1_+$, where $w(t) = e(x(t),t)$.

Then we have

Theorem 2.2 Given dynamical system (4)-(6), if Assumptions 2.2 are met, then feedback control

$$\tilde{p}(\cdot) : R^n \times R^1_+ \to R^p$$

given by

$$\tilde{p}(x,t) = -\frac{B'(x,t)P(t)x}{\|B'(x,t)P(t)x\|} \max_{w \in W} \|E(x,t,w)x\| \quad \forall\ (x,t) \notin N$$

$$\tilde{p}(x,t) = u \text{ with } \|u\| \leq \max_{w \in W} \|E(x,t,w)x\| \qquad \forall\ (x,t) \in N$$

where

$$N \triangleq \{(x,t) \mid B'(x,t)P(t)x = 0\}$$

and

$$\dot{P}(t) + P(t)A(t) + A'(t)P(t) + Q(t) = 0, \quad \lim_{t\to\infty} P(t) = 0$$

guarantees uniform asymptotic stability in the large of $\{0\}$ for all admissible uncertainties $e(\cdot)$.

Proof The proof is entirely analogous to that of Theorem 2.1; we shall not repeat it here.

Remarks

1. If matrices $A(t) = A$ = constant and $Q(t) = Q$ = constant, then $P(t) = P$ = constant such that [16]

$$PA + A'P + Q = 0$$

2. If matrix $A(\cdot)$ is not stable, but $\{A(\cdot), B(\cdot)\}$ is stabilizable, then the results of this paper apply to the stabilized nominal systems; e.g., see [25].

3. If the system is subject to both parameter uncertainty and input disturbance, the magnitude of $\tilde{p}(x,t)$ is

$$\max_{w \in W} \|E(x,t,w)x\| + \|D(x,t)\rho(x,t)\|$$

4. The results of this paper can be extended to nonlinear systems which are linear in the input [24]; that is,

$$\dot{x}(t) = f_o(x(t),t) + \delta f(x(t),t) + B(x(t),t)u + C(x(t),t)v$$

where

$$\dot{x}(t) = f_o(x(t),t) + B(x(t),t)u$$

is the nominal system, and

$$\delta f(x(t),t) = \text{model uncertainty}$$
$$v(t) = \text{input disturbance}$$

In [24], stability results are given for both norm bounds and component bounds.

§3. APPLICATION

In this section we demonstrate the use of stabilizing control $\tilde{p}(\cdot)$ for a system subject to input disturbance. In particular, we consider the problem of longitudinal motion of an aircraft in a gusty wind. Let

$\alpha \triangleq$ angle of attack (relative to unperturbed air)

$\theta \triangleq$ orientation of aircraft (relative to inertial line)

$q \triangleq \dot{\theta}$

$\delta_E \triangleq$ (active) elevator control angle

$\delta_A \triangleq$ (active) aileron control angle

$W \triangleq$ wind gust speed

Approximate equations of motion are of the form

$$\begin{bmatrix} \dot{\alpha} \\ \dot{q} \end{bmatrix} = A \begin{bmatrix} \alpha \\ q \end{bmatrix} + B \begin{bmatrix} \delta_E \\ \delta_A \end{bmatrix} + c\,W$$

where A, B are constant matrices and c is a constant vector.

Analog computations were carried out for

$$A = \begin{bmatrix} -2.03 & 1.00 \\ -8.57 & -2.75 \end{bmatrix}$$

$$B = \begin{bmatrix} -0.021 & -0.156 \\ -1.82 & -0.550 \end{bmatrix}$$

$$c = \begin{bmatrix} 0.008 \\ 0.034 \end{bmatrix}$$

$$Q = I$$

To avoid noisy state measurements near the origin, a modified control was employed; namely, for states in a (square) neighborhood N of $\{0\}$, null control was used.

Figures 1, 2, and 3 show the behavior of the system subject to a constant disturbance; namely, W = constant. In Figure 1, state plane trajectories are given for three cases:

1a. Uncontrolled system with initial state outside N.

1b. Controlled system with initial state outside N.

1c. Controlled system with initial state $\{0\}$.

The state of the disturbed but uncontrolled system lends to a steady state other than the origin, as shown in 1a. For 1b the state of the controlled system moves into neighborhood N and once inside does not leave it. Similarly, for 1c the state does not leave N. Figure 2 shows the time-histories of the state variables for case 1c. Figure 3 portrays the control histories for the latter case; as expected, chattering occurs at the boundary of N.

Figures 4 through 7 show the behavior of the system subject to sinusoidal disturbance, $W = W_o \sin t$ with W_o = constant. Figure 4 presents state space trajectories for the uncontrolled system in 4a and for the controlled system in 4b. The time-histories of the state variables for case 4b are shown in Figure 5, and the control histories in Figures 6 and 7. As shown in 4a, the uncontrolled state goes into a limit cycle, whereas 4b shows the controlled state moving into neighborhood N where it remains. Chattering of the controls occurs again once the state is in N.

ACKNOWLEDGMENTS

This paper is based on research supported by ONR. Continuing discussions of this problem with Dr. S. Gutman are gratefully acknowledged; I am also indebted to him for providing the numerical example.

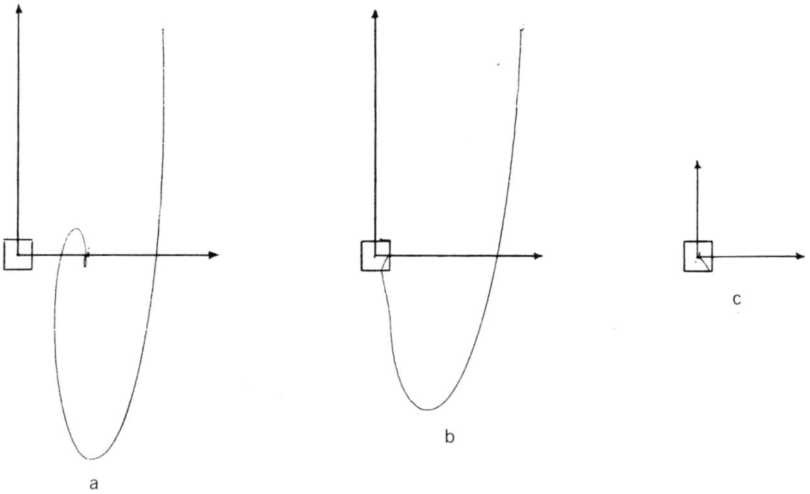

FIGURE 1

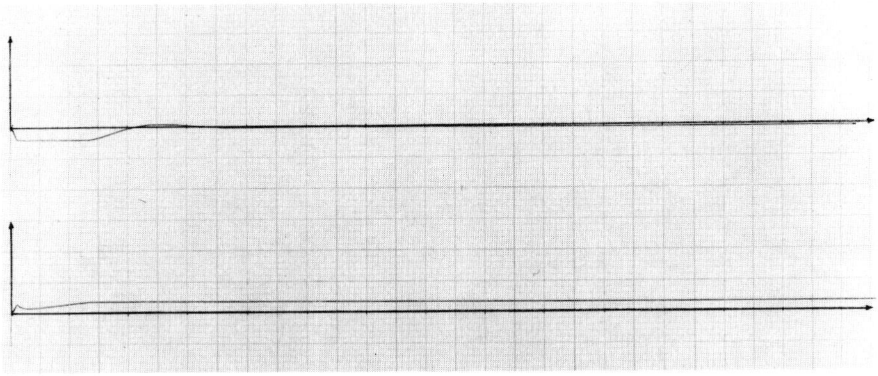

FIGURE 2

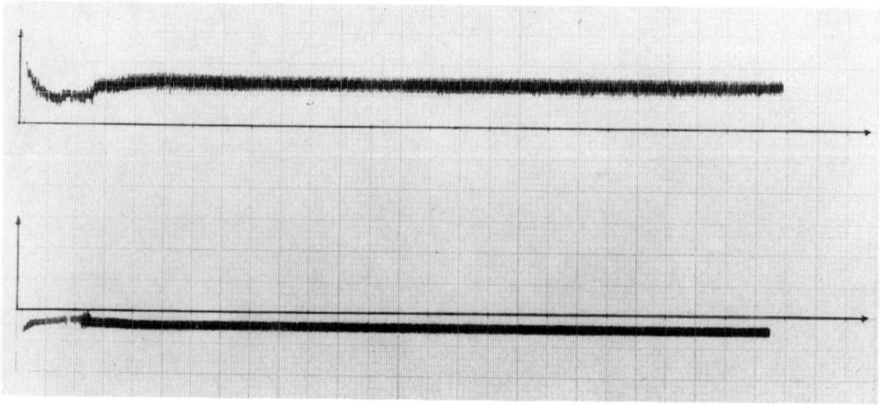

FIGURE 3

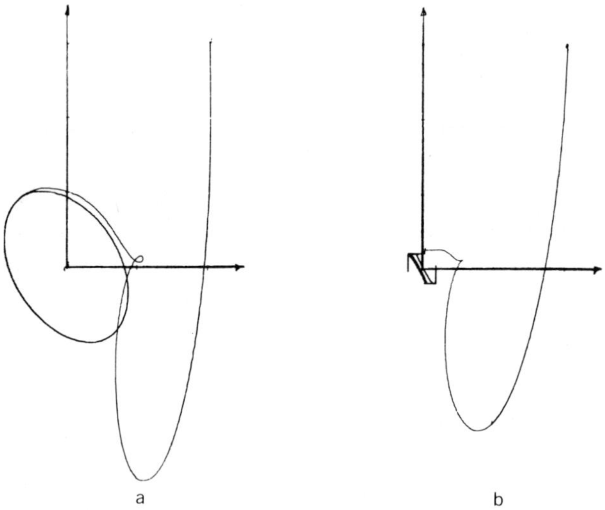

a b

FIGURE 4

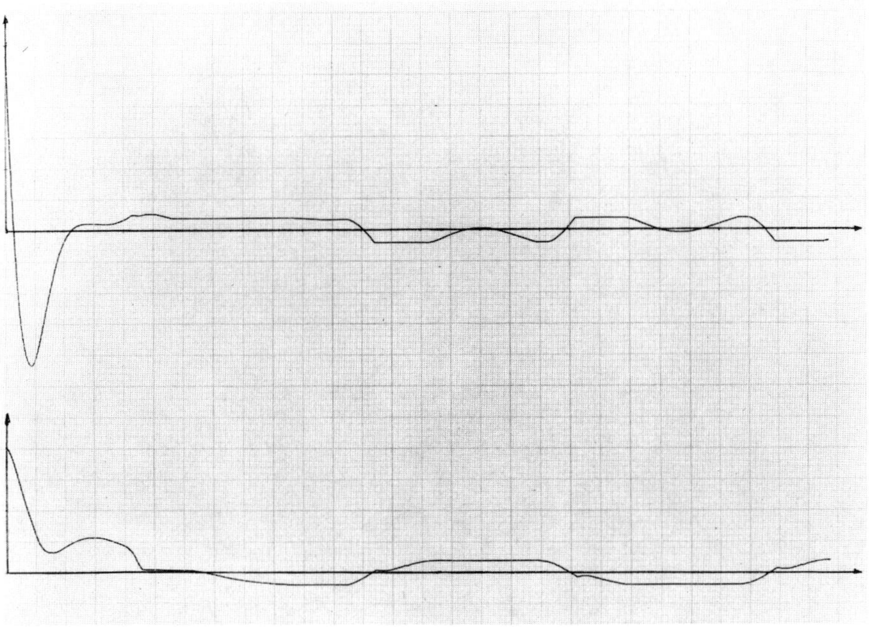

FIGURE 5

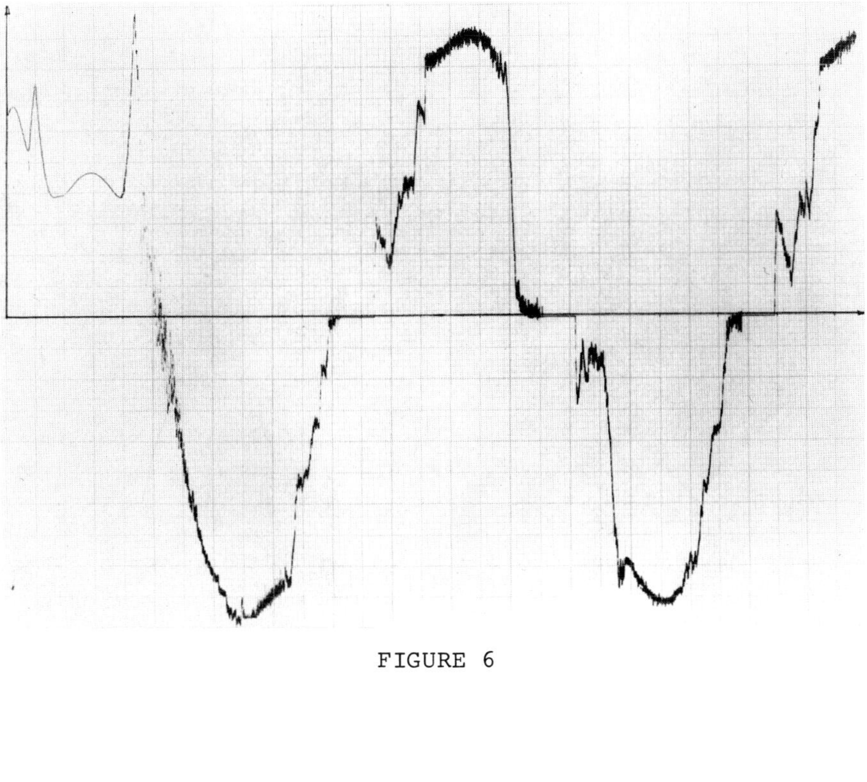

FIGURE 6

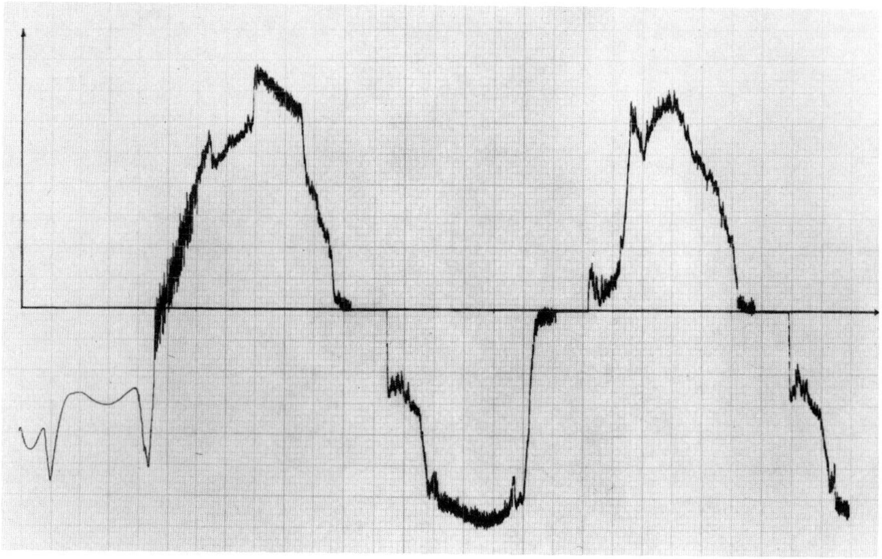

FIGURE 7

REFERENCES

1. Ragade, R. K. and I. G. Sarma, A Game-theoretic Approach to Optimal Control in the Presence of Uncertainty, IEEE Trans. on Automatic Control, vol. AC-12, no. 4, 1967.

2. Sarma, I. G. and R. K. Ragade, Some Considerations in Formulating Optimal Control Problems as Differential Games, Intl. J. of Control, vol. 4, 1966, p. 265 f.

3. Bertsekas, D. P. and I. B. Rhodes, Sufficiently Informative Functions and the Minimax Feedback Control of Uncertain Dynamic Systems, IEEE Trans. on Automatic Control.

4. Chang, S. S. L. and T. K. C. Peng, Adaptive Guaranteed Cost-Control of Systems with Uncertain Parameters, IEEE Trans. on Automatic Control, vol. AC-17, no. 4, 1972.

5. Speyer, J. L. and U. Shaked, Minimax Design for a Class of Linear Quadratic Problems with Parameter Uncertainty, IEEE Trans. on Automatic Control, vol. AC-19, no. 2, 1974.

6. Menga, G. and P. Dorato, Observer-feedback Design for Linear Systems with Large Parameter Uncertainty, Proc. IEEE Conf. on Decision and Control, Phoenix, 1974, pp. 872 f.

7. Davison, E. J., The Output Control of Linear Time Invariant Multivariable Systems with Unmeasurable Arbitrary Disturbances, IEEE Trans. on Automatic Control, vol AC-17, no. 5, 1972.

8. Gutman, S. and G. Leitmann, Stabilizing Control for Linear Systems with Bounded Parameter and Input Uncertainty, Proc. 7th IFIP Conf. on Optimization Techniques, Nice, Springer Verlag, 1975.

9. Taran, V. A., Improving the Dynamic Properties of Automatic Control Systems by Means of Nonlinear Corrections and Variable Structure, Automation and Remote Control, vol. 25, no. 1, 1964.

10. Buyakas, V. I., Optimal Control by Systems with Variable Structure, Automation and Remote Control, vol. 25, no. 4, 1966.

11. Utkin, V. I., On the Compensation of the Response to Disturbances, Technical Cybernetics, vol. 1, 1965, pp. 155 f.

12. Drazenovic, B., The Invariance Conditions in Variable Structure Systems, Automatica, vol. 5, 1969, p. 287 f.

13. Blaquière, A., F. Gérard, and G. Leitmann, Quantitative and Qualitative Games, Academic Press, N. Y., 1966.

14. Isaacs, R., Differential Games, Wiley and Sons, N. Y., 1965.

15. Stalford, H. and G. Leitmann, Sufficient Conditions for Optimality in Two-Person Zero-Sum Differential Games with State and Strategy Constraints, J. of Math. Anal. and Appl., vol. 33, no. 3, 1971.

16. Kalman, R. E. and J. E. Bertram, Control System Analysis and Design via the Second Method of Lyapunov, J. of Basic Engineering, vol. 82, no. 2, 1960.

17. Filippov, A. G., Application of the Theory of Differential Equations with Discontinuous Right-hand Sides to Nonlinear Problems in Automatic Control, Proc. First IFAC Cong., 1960, p. 923 f.

18. André, J. and P. Seibert, Über stückweise lineare Differentialgleichungen die bei Regelungsproblemen auftreten, I II, Arch. Math., vol. 7, 1956, pp. 148 f and 157 f.

19. André, J. and P. Seibert, After Endpoint Motions of General Discontinuous Control Systems and Their Stability Properties, Proc. First IFAC Cong., 1960, p. 919 f.

20. Alimov, Y. I., On the Application of Lyapunov's Direct Method to Differential Equations with Ambiguous Right Sides, Automation and Remote Control, vol. 22, no. 7, 1961.

21. Roxin, E., On Generalized Dynamical Systems Defined by a Contingent Equation, J. of Differential Equations, vol. 1, 1965, p. 188 f.

22. Roxin, E., On Asymptotic Stability in Control Systems, Rend. Circ. Mat. di Palermo, Serie II, Tomo XV, 1966.

23. Gutman, S., Differential Games and Asymptotic Behavior of Linear Dynamical Systems in the Presence of Bounded Uncertainty, Ph.D. dissertation, University of California, Berkeley, 1975.

24. Gutman, S., Uncertain Dynamical Systems - A Differential Game Approach, NASA TMX-73, 135, April 1976.

25. Gutman, S. and G. Leitmann, On a Class of Linear Differential Games, J. of Optimization Theory and Appl., vol. 17, nos. 5/6, 1975.

STOCHASTIC CONTROL PROBLEMS IN FISHERY MANAGEMENT

William J. Palm

University of Rhode Island
Kingston, Rhode Island

§1. INTRODUCTION

The need for prudent management of our natural resources is obvious and it is of special importance to our fisheries, some of which are endangered by overfishing. These fish resources are very important to the U. S. food supply and economy, as well as to those of other nations. As a result of the on-going United Nations Law of the Sea Conference, it seems likely that the U. S. will be responsible for managing the fish resources within a zone extending 200 miles out from our coasts. If such an international agreement fails to materialize, Congress has made provisions to establish such a zone unilaterally. In any event, it will be necessary for the federal government or an international agency to have an analytically-based management scheme for regulating the amount of fishing applied to each species in this 200 mile zone. Regulatory mechanisms include catch quotas and limits on fishing effort (number and size of boats permitted entry into the fishery, number of hooks or other gear used, etc.). These decisions on quotas and limits are more easily and objectively made if there are available optimization methods and models of the fishery which include biological and economic factors.

The fishery management problem can be viewed as a problem of optimization, estimation and control by considering the catch quotas or fishing effort limits to be the control variables of the system. Thus, the methods of system theory developed for engineering design also can have direct application to this resource management problem. Population fluctuations due to environmental influences and to intrinsic growth processes require that the control variables (quotas or limits) be adjusted in a feedback manner. That is, a fishery is a dynamic system which requires feedback control for optimal regulation. Estimation techniques are required for developing a model of the system, for parameter identification, and for estimating the states of the system for feedback purposes. Optimization is necessary to determine the feedback control function which allows maximum utilization of the resource without jeopardizing its future use.

A dynamic model of the fish population, the effect of fishing, and related economic aspects is needed for the control analysis. There is much misunderstanding of these model requirements among people who are not biologists. The data availability in fisheries science is too limited generally to support complex models which include physiological factors, or competition and predation between species. Thus, it becomes important to choose a control model structure which can be supported by available data. Usually data is available on the yearly catch and the amount of fishing effort (such as a number of boat-days) expended to obtain that catch. A model which can be supported by catch and effort data is reviewed below.

Because of these restrictions, a relatively simple model must be used. Thus, model coefficients which are treated as constants may actually be variables in themselves, with a stochastic component. These effects can be modeled as "noise" and appropriate stochastic control methods can be applied. Also, the simple model structure usually prevents an extremely accurate fit to the data, and thus there can be a large amount of parameter uncertainty. This also must be taken into consideration in any control analysis.

The purpose of this paper is to discuss the parameter estimation problem and its effect on the control analysis. A review of current methods is given. An attempt was made with modern estimation methods (the extended Kalman filter), but was not successful. These results are discussed using the Pacific Yellow-fin Tuna fishery as an example. Finally, formulation and analysis of the control problem is presented with emphasis on reduction of sensitivity to parameter uncertainties.

§2. THE FISHERY MODEL

Here we summarize the differential equation model of fish population dynamics which was used in the study. Equivalent difference equation models can be developed by using appropriate finite difference approximations to the time derivative. Since no seasonal effects are included, the unit of time for this model is one year. The principal advantage of this single-species model is that it requires only data on catch and fishing "effort" for parameter estimation. That is, physiogical data, which is difficult to obtain, is not required. Of course, the models' predictive capability is correspondingly limited to prediction of yearly catch, given fishing effort as the input.

The general model form is based on conservation of mass, and is the following:

$$\frac{dP}{dt} = G - F \tag{1}$$

where:

P = size of the "fishable stock" (that portion of the population subject to capture) in terms of biomass or numbers of individuals;

t = time;

G = growth rate (includes natural mortality, growth due to increase in organism size, and growth due to entry of new fish into the fishable stock);

F = fishing mortality rate.

A large variety of model forms result when functional forms for G and F are chosen. The most commonly used form is the "Schaefer" model, in which

$$G = (a - bP)P \tag{2}$$

There are only two common forms proposed for the fishing mortality rate, F. The first simply treats F as a given constant, and has been used in several control analyses. However, this is generally not accepted because fishing success depends on the size P of the fishable stock, the amount of effort expended in fishing, and the efficiency of the fishing method. These three factors are included in the following commonly used form:

$$F = qfP \quad (3)$$

where f is the "fishing effort," which may be expressed as the number of standard units of gear employed (e.g., number of hooks or lobster pots), or as the amount of time spent fishing (e.g., boat-days). The constant of proportionality, q, is the "coefficient of catchability," and is a measure of fishing efficiency.

Other, more complex fishery models can be proposed, but one must constantly keep in mind the general lack of data necessary to support such models. In contrast to the situation for land-based agriculture, there exists a substantial scarcity of data on the sea and its inhabitants. As an example, for many if not all fishing grounds, there does not exist an unbroken series of temperature measurements. Also, the growth, feeding and behavioral characteristics of commercial fish in their natural environment is not well understood. Thus, there is little basis on which to build models incorporating fish physiology, and almost no basis for building predator-prey type models.

The model used in this study is given by (1), with G and F given by (2) and (3). Thus

$$\frac{dP}{dt} = (a - bP)P - qfP \quad (4)$$

If the fishing effort f is held constant at the value f_o, the corresponding equilibrium value of P, denoted P_o, is found from (1) by setting the derivative equal to zero. The "equilibrium yield" or catch in one time period (a year) is thus:

$$C_o = qf_oP_o \quad (5)$$

A commonly used criterion for fishery management is the "maximum equilibrium yield" criterion. The value of f_o which maximizes C_o is denoted by f_o^*, and is found from (5) to be:

$$f_o^* = \frac{a}{2q} \tag{6}$$

The corresponding value of P_o is:

$$P_o^* = \frac{a}{2b} \tag{7}$$

These relations will be of use later.

§3. PARAMETER ESTIMATION

The assumed data base for parameter estimation for our models is the catch C and fishing effort f for each year. Note that we are not given any direct information on the biomass P of the catchable stock. Due to lack of data on P, a change of model variables is made. Since it is assumed that catch C is equal to qfP, we define the new variable to be catch per unit effort, N. That is,

$$N = \frac{C}{f} = qP \tag{8}$$

The model is then expressed in terms of N and f, and the data is recomputed to give N and f for each year. This transformation gives a model in terms of variables N and f, for which data is available.

For example, the so-called Schaefer model is:

$$\frac{dP}{dt} = (a - bP)P - qfP \tag{9}$$

Use of the transformation, N = qP, gives:

$$\frac{dN}{dt} = (a - b_1N)N - qfN \tag{10}$$

where $b_1 = b/q$. Equation (10) is used to estimate a, q, and b_1, from which b is recovered.

Current Methods

There have been several methods used by fisheries scientists for parameter estimation. These are reviewed by Pella and Tomlinson (1969). The first method used avoided dealing with differential equations such as (10) by assuming the population makes transitions between a series of steady-states. By identifying enough of these states, enough algebraic equations corresponding to an equilibrium state of equation (10) can be obtained to allow calculation of the unknowns a, b_1, and q. We shall refer to this as the quasi-steady state method (QSS).

If the model is simple enough, such as (10), we can integrate it over a given year, with f a given constant. With this formula, a search method can be employed to find the parameter set which minimizes the sum of the squares of the deviations between predicted and actual catches (the MSSCD method).

Also, equation (10) may be approximated by a finite difference equation. The following approximation has been used up to now:

$$\frac{1}{N}\frac{dN}{dt} = \frac{N_{i+1} - N_{i-1}}{2N_i} \tag{11}$$

where N_i is the value of N in year i. Dividing both sides of (10) by N, and using (11) gives:

$$\frac{N_{i+1} - N_{i-1}}{2N_i} = a - b_1 N_i - qf_i \tag{12}$$

N_i and f_i are known, and the left-hand side of (12) can be computed from the data. Thus (12) is in the form of a linear regression equation:

$$Y_i = c_o + c_1 N_i + c_2 f_i \tag{13}$$

Least-squares regression can then be used to estimate c_o, c_1, and c_2, from which we recover a, b_1, and q. We shall refer to this as the LSR1 method.

Equation (11) is a crude approximation for the relative derivative, and it is proposed here that the following is a better approximation:

$$\frac{1}{N}\frac{dN}{dt} = 2\,\frac{N_{i+1} - N_i}{N_{i+1} + N_i} \tag{14}$$

In a manner similar to that above, we can use (14) to convert the original differential equation into a difference equation to which least-squares regression may be applied. We refer to this as the LSR2 method.

Pella and Tomlinson (1969) compared the first three methods for data on the Pacific Yellow-fin Tuna fishery from 1935 to 1967. I have applied the LSR2 method to the same data. The results are shown in the following table.

Method	a	$b \times 10^4$	$q \times 10^3$	f_o^*	C_o^*
QSS	3.93	213.60	62.9	31.2	180.8
MSSCD	64.5	0.06	990.0	32.6	185.8
LSR1	0.212	0.90	5.4	19.6	122.8
LSR2	1.005	13.27	13.9	36.1	190.0

Here f_o^* is the maximum equilibrium yield value of f, in thousands of boat-days, and C_o^* is the maximum equilibrium yield in millions of pounds.

Clearly, all four methods produce widely disparate estimates for the parameters. However, three of the methods produce values of f_o^* and C_o^* which are surprisingly close. The failure of the LSR1 method to do so seems to cast more doubt on the validity of the approximation (11).

Simulations of the model using the parameter values given in the above table, with the data of f as input, show that the model is capable of following the general trend of the catch data. However, for some years there is a large deviation between predicted and actual catch. The conclusion seems to be that the above methods (QSS, MSSCD, and LSR2) are useful for computing f_o^* and C_o^*, but more parameter estimation accuracy is needed if we are to have confidence in the yearly predictions.

Application of Modern Estimation Techniques

In light of the above results, a brief investigation was made into the applicability of modern estimation techniques. The extended Kalman filter [Gelb, 1974] has been used with state augmentation to estimate the parameters of dynamic models. Since it was suspected that the parameters a, b, and q might not be constant in the real situation, an artificial data set was created by simulating equation (9) for given constant values of a, b, and q, and for a given series of values for f. The extended Kalman filter was then applied to estimate a, b, and q. The measurement equation was the catch equation: $C = qfP$.

The attempt was not successful. At best, only two of the three coefficient estimates could be made to converge to their true values. Phenomena, such as negative error covariances due to small measurement noise, which have been noted by other observers, were seen. The limitation of only one measurement per year seemed to be a factor in the filter's performance. It is possible that better approximations of the model and measurement nonlinearities, such as is used with the iterated, second-order, and statistically linearized Kalman filters [Gelb, 1974], would improve the situation. Also suggested is Invariant Imbedding as a possible method. Finally, the Gaussian, white-noise assumptions employed might not be realistic.

§4. PREVIOUS CONTROL RESULTS

Goh [1969, 1973] and Cliff and Vincent [1973] attacked the problem of maximizing yield directly. Their models differ somewhat but their results can be generally summarized as follows. The yield over the time interval (0, T) is the integral of the rate of catch, qfP, where P is the size of the population subject to capture, f is the fishing effort (our control variable), and q is a proportionality constant. Thus, the function f was found which would maximize the yield over the planning interval (0, T), where

$$\text{Yield} = \int_0^T qfPdt \tag{15}$$

To avoid zero population at time T, let $T \to \infty$. This results in the following singular, bang-bang control law:

$$f = \begin{cases} f_m & \text{if } P > P^* \\ f^* & \text{if } P = P^* \\ 0 & \text{if } P < P^* \end{cases} \tag{16}$$

where f_m is the maximum available fishing effort, and (f^*, P^*) are the values of f and P which maximize yield under equilibrium conditions. Application of the control law (2) will bring P to its optimal equilibrium value P^*.

There are two difficulties with the above formulation. It is socially undesirable to have the fishing effort switching between three distinct levels. The requirement of stopping fishing whenever P drops below P^* seems unnecessarily severe. Also, this singular control formulation with its necessary conditions for optimality is extremely difficult to apply to models which have more complicated nonlinearities. Thus another formulation was needed.

Before proceeding to the new formulation, we mention some other studies. Clark [1973] and Clark et al. [1973] analyze the problem of optimal reduction of effort in an overexploited fishery ($f \gg f^*$). This is a different problem than the one treated here. We assume that the fishery is already managed well enough so as not to require massive changes in fishing effort, with their resultant economic dislocations. Related work which emphasizes the role of economics rather than population dynamics is described by Quirk and Smith [1970], Mann [1970], Booth [1972], and Jacquette [1974]. Other treatments of the control of natural populations have been concerned primarily with pest management. Their results are generally not applicable to the fishery problem because of differing performance objectives and model structures.

§5. NEW CONTROL FORMULATION

The need to have a realistic control formulation, without unnecessary mathematical complexity, resulted in the following approach to the problem. We denote by P^* and f^* the values of P and f which produce maximum benefit according to whatever criterion one wishes to choose. The most common criterion is the maximum equilibrium yield. There has recently been some doubt cast on the wisdom of this choice, and therefore we do not restrict ourselves to it. Our only requirement for the present is that (P^*, f^*) be an equilibrium solution of the fishery model. The values of P^* and f^* would be the result of a static optimization procedure depending on what criterion was chosen. This equilibrium requirement can be relaxed in a manner similar to that used in the tracking problem of linear-quadratic control theory.

The new control formulation was developed as follows. If P, the size of the population subject to capture, is greater than P^*, it seems reasonable to allow the fishing effort f to be equal to the maximum available, f_m (let everyone fish for this species as much as they want). Of course, if $P = P^*$, then set $f = f^*$ to maintain the optimum condition. Finally, if $P < P^*$, we would like to choose f to return P to the optimal value P^*. We would like this return to be as quick as possible but without inducing overshoot and its resultant oscillations. This is reminiscent of the classical regulator design approach, in which the dominant time constant is made as small as possible, while keeping the damping ratio ζ of the dominant root equal to unity. Recall that the generalized step-input response of a second-order system overshoots and oscillates about the final value when $\zeta < 1$; and sluggishly approaches the final value exponentially when $\zeta > 1$. In this first attempt at a solution, P and f were assumed to be close enough to P* and f* so that a linearized analysis could be used. Thus, the following control formulation is suggested:

a. If $P > P^*$, set $f = f_m$.
b. If $P = P^*$, set $f = f^*$.
c. If $P < P^*$, choose f to minimize the system's dominant time constant and to give a dominant damping ratio of unity.

The above formulation represents an indirect approach to yield maximization [if (P*, f*) represents the maximum equilibrium yield solution], as opposed to the direct maximization approach mentioned earlier. Simulations of this formulation showed that it gives a yield which is very close (98%) to the maximum yield obtained with the direct maximization methods of Goh, Cliff, and Vincent. As management models and problems become more complex, utilization of such indirect approaches to optimization should receive more attention than they have in the past. With the increasing applications of systems theory to economics, there has unfortunately been a tendency toward direct optimization in complex problems where it has proven to be unwieldy. Such brute force methods are limited in the generality of their conclusions, as well as hampered unnecessarily by difficulties with numerical methods.

The above formulation differs from that presented by Palm [1975a, 1975b] in which a performance index was chosen to return P to P*, for any value of P. This criterion leads to the Linear-Quadratic control approach.

§6. CONTROL ANALYSIS

For the control formulation selected for this study, the fishing effort f is chosen to return the catchable stock biomass P to its optimal value P* as quickly as possible but without oscillation, when $P \le P^*$. Mathematically, we achieve this condition by selecting a feedback control law for f which minimizes the system's dominant time constant τ, while keeping the dominant damping ratio ζ equal to unity.

To synthesize a control law, the model (9) is linearized about the desired equilibrium (P*, f*) to obtain:

$$\frac{dx}{dt} = -a_1 x + b_1 u \tag{17}$$

$$x = P - P^* \tag{18}$$

$$u = f - f^* \tag{19}$$

$$a_1 = bP^* \tag{20}$$

$$b_1 = qP^* \tag{21}$$

Following the classical control synthesis procedure, we select a feedback control law and test its performance. An obvious first choice is the proportional control law:

$$u = \frac{-k_1}{b_1} x \tag{22}$$

where k_1 is the proportional control "feedback gain." Substitution of (22) into (17) gives a differential equation with the following characteristic equation:

$$s + a_1 + k_1 = 0 \tag{23}$$

where s is the Laplace transform variable. The system's time constant is:

$$\tau = \frac{1}{a_1 + k_1} \tag{24}$$

To minimize τ, we choose k_1 as large as possible. However, for a time scale of one year, it makes sense to choose k_1 to give a time constant of the order of one year. This puts a realistic limit on the magnitude of k_1. Setting τ equal to unity, and solving (24) for k_1 gives:

$$k_1 = 1 - a_1 \tag{25}$$

This is the nominal value of k_1. However, the true system dynamics will differ from those of the model (17) since there is (at least) estimation error in the model's coefficients. In particular, suppose P* is the maximum equilibrium yield value of P. Then, from (7), a_1 is

$$a_1 = \frac{a}{2} \tag{26}$$

Taking the value of "a" from the LSR2 method, we obtain 0.5 for the nominal value of k_1. From the table of coefficients for the Yellow-fin Tuna, we see that the estimates of the coefficient "a" vary from 0.212 to 64.5. Thus the value of a_1 might be in the interval (0.1, 32.25), and the actual time constant, given by (24) with $k_1 = 0.5$, might lie in the interval (0.03, 1.7) years. Thus, it is possible for the actual system time constant to be greater than the desired value (1.7 years vs 1 year).

Since we desire to return the system to the (f*, P*) equilibrium point as fast as possible, it is of interest to investigate how the control law might be designed so as to reduce its sensitivity to parameter uncertainties. Again following the classical approach, we try a proportional-plus-integral control law:

$$u = -\frac{k_1}{b_1}x - \frac{k_2}{b_1}\int_0^t x(t)dt \tag{27}$$

The characteristic equation for the resulting closed-loop system is:

$$s^2 + (k_1 + a_1)s + k_2 = 0 \tag{28}$$

With the requirement of no oscillations with minimum time constant, we set both roots equal to each other. This gives:

$$k_1 = -a_1 + \frac{2}{\tau} \tag{29}$$

$$k_2 = \frac{1}{\tau^2}$$

Again choosing τ equal to unity with a_1 equal to 0.5, we obtain $k_1 = 1.5$ and $k_2 = 1$, for the nominal values. If the true value of a_1 is 0.1, then these gain values give for the roots of (28):

$$s = -0.8 \pm 0.6i \tag{31}$$

and the dominant time constant is 1.25 years, with a damping ratio of 0.8. Thus the system will display damped oscillations.

This oversimplified example shows that it is possible to design a control scheme to reduce sensitivity to parameter uncertainties. However, something might have to be sacrificed to do this, such as acceptance of oscillatory behavior.

§7. CONCLUSION

At present, simple models of fishery dynamics must be used because the scarcity of appropriate data prevents the development of more realistic and precise models. The effect of the resultant large parameter uncertainties can be reduced somewhat by the use of a proper control law. The emphasis of future applications of systems theory to this area should be in the application of modern estimation and identification techniques in order to reduce these parameter uncertainties.

ACKNOWLEDGMENT

This work was supported by NSF under Grant ENG74-16358 and by a University of Rhode Island Faculty Fellowship.

BIBLIOGRAPHY

1. Booth, D. E., A Model for Optimal Salmon Management, The Fishery Bulletin, 70 (1972), pp. 497-506.

2. Clark, C. W., The Economics of Overexploitation, Science, 181 (1973), pp. 630-634.

3. Clark, C. W., G. Edwards, and M. Friedlander, Beverton-Holt Model of a Commercial Fishery: Optimal Dynamics, J. Fish. Res. Bd. Can., 30 (1973), pp. 1629-1640.

4. Cliff, E. M. and T. L. Vincent, An Optimal Policy for a Fish Harvest, J. Optimization Thy. and Appl., 12 (1973), pp. 485-596.

5. Gelb, A. (ed.), Applied Optimal Estimation, MIT Press, Cambridge, Mass. (1974), 374 p.

6. Goh, B. S., Optimal Control of a Fish Resource, Malayan Scientist, 5 (1969), pp. 65-70.

7. Goh, B. S., Optimal Control of Renewable Resources and Populations, presented at 6th Hawaii Int. Conf. on System Sciences (1973).

8. Jacquette, D. L., A Discrete-Time Population Control Model with Setup Cost, Operations Research, 22 (1974), pp. 298-304.

9. Mann, S. H., Mathematical Theory for the Harvest of Natural Animal Populations When Birth Rates are Dependent on Total Population Size, Math. Biosc., 7 (1970), pp. 97-110.

10. Palm, W. J., An Application of Control Theory to Population Dynamics, in Differential Games and Control Theory, E. O. Roxin, P. T. Liu, and R. L. Sternberg, eds., Marcel Dekker, Inc., New York (1975a).

11. Palm. W. J., Fishery Regulation via Optimal Control Theory, Fishery Bulletin, 73 (1975b), pp. 830-837.

12. Pella, J. J. and P. K. Tomlinson, A Generalized Stock Production Model, Inter-Am. Trop. Tuna Comm., Bull. 13 1969), pp. 419-496.

13. Quirk, J. P. and V. L. Smith, in Economics of Fisheries Management (A. D. Scott, ed.), H. R. Macmillan Lectures in Fisheries, University of British Columbia (1970), pp. 3-32.

FERMAT'S PRINCIPLE IN A STOCHASTIC MEDIUM

Edmond Ghandour

University of Tel-Aviv
Ramat-Aviv, Israel

ABSTRACT

We consider a model medium whose mean-value speed of propagation is perturbed by white-noise. Upon stating an appropriate stochastic Fermat's Principle and employing notions from stochastic optimal control, we obtain the appropriate perturbed Eikonal equation, being a parabolic differential equation for the average traversal time of a wavefront. When the stochastic perturbations are small, the problem reduces to one of singular perturbation of the Hamilton-Jacobi equation. We discuss some questions related to physical scales of interaction common to all random propagation problems. Using a result of Fleming [1] we solve the perturbed Hamilton-Jacobi equation for the approximate averaged traversal time and mean stochastic-path. This also allows us to consider the problem of tracing stochastic rays in an ocean waveguide and examine some aspects associated with the formulation of an approximate effective speed of propagation for the medium in question.

§1. INTRODUCTION

One of the basic problems of geometrical optics which involves the study of the laws of wave propagation in the limiting small wavelength case $\lambda \to 0$, is to determine the *rays*, i.e., trajectories whose tangent at any point is in the same direction as the direction of propagation of sound. For mediums which are characterized deterministically, this classical problem can be treated in either one of two ways:

(i) by examining asymptotic solutions to the wave equation, or

(ii) in the framework of control theory by studying the Pontryagin problem, namely, determine the characteristics associated with the Hamilton-Jacobi equation for the traversal time of propagation.

We shall use the latter approach, generalized appropriately for a random medium. By an inhomogeneous isotropic time-dependent random medium, we shall mean an ensemble of media $M(\omega)$, $\omega \in \Omega$, Ω the probability space, in which a probability measure $P(\omega)$ is defined. For each ω, the properties of the medium are characterized by $v(\underline{x},t;\ \omega)$ or $n(\underline{x},t;\ \omega)$, defining the velocity of propagation and index of refraction, respectively. For fixed point in space and time, $M(\omega)$ is a random medium if v and n are measurable functions of ω. Henceforth, we shall take as our underlying probability space the triplet (Ω,B,P), where B is a σ-algebra of subsets of Ω and P a probability measure on B.

There are various phenomenological questions as to how one is to model a random medium. We shall consider, as our model, a medium whose mean velocity of propagation is perturbed by an additive white-noise term, i.e., we shall characterize the displacements along rays by Ito type equations. We consider this kind of a medium because in general, it is in the white-noise limit that the stochastic effects become most significant. We shall demonstrate this point later on as well as discuss the appropriate physical scales of interaction involved in wave propagation in a random medium. An important aspect in being able to study the effects on propagation of the random perturbations for long time intervals as the measure of the perturbations is decreased is an estimate given by Fleming [1] for the deviations of the perturbed trajectories from the unperturbed geodesic path of propagation, and therefore, also an estimate for the exit time of a distrubance from an appropriate compact domain. The object here is to state an appropriate Fermat's principle for the random medium which would enable us to determine the stochastic rays and hence also the average perturbed path of propagation. The results can then be suitably applied to various types of problems, e.g., propagation in a wave-guide.

We rely in this presentation on previous work by the author [2,3,4], and so, since some of the results have been given there in more detail, when appropriate we shall give here only a brief summary. The work of Fleming [1] is important to us here in that it sets a rigorous mathematical framework for treating the perturbed Hamilton-Jacobi equation, results of which have been used in the calculation.

§2. FERMAT'S PRINCIPLE EXTENDED TO STOCHASTIC MEDIUM

Before discussing the random problem, let us recall the classical deterministic statement. Suppose the medium in which the disturbance propagates occupies the bounded domain G, $G \subset R^3$. In general, we shall consider points on the trajectories of propagation to be in $G_o \subset G$, G_o an open subset, and such that G_o is sufficiently far from the boundary ∂G, thus, the nature of the boundary of G is not that important. Assume that the displacement of a ray at any point $\eta \in G_o$, $\eta(t) = (\eta_1(t), \eta_2(t), \eta_3(t))$, and time t evolves according to the system

$$d\eta(t) = f(\eta,t,u)dt = v(\eta,t)udt \tag{1}$$

where v denotes the velocity of propagation while u is a unit control vector defined by the *spherical coordinates* θ and ϕ, namely,

$$u = (\sin\theta \cos\phi, \sin\theta \sin\phi, \cos\theta) \tag{2}$$

and at an initial time s we have $\eta(s) = x$. We take $(s,x) \in A$, $A = (t_o,t^*) \times G_o$, and assume $u \in V$, V an admissible class of controls. Our system is assumed completely observable, that is, we can write $u = U(t,\eta)$ as in (2), where $U \in V$. f is assumed sufficiently smooth and is restricted in such a way that a unique solution exists to the equation (1). Let the curve γ represent the path of a ray connecting the generic point (s,x) and a fixed point (t_1,x_1), t_1 finite. Define the traversal time taken to go from x at s to x_1 on γ by

$$T^u(s,x) = \int_s^{t_1} dt^u \tag{3}$$

where dt^u is given by (1) and we have indicated the dependence of the propagation time upon the control vector u. The classical Fermat's principle can be stated as follows: γ is the path of a ray of light iff $T^u(s,x) \geq T^{u^*}(s,x)$ for any other curve in a close neighborhood of γ, where u* is the minimizing control. Mathematically, this statement becomes $T(s,x) =$

$T^{u*}(s,x) = \inf_{u \in V} T^{u}(s,x)$, the principle of a path of least time. In A, $T(s,x)$ satisfies the Hamilton-Jacobi equation

$$T_s + H(s,x,T_x) = 0 \tag{4}$$

where

$$H(s,x,T_x) = \min_{u \in V}[1 + T_x' f(x,s,u)] \tag{5}$$

and T satisfies homogeneous end condition on the terminal point. T_x represents the gradient vector ∇T. Equation (4) describes the way the wave-front evolves in time, and this problem can be reduced to a family of initial value problems defining the characteristics. Equation (4) also depicts the *ray property*, namely, the orthogonality of the wave-fronts and the light rays for isotropic media. The optimal control minimizing (5) can be readily shown to be given by

$$\tan\theta^* = \frac{(T_x^2 + T_y^2)^{1/2}}{T_z}, \qquad \tan\phi^* = \frac{T_y}{T_x} \tag{6}$$

thus determining the direction of propagation and hence also the free-space geodesic path of propagation. For a medium in which the velocity of propagation depends only on position, i.e., $v = v(\eta)$, the substitution of (6) into (4) implies the *Eikonal* equation $|\nabla T|^2 = v^{-2}$.

The extension to the random medium is as follows. Perturb the system (1) in such a way that the displacements of a ray in a random medium are characterized by

$$d\eta(t) = f(\eta,t,u)dt + \sqrt{\varepsilon}\,\sigma(\eta,t)dw \tag{7}$$

where f is as in (1) linear in the unit control vector u, ε is a small parameter, and again u is defined by (2). w is a three-dimensionsl Brownian motion process with $E(w_t-w_s)(w_t-w_s)^T = I|t-s|$, $(\ )^T$ denoting the transpose and I the identity matrix. Again, our system has been assumed to be completely observable, $\eta(s) = x$, and σ not-degenerate 3 x 3 matrix from which we define the matrix a with elements $a_{ij} = \sum_{k=1}^{3} \sigma_{ik}\sigma_{jk}$. Now, $u \in Y$, Y an admissible control class, is a Markov control function. Under suitable smoothness of $f(\eta,t,\cdot)$ and $\sigma(\eta,t)$ and a class Y guaranteeing the existence of a unique solution to (7), $\eta(t)$ will be a Markov process and with it will be associated a second order diffusion operator

$$L^{\varepsilon} = \frac{1}{2}\,\varepsilon \sum_{i,j=1}^{3} a_{ij} \frac{\partial^2}{\partial\eta_i\partial\eta_j} + v \sum_{i=1}^{3} u_i \frac{\partial}{\partial\eta_i} = L_2 + L_1 \qquad (8)$$

Now, consider the class Ψ defined as all continuous random trajectories in G_o passing through (s,x) and terminating with probability one not in x_1 but instead in a neighborhood $S \subset G_o$ of x_1, defined as the *target set* S. Since the medium is random, the waves propagating along curves in Ψ will arrive at S in random times. Consequently, we let τ_S denote the random *time of arrival* of a disturbance at S. Naturally, the excitation time of points on S which serve as a source of further propagation of disturbances is random. We define next $T^u(s,x)$ as the *average traversal time* of a disturbance in going from (s,x) along the random trajectories in the class Ψ to S, this for the unit control vector u. Let the curve $\nu \in \Psi$ represent the stochastic ray. Fermat's principle for the random medium can then be stated as follows: the paths in Ψ which minimize the average traversal time and such that $T^u(s,x) > T^{u^*}(s,x)$ for all curves neighboring ν define the stochastic rays.

Mathematically, this is equivalent to

$$T(s,x) = T^{u^*}(s,x) = \inf_{u \in Y} E_{s,x}\left[\int_s^{\tau_S} dt^u\right] \tag{9}$$

where $E_{s,x}$ is the expectation over the trajectories in Ψ, and dt^u is constrained by (7).

Under suitable conditions on f, σ and Y (see [5]), the function $T(s,x)$ satisfies in G_o the equation

$$T_s + L_2 T + H(s,x,T_x) = 0, \quad T(s,x) = 0 \text{ for } x \in \partial S \tag{10}$$

where u^* yields $H = 1 + T_x f = \min$, and L_2 is as in (8). Because σ is <u>not</u> a function of the control, it follows that the minimizing unit direction vector has exactly the same form as in (6), with (T_x, T_y, T_z) now solutions of (10). The task now becomes to reduce problem (10) to one which is mathematically tractable and then determine the stochastic rays, and hence, the perturbations from the rays γ induced by the presence of the random inhomogeneities.

The problem is reduced in complexity if we concentrate on finding the solutions to (10) in regions where the solution to (4) is sufficiently smooth. Following the terminology of Fleming [1], these regions will be called regions of strong regularity $N \subset G_o$. We shall assume N to contain γ, and this will be so provided; (s,x) is a regular point and not conjugate [6]. It now becomes feasible to seek an estimate for the containment probability that the optimal stochastic trajectory will remain in N and exit on ∂S. This and its consequences will be discussed in Section 4. We also note that physically N can be considered as making up the ray-tube. From now on, we shall designate by $(\ )^o$ and $(\ )^\varepsilon$ relevant quantities associated with the deterministic and stochastic problems, respectively. Thus, $\gamma^{o}(s,x)$ and $\gamma^{\varepsilon}(s,x)$ refer to γ and ν, respectively.

§3. PHYSICAL SCALES OF INTERACTION

If instead of an additive-white-noise in (7) we consider the mean velocity of propagation as perturbed by a real three-dimensional stationary random process, then we can associate with the inhomogeneities of the medium a correlation length ℓ. Physically, we can then recognize the existence of the two most significant scales of interaction of waves: (i) physical processes on the scale of ℓ, and (ii) interactions in the random medium on a scale $>> \ell$. In this latter case, if a scattered wave propagates for distances much larger than ℓ, then the interaction of the random wave with the random medium will be statistically independent. It then follows that in order to pursue a correct asymptotic analysis as $\varepsilon \to 0$, we must take account of stochastic effects on time scales $\sim \frac{1}{\varepsilon}$, i.e., after a number of correlation lengths, and this is precisely the white-noise limit. To see this, it is sufficient to consider the following autonomous system:

$$\dot{\eta}(t) = v(\eta) + \sqrt{\varepsilon}\, \rho(t) \tag{11}$$

where ρ is a stationary process with zero mean, and

$$E\{\rho(t)\rho(t+s)\} = R_{\rho}(s), \; e = \int_{-\infty}^{\infty} R_{\rho}(s)ds < \infty \tag{12}$$

If we now let $\tau = \varepsilon t$ and define $y^{\varepsilon}(\tau) = \eta(\frac{\tau}{\varepsilon})$, then (11) becomes

$$\frac{dy^{\varepsilon}(\tau)}{d\tau} = \frac{1}{\varepsilon} v(y^{\varepsilon}(\tau)) + \xi^{\varepsilon}(\tau) \tag{13}$$

where $\xi^{\varepsilon}(\tau) = 1/\sqrt{\varepsilon}\ \rho(\tau/\varepsilon)$. This scaling implies that

$$R_{\xi^{\varepsilon}}(\tau) = E\{\xi^{\varepsilon}(s)\xi^{\varepsilon}(s + \tau)\} = \frac{1}{\varepsilon} R_{\rho}(\frac{\tau}{\varepsilon}),$$

and hence,

$$\lim_{\varepsilon \to 0} R_{\xi^{\varepsilon}}(\tau) = e\delta(\tau)$$

so $\xi^{\varepsilon}(\tau)$ approaches a white-noise. Clearly, the scaling $\hat{\tau} = \sqrt{\varepsilon}t$ will not give us this limit. For Brownian motion in (7) this follows immediately from the similarity property, i.e., $\sqrt{\varepsilon}\ w(\frac{t}{\varepsilon})$ is similar to $w(t)$, where by similarity we mean that both processes induce the same probability space. The importance of the white-noise limit as well as the possibility of using the dynamic programming formalism are what motivated us to model our random medium by the system (7).

§4. SOME CONSEQUENCES AND APPLICATIONS

We return to the question of the containment probability that $\gamma^{\varepsilon}(s,x)$, the ray path in the perturbed problem, will remain in N and exit an ∂S. Define the exit place of a random trajectory by $e_N^{\varepsilon} = (\tau_N, \eta^{\varepsilon}(\tau_N))$, where τ_N is the exit time from $N \subset A$ with $\tau_N < \tau_S$ and $(s,x) \in N'$, N' a bounded set with $\bar{N}' \subset N$, $\bar{N}' = N' \cup \partial N'$. Let Σ denote an open cylindrical set about $\gamma^{o}(s,x)$, $\Sigma \subset N'$, with its boundary in the neighborhood of the terminal point coinciding with ∂S, ∂S on the wavefront. Then, in the context of our problem, Fleming [1] has proven the following:

<u>Theorem</u> There exist positive α (depending on $t^* - t_o$ and a bound for $|f_x|$), β and ε^* such that

$$1 - \alpha \sqrt{\varepsilon}\ e^{-\beta/\varepsilon} \leq \Pr\{\gamma^{\varepsilon}(s,x) \in \Sigma\} \leq 1 \tag{14}$$

with $0 < \varepsilon \leq \varepsilon^*$ and $(s,x) \in \Sigma$. Clearly, by this is implied that $\Pr(e_N^{\varepsilon} \in \partial N - \partial S) \leq \alpha\sqrt{\varepsilon}\ \ e^{-\beta/\varepsilon}$. Ventsel and Freidlin [7] have also proven a related result for the deviation of η^{ε} from η^{o}, the main term in the probability having the form,

$$\exp - \frac{1}{2\varepsilon} I(\eta^{o})$$

where $I(\eta^{o})$ is a nonnegative functional. Qualitatively similar results for a system of the type (11) was also shown to be valid by Freidlin [8].

Turning now to the physical consequences of the above theorem, we can consider N as making up the ray-tube of the deterministic problem. Clearly Σ will be a smaller ray-tube which we can conceive as being formed by all rays which begin at time s from an element dA of a wave-front and at time t_1 intersect another wave-front element, which for convenience we take as ∂S. We take $x \in dA$ and $x_1 \in \partial S$. Let γ^{o} be the deterministic ray joining x and x_1, and let $y(t) \in \Psi'$, Ψ' a subset of Ψ all of whose trajectories do not exit Σ but an ∂S. Let $z(t) \in \Psi''$, $\Psi'' \subset \Psi$, be trajectories which exit Σ before reaching ∂S (see Figure 1). Then, on an appropriate propagation

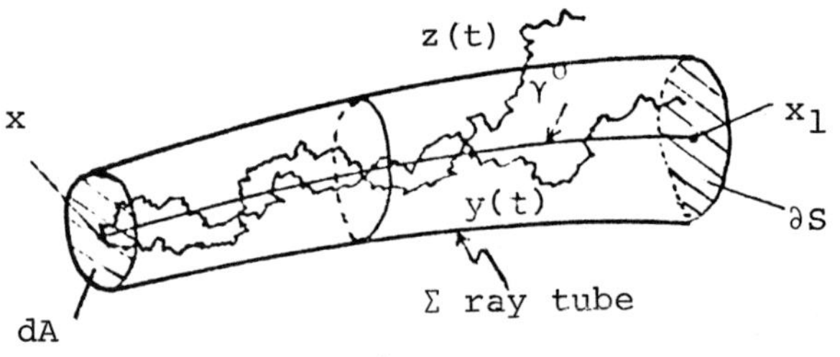

Figure 1

time scale, $y(t)$ will reach ∂S with the probability in (14). Obviously, the $y(t)$ which minimize $T^u(s,x)$ make up the stochastic rays.

Fleming [1] has also obtained results about the uniform convergence of the perturbed control vector and the solution of (10) as $\varepsilon \to 0$ for $(s,x) \in N'$. Of interest to us is the validity of the expression

$$T^\varepsilon = T^o + \varepsilon\mu + o(\varepsilon) \tag{15}$$

which enables us to cast the perturbed problem into an Eikonal form. To simplify the ensuing discussion, from now on we shall consider (7) in its autonomous form, and take σ to be the identity matrix. All the results discussed above can be modified appropriately for this case with the estimate (14) valid, now in a subdomain of an autonomous region of strong regularity. Problem (10) now becomes of the elliptic type

$$\frac{1}{2}\,\varepsilon\Delta_x T^\varepsilon + H(x,T_x^\varepsilon) = 0, \quad T^\varepsilon(x) = 0 \text{ on } \partial S \tag{16}$$

where $H(x,T_x^\varepsilon) = 1 + v(x)\left(|T_x^\varepsilon|\right)$ and we may take $t_o = 0$. Expression (6) for the optimal control is still valid in this case. We remark upon one additional point that one may calculate an estimate of the type (15), i.e., $P_x(T^\varepsilon - T^o \le \delta)$, by solving the initial-boundary value problem in a suitable subdomain $\Sigma \subset G_o$, namely,

$$\frac{\partial q^\varepsilon}{\partial t} = L^\varepsilon q^\varepsilon, \qquad x \in \Sigma,\ t > 0$$

$$q^\varepsilon(x,0) = 0 \text{ for } x \in \Sigma$$

and

$$q^{\varepsilon}(x,t) = 1 \qquad \text{for } x \in \partial S, t > 0 \tag{17}$$

where L^{ε} is as in (8) with $u = u^*$, thus a coupled problem to the equation for T^{ε}. The unique solution to (17) is $q^{\varepsilon}(x,t) = P_x(\tau^{\varepsilon} \leq t)$, where τ^{ε} is the exit time of the optimal path $\eta^{\varepsilon}(t)$ from Σ through ∂S.

We now state some results that can be obtained by solving (16) through the use of (15) and the appropriate characteristic equations [2-4].

A. $$\mu = -\frac{1}{2}\int_{\gamma^o} \frac{\Delta_x T^o(x^o(s))}{v(x^o(s))}\, ds = -\frac{1}{2}\int_0^{T^o} \Delta_x T^o(\eta^o(t)dt \tag{18}$$

where $x^o(s)$ is $\gamma^o(x)$ and $T^o(x) = \int_{\gamma^o} ds/v(x^o(s))$.

B. Equation (16) can be recast into an Eikonal form by writing

$$|T_x^{\varepsilon}|^2 = g^{-2}(x), \qquad T^{\varepsilon}(x) = 0, \qquad x \in \partial S \tag{19}$$

where g is an approximate effective speed of propagation given by

$$g^{-2} = v^{-2}\left(1 + \frac{1}{2}\,\varepsilon\Delta_x T^o + \frac{1}{2}\,\varepsilon^2\Delta_x\mu\right)^2$$

with T^o and μ as in (18). The error introduced into the solution T^{ε} from (19) is $0(\varepsilon^2)$, since

$$\left|g^{-2} - v^{-2}\left(1 + \frac{1}{2}\,\varepsilon\Delta_x T^{\varepsilon}\right)^2\right| = o(\varepsilon^3).$$

From (19), through the reduction to the characteristic system, we can obtain the ray equation, and therefore, also the approximate perturbed path of propagation $\gamma^{\varepsilon}(x)$. In addition, it can be shown that the normals ℓ^o and ℓ^{ε} to the wave fronts T^o = constant and T^{ε} = constant, respectively, have approximate inclinations given by

$$\ell^{o} \cdot \ell^{\varepsilon} = \frac{[1 + \varepsilon v \ell^{o} \cdot \nabla \mu]}{[1 + \frac{1}{2} \varepsilon \Delta_{x} T^{o}]} + o(\varepsilon) \tag{20}$$

thus indicating that the foregoing results are valid for small perturbations.

With the effective speed of propagation g we can associate an equivalent medium, in the sense of inducing the same trajectories of propagation, i.e., same extremals. In fact, the stochastic rays induced by g are the extremals of the minimum problem $T = \min \int_{o}^{s} g^{-1} ds$, where the minimum is over all curves joining x and x_1. Here we also get the equivalence of the optimal policies. In general, the effect of the randomness is to decrease the local group velocity.

C. For a two-dimensional stratified medium for which $v = v(y)$ only, we derive a modified Snell's law for the random medium considered, in particular, for g as in part B, we have that

$$\frac{\cos \phi^{\varepsilon}}{\cos \phi_{o}^{\varepsilon}} = \frac{g}{g\big|_{o}}, \quad g = v(y) \left[1 + \frac{1}{2} \varepsilon \frac{\partial^{2} T^{o}}{\partial y^{2}} + \frac{1}{2} \varepsilon^{2} \frac{\partial^{2} \mu}{\partial y^{2}} \right]^{-1} \tag{21}$$

where ϕ^{ε} is the angle the stochastic ray makes with the x-axis and ϕ_{o}^{ε} refers to an initial angle of the ray. For a wave-guide problem with a source at the origin for which v varies only in a layer $|y| < y^*$, where $v < v^*$ and $v = v^*$ outside this layer, with $\min v = v_o$ at $y = 0$ and $v'(0) = 0$, $v''(0) \neq 0$, and v increasing to v^* smoothly with higher derivatives vanishing on $|y| = y^*$, we can show that the stochastic rays will penetrate the v^* region and continue in straight paths if $\phi_{o}^{\varepsilon} > \cos^{-1} (v_o/v^* \lambda_o)$, while if $\phi_{o}^{\varepsilon} < \cos^{-1}(v_o/v^* \lambda_o)$, then the stochastic rays will remain in the channel,

$$\lambda_o = \left(1 + \frac{1}{2}\varepsilon^2 \frac{\partial^2\mu}{\partial y^2}\right)\Bigg|_{y=0}$$

Inside the channel the stochastic rays execute a closely approximated oscillatory motion about the x-axis with angle

$$\phi^\varepsilon = \cos^{-1} \frac{v(y)}{v_o} \lambda_o\left(1 + \frac{1}{2}\varepsilon \frac{\partial^2 T^o}{\partial y^2} + \frac{1}{2}\varepsilon^2 \frac{\partial^2\mu}{\partial y^2}\right)\cos \phi_o^\varepsilon \tag{22}$$

In the limiting case for which $\varepsilon \to 0$, these results approach the deterministic solution calculated by Blum and Cohen [9], and qualitatively agree with experimental investigations.

REFERENCES

1. Fleming, W. H., Stochastic Control for Small Noise Intensities, SIAM J. Control, vol. 9, 1971.
2. Ghandour, E., Propagation of Disturbances in a Random Medium, J. Inst. Math. Applic., to appear.
3. Ghandour, E., Stochastic Rays in a Waveguide, J. Optical Society of America, vol. 66, no. 9, 1976.
4. Ghandour, E., On an Approximate Effective Speed of Propagation, J. Franklin Institute, to appear.
5. Kushner, H. J., Stochastic Stability and Control, Academic Press, New York, 1967.
6. Fleming, W. H., The Cauchy Problem for a Nonlinear First Order PDE, J. Diff. Eqn., vol. 5, 1969.
7. Ventsel, A. D. and M. I. Freidlin, On Small Random Perturbations of Dynamical System, Russian Math. Surveys, vol. 25, 1970.
8. Freidlin, M. I., Probabilities of Large Deviations for Randomly Perturbed Dynamic Systems and Stochastic

Stability, Theory of Prob. Applic., vol. 18, 1973.

9. Blum, J. W. and D. S. Cohen, Acoustic Wave Propagation in an Underwater Sound Channel, J. Inst. Math. Applic., vol. 8, 1971.

INFORMATION STRUCTURES IN DIFFERENTIAL GAMES

Geert Jan Olsder

Twente University of Technology
Enschede, The Netherlands

§1. INTRODUCTION

In order to state a nonzero-sum differential game, the following ingredients are necessary: a model or system, cost functions for each player, a solution concept which in this paper is the Nash-solution concept, and an information structure (i.s.). Loosely speaking, an i.s. states in which way the time dependent controls are allowed to depend on each other, the current state and/or past states of the system. By means of a simple example (a two-stage differential game) 20 different i.s. are given, among which are the well known open-loop, closed-loop, and Stackelberg structures.

It is not by accident that the Nash-equilibrium concept has been chosen in order to show some features of several i.s. For the Pareto solution concept [2], it can be shown that the differential game is equivalent to an optimal control problem in which case the concept of i.s. does not make much sense. Although with respect to deterministic optimal control problems different i.s. exist, they all give the same answers. (With respect to stochastic optimal control problems, it is well known that there is a difference in answers, depending on whether one considers open-loop or closed-loop solutions.)

For the min-max solution concept the numerical values of the cost functions are independent of the i.s. used.

The differential games considered are of the linear-quadratic type. In some sections the linear model will be deterministic; in others there will be additional noise. The observations are assumed to be noiseless; i.e., if a player measures the state vector, this measurement will be exact.

It is assumed that the i.s. is fixed before the game starts and is known to the players. The i.s. states in what order the decisions have to be made and on what (available) information a decision may depend. For a deterministic game, the result is already determined before the actual game starts. Therefore, an i.s. can be considered as a projection of the time-axis on the initial time; it states in which order the projections of several decisions and state variables take place.

In which way does a structure arise? It can for instance be dictated by some governmental law. Another possibility is that the costs of constructing a device, for instance, for measuring the state of the system at each instant of time are too high and, therefore, one sticks to open-loop. To enact a law by the government can be interpreted as a differential game in itself, with the government and the people as players. The government announces its decision (i.e., the law) and the people try to interpret this as much as possible in their own way. This particular i.s., where one player has to announce his strategy before the other player makes his decision, is called the Stackelberg structure.

A basic reference to this paper is [1], where a start has

been made in considering various i.s. In [1] three i.s. have been considered with respect to a two-person game; one in which both players play open-loop, another in which both play closed-loop, and the third (open/closed-loop) in which one player plays open- and the other closed-loop. The name attached to this latter case, however, is confusing as will be shown in this paper.*

In Section 2 the statement of the general problem is given and for five i.s. the corresponding solution has been given in Section 4. In Section 3 eleven different i.s. are shown by means of a simple example. In Section 5 some remarks about the general M-player game are made. In Sections 6 and 7, two particular i.s. have been dealt with. In Section 6 one of the players has a time-delay in his observations, and in Section 7 one of the players has to announce his strategy $k(> 0)$ steps ahead of time.

§2. STATEMENT OF THE GENERAL PROBLEM

We are given a system of which the evolution of the state x over the index set $\theta \triangleq \{0, 1, \cdots, N-1\}$ is governed by

$$x(i+1) = A(i)x(i) + B_1(i)u_1(i) + B_2(i)u_2(i)$$

$$x(0) = x_0 \tag{2.1}$$

where x, u_1, and u_2 are vectors with n, r_1, and r_2 components, respectively, and where A, B_1, and B_2 are matrices of appropriate sizes. The vectors $u_j(i)$, $i \in \theta$, are chosen by Player P_j, $j = 1, 2$, in such a way as to minimize the cost function $J_j(N,0)$, where

$$J_j(N,m) = \sum_{i=m}^{N-1} \Big[x^T(i+1)Q_j(i+1)x(i+1) + u_1^T(i)R_{j1}(i)u_1(i)$$

*See footnote in subsection 3.3.

$$+ u_2^T(i)R_{j2}(i)u_2(i)] \tag{2.2}$$

The square matrices Q_i and R_{ij} have appropriate sizes and satisfy $R_{11}(j)$, $R_{22}(j) > 0$, $Q_i(j) \geq 0$, $R_{k\ell}(j) \geq 0$ with $k \neq \ell$, $\forall\ j \in \theta$. The symbol T denotes transpose.

For the determination of $u_1(i)$ player P_1 has access to some information about the value of the state vector $x(i)$ and/or some past values of the state vector, i.e., $x(j)$, $j = 0, \cdots, i-1$. Besides P_1 may have access to the functional form of the controls $u_2(j)$ if P_2 announces these functional forms beforehand. It is assumed that this information is exact; there is no measurement noise. The information of P_1 at stage i will be denoted by η_1^i. The information η_2^i is defined similarly. Each η_j^i will generate an information space Z_j^i. At stage i, player P_j will pick a control law $\gamma_j^i \in \Gamma_j^i$, where Γ_j^i is the space of all admissible control laws for player j. The class of all Γ_j^i can, for instance, be the space of all Borel-measurable functions mapping Z_j^i into R^{r_i}, the r_i-dimensional Euclidean space. If player P_1 chooses the control law sequence $u_1(i) = \gamma_1^i \in \Gamma_1^i$, $i = 0, \cdots, N-1$, and if P_2 chooses $u_2(i) = \gamma_2^i \in \Gamma_2^i$, then the value of the cost function will be denoted by $\tilde{J}_j(\{\gamma_1\},\{\gamma_2\}) \triangleq J_j(M,0)$, where $\gamma_j \triangleq \{\gamma_j^0, \cdots, \gamma_j^{N-1}\}$.

With these definitions, a set $(\{\gamma_1^*\}, \{\gamma_2^*\})$ is said to constitute a Nash equilibrium solution for the two-person differential game described above if

$$\tilde{J}_1(\{\gamma_1^*\},\{\gamma_2^*\}) \leq \tilde{J}_1(\{\gamma_1\},\{\gamma_2^*\}) \tag{2.3}$$

$$\tilde{J}_2(\{\gamma_1^*\},\{\gamma_2^*\}) \leq \tilde{J}_2(\{\gamma_1^*\},\{\gamma_2\}) \tag{2.4}$$

for all γ_1 with components $\gamma_1^i \in \Gamma_1^i$ and for all γ_2 with components $\gamma_2^i \in \Gamma_2^i$. This Nash equilibrium solution will be referred to as a global Nash solution hereafter in order to distinguish it from another Nash solution to be defined now.

Denote the control sequence $\{\gamma_j\}$ with the kth component missing by $\{\delta_j^k\}$, i.e., $\{\delta_j^k\} = \{\gamma_j^0, \gamma_j^1, \cdots, \gamma_j^{k-1}, \gamma_j^{k+1}, \cdots, \gamma_j^{N-1}\}$. The set $(\{\gamma_1^*\}, \{\gamma_2^*\})$ is said to constitute a stagewise Nash solution if for all $\gamma_j^k \in \Gamma_j^k$, $j = 1, 2$, $k = 0, \cdots, N-1$, the following inequalities hold:

$$J_1(\{\gamma_1^*\},\{\gamma_2^*\}) \le J_1(\{(\delta_1^k)^*\}\ \gamma_1^k,\{\gamma_2^*\})$$

$$J_2(\{\gamma_1^*\},\{\gamma_2^*\}) \le J_2(\{\gamma_1^*\},\{(\delta_2^k)^*\},\ \gamma_2^k)$$

It is easily seen that the stagewise Nash solution concept is weaker than the global Nash solution concept. Quite often the stagewise Nash solution concept is extremely tractable from a computational point of view. If, moreover, the stagewise Nash solution is unique and it is known that a global Nash solution exists [6], then the stagewise Nash-solution equals the global Nash solution.

§3. DIFFERENT INFORMATION STRUCTURES AND AND ELUCIDATION BY MEANS OF AN EXAMPLE

In this section we will elucidate the influence of different information structures on the Nash-solution of the following two-person two-stage nonzero-sum game:

$$x(i+1) = x(i) + u_1(i) + u_2(i),\ i = 0, 1, x(0) = x_0 \quad (3.1)$$

$$J_1(2,0) = 2x^2(1) + 2x^2(2) + u_1^2(0) + u_1^2(1) \quad (3.2)$$

$$J_2(2,0) = x^2(1) + x^2(2) + u_2^2(0) + u_2^2(1) \quad (3.3)$$

In this section x, u_1, and u_2 are scalars. In each of the following subsections a particular i.s. will be dealt with.

The solutions to be presented will always be the stagewise Nash solutions. It turns out that for all i.s. presented for which the solution exists, it is globally optimal as well. This will not be mentioned explicitly each time. The idea of the proof is that of keeping fixed the strategy found for one player. Then the other player faces a standard optimal control problem of which the solution has been found, and of which the global optimality must be proved.

3.1 Open-Loop Information Structure

The information η_j^i equals the initial condition x_0 for both players and both stages. Hence, by definition $u_j(i) = \gamma_j^i(x_0)$; $j = 1, 2, i \in \theta$. In order to obtain the Nash solution of the problem described by equations (3.1-3.3), equation (3.1) is substituted into (3.2 and (3.3) in such a way that $J_1(2,0)$ and $J_2(2,0)$ become functions of x_0, $u_1(i)$ and $u_2(i)$ only:

$$J_1 \triangleq 2\{x_0 + u_1(0) + u_2(0)\}^2 + 2\{x_0 + u_1(0) + u_2(0) + u_1(1) + u_2(1)\}^2 + u_1^2(0) + u_1^2(1) \tag{3.4}$$

$$J_2 \triangleq \{x_0 + u_1(0) + u_2(0)\}^2 + \{x_0 + u_1(0) + u_2(0) + u_1(1) + u_2(1)\}^2 + u_2^2(0) + u_2^2(1) \tag{3.5}$$

Differentiation of (3.4) and (3.5) with respect to $u_1(i)$, $i = 0, 1$ and $u_2(i)$, $i = 0, 1$, respectively, and equating the results equal to zero yields four equations with four unknowns $u_1(i)$, $u_2(i)$ of which the solution is:

$$u_1(0) = -\frac{10}{19}x_0; \quad u_1(1) = -\frac{2}{19}x_0;$$

$$u_2(0) = -\frac{5}{19}x_0; \quad u_2(1) = -\frac{1}{19}x_0 \tag{3.6}$$

3.2 Closed-loop Information Structure

The information η_j^i now equals the state vector x(i) for both players, i.e., $u_j(i) = \gamma_j^i(x(i))$, j = 1, 2, i = 0,1. Properly speaking, this i.s. should be called closed-loop no memory, in order to distinguish this structure from the kind in which $u_j(0) = \gamma_j^0(x_0)$, $u_j(1) = \gamma_j^1(x_0, x(1))$.

In [1] it has been shown that for the general game, as described in Section 2, the closed-loop no-memory solution is unique provided it exists, and that the closed-loop solution with memory is not unique. The corresponding trajectories in state space are not unique either. If noise is added to the state equations, the closed-loop solution with memory becomes unique again.

The solution (closed-loop, no memory) is obtained by applying the dynamic programming principle. At stage 1 the system is at state x(1) and at that stage $u_j(1) = \gamma_j^1(x(1))$, j = 1,2, must be applied. The choice of $u_1(1)$ and $u_2(1)$ will only influence the following part of the cost functions:

$$J_1(2,1) = 2x^2(2) + u_1^2(1) = 2(x(1) + u_1(1) + u_2(1))^2 + u_1^2(1)$$

$$J_2(2,1) = x^2(2) + u_2^2(1) = (x(1) + u_1(1) + u_2(1))^2 + u_2^2(1)$$

Minimization of $J_1(2,1)$ and $J_2(2,1)$ with respect to $u_1(1)$

and $u_2(1)$, respectively, gives two equations of which the solution is

$$u_1(1) = -\frac{1}{2}x(1),\ u_2(1) = -\frac{1}{4}x(1) \tag{3.7}$$

In order to obtain $u_1(0)$ and $u_2(0)$, equations (3.7) will be substituted into $J_i(2,0)$. Minimization of these $J_i(2,0)$ with respect to $u_i(0)$, $i = 1,2$, yields

$$u_1(0) = -\frac{19}{36}x_0;\quad u_2(0) = -\frac{1}{4}x_0 \tag{3.8}$$

Equations (3.7) and (3.8) constitute the Nash solution of this differential game with the closed-loop i.s. There was no reason to believe that the outcomes of the games for open-loop and closed-loop structure should be equal, because the equilibrium conditions (and consequently, the state vectors which satisfy them) are not the same in open-loop and closed-loop problems, as is for instance clearly explained in [2].

Remark Sometimes the statement is made that every closed-loop solution is an open-loop solution. This statement has to be interpreted with care. If one adopts the definitions of open- and closed-loop as given in this paper, this statement is not true as is easily shown by comparing the answers in subsections 3.1 and 3.2.

The solution

$$u_1(0) = -\frac{10}{19}x_0;\quad u_2(0) = -\frac{5}{19}x_0;$$

$$u_1(1) = -\frac{1}{2}x(1);\quad u_2(1) = -\frac{1}{4}x(1) \tag{3.9}$$

which has a closed-loop appearance, gives rise to the same path in state space as the open-loop solution (3.6) does.

In the same way, the solution

$$u_1(0) = -\frac{19}{36} x_0; \quad u_2(0) = -\frac{1}{4} x_0;$$

$$u_1(1) = -\frac{1}{9} x_0; \quad u_2(1) = -\frac{1}{18} x_0 \tag{3.10}$$

which has an open-loop appearance, generates the same path in state space as the solution (3.7, 3.8) does. In order to distinguish all these solutions one could call solution (3.6) the open-loop, closed-eye solution; solution (3.9) could be called open-loop, open-eye; solution (3.7) and (3.8) could be called closed-loop, open-eye and ultimately, solution (3.10) could be called closed-loop, closed-eye.

3.3 Open- versus Closed-loop Information Structure

In subsections 3.3-3.8 the roles of the players can be interchanged which will give rise to different numerical solutions. These duplicate problems will not be mentioned further.

In this subsection it is assumed that P_1 plays open-loop and P_2 closed-loop. The functional form of the controls is $u_1(i) = \gamma_1^i(x_0)$; $i \in \theta$ and $u_2(0) = \gamma_2^0(x_0)$, $u_2(1) = \gamma_2^1(x_0, x(1))$. Note that the control $u_2(1)$ is not memoryless. A control of the form $u_2(1) = \gamma_2^1(x(1))$ does not make sense as will be seen soon.

Suppose for the time being that $\gamma_1^i(x_0)$, $i = 1, 2$, are known, then P_2 faces the closed-loop optimal control problem; minimize formula (3.3) subject to $x(i+1) = x(i) + \gamma_1^i(x_0) + u_2(i)$, $i = 0, 1$; $x(0) = x_0$. The solution is:

$$u_2(1) = -\frac{1}{2} (x(1) + \gamma_1^1(x_0)) \tag{3.11}$$

$$u_2(0) = -\frac{1}{5}(3x(0) + 3\gamma_1^0(x_0) + \gamma_1^1(x_0)). \qquad (3.12)$$

On the other hand, P_1 assumes that $\gamma_2^0(x_0)$ and $\gamma_2^1(x_0, x(1))$ are given and then P_1 faces the following open-loop problem: minimize formula (3.2) subject to $x(i+1) = x(i) + u_1(i) + \gamma_2^i$, $i = 0, 1$; $x(0) = x_0$. The two state equations are substituted into formula (3.2) and then P_1 faces the unconstrained minimization of

$$2\{x_0 + u_1(0) + \gamma_2^0(x_0)\}^2 + 2\{x_0 + u_1(0) + \gamma_2^0(x_0)$$
$$+ u_1(1) + \gamma_2^1(x_0, x_0 + u_1(0) + \gamma_2^0(x_0))\}^2$$
$$+ u_1^2(0) + u_1^2(1)$$

Differentiation with respect to $u_1(0)$ and $u_1(1)$ gives:

$$2\{x_0 + u_1(0) + \gamma_2^0(x_0)\} + \{x_0 + u_1(0) + \gamma_2^0(x_0) + u_1(1)$$
$$+ \gamma_2^1(x_0, x_0 + u(0) + \gamma_2^0(x_0))\} + u_1(0) = 0 \qquad (3.13)$$

where we used the fact that $\partial\gamma_2^1/\partial x(1) = -\frac{1}{2}$ according to equation (3.11) and $\partial x(1)/\partial u_1(0) = 1$ according to equation (3.1), and

$$2\{x_0 + u_1(0) + \gamma_2^0(x_0) + u_1(1) + \gamma_2^1(x_0, x_0 + u(0)$$
$$+ \gamma_2^0(x_0))\} + u_1(1) = 0 \qquad (3.14)$$

Equations (3.11-3.14) constitute four equations with four unknowns. The solution is:

$$u_1(0) = \gamma_1^0(x_0) = -\frac{1}{2}x_0; \quad u_1(1) = \gamma_1^1(x_0) = -\frac{1}{9}x_0;$$
$$u_2(0) = \gamma_2^0(x_0) = -\frac{5}{18}x_0; \quad u_2(1) = \frac{1}{18}x_0 - \frac{1}{2}x(1)$$

Remark In references [1,5] Başar also defines open-versus-closed-loop solutions. Those definitions, however, are different from ours. In our approach Başar's definition corresponds to the Stackelberg open-versus-closed-loop (leader = open-loop) structure, to be dealt with in subsection 3.7.*

3.4 Stackelberg Open-loop Information Structure

In Stackelberg games, for an extensive treatment see [3], one player, called the leader, announces his strategy first. As soon as the other player, called the follower, knows the strategy of the leader, he will choose his own. The essence is that the follower's strategy will depend on the leader's strategy.

Let's assume P_2 is the leader. In this subsection we restrict ourselves to the open-loop case. Hence $u_2(0) = \gamma_2^0(x_0)$, $u_2(1) = \gamma_2^1(x_0)$, $u_1(0) = \gamma_1^0(x_0, u_2(0), u_2(1))$, $u_1(1) = \gamma_1^1(x_0, u_2(0), u_2(1))$. Once P_1 knows γ_2^0 and γ_2^1, he faces the ordinary optimal control problem: minimize $J_1(2,0)$ in equation (3.2) subject to $x(1) = x_0 + u_1(0) + \gamma_2^0(x_0)$, $x(2) = x(1) + u_1(1) + \gamma_2^1(x_0)$. A straightforward calculation shows that the minimizing functions are

$$u_1(0) = -\frac{1}{11}(2x_0 - 2u_2(0) - 6u_2(1))$$

$$u_1(1) = -\frac{1}{11}(8x_0 - 8u_2(0) - 2u_2(1)) \qquad (3.15)$$

Player P_2, the leader, will realize that P_1 will choose his strategy according to equations (3.15). Hence P_2 faces the control problem: Minimize $J_2(2,0)$ subject to $x(1) = x_0 + u_1(0) + u_2(0)$, $x(2) = x(1) + u_1(1) + u_2(1)$, where $u_1(0)$ and $u_1(1)$ are given in (3.16). The solution can again be obtained

---

*Dr. Başar has pointed out to me that this remark is incorrect.

in a straightforward way:

$$u_2(0) = \gamma_2^0(x_0) = -\frac{11}{145}x_0$$

$$u_2(1) = \gamma_2^1(x_0) = \frac{3}{145}x_0 \qquad (3.16)$$

Equations (3.15) and (3.16) constitute the optimal solution.

3.5 Stackelberg Feedback Structure

One player, say P_2, is again the leader and the other one the follower. The strategies now depend on the current state. As far as the leadership of P_2 is concerned, he can announce both strategies $u_2(0) = \gamma_2^0(x_0)$, $u_2(1) = \gamma_2^1(x(1))$ ahead of time (Stackelberg closed-loop to be discussed in subsection 3.6) or he announces his strategies one at a time, i.e., at stage 0 he announces $u_2(0)$ after which P_1 has to decide about $u_1(0)$ and at stage 1 P_2 announces $u_2(1)$ after which P_1 decides about $u_1(1)$. The latter case is called the Stackelberg feedback structure, to be discussed now. Here $u_2(0) = \gamma_2^0(x_0)$, $u_1(0) = \gamma_1^0(x_0,u_2(0))$, $u_2(1) = \gamma_2^1(x(1))$, $u_1(1) = \gamma_2^1(x(1),u_2(1))$.

The solution, obtained by applying the dynamic programming argument, is

$$u_1(1) = -\frac{2}{3}\{x(1) + \gamma_2^1(x(1))\}$$

$$u_2(1) = \gamma_2^1(x(1)) = -\frac{1}{10}x(1)$$

$$u_1(0) = \gamma_1^0(x_0,\ u_2(0)) = -\frac{254}{354}(x_0 + \gamma_2^0))$$

$$u_2(0) = \gamma_2^0(x_0) = -\frac{11000}{136316}x_0$$

3.6 Stackelberg Closed-loop Information Structure

The Stackelberg closed-loop solution has been defined in the previous subsection. Suppose again that P_2 is the leader, who now announces $u_2(0) = \gamma_2^0(x_0)$ and $u_2(1) = \gamma_2^1(x_1))$ beforehand. The functional form of $u_1(0)$ and $u_1(1)$ is: $u_1(0) = \gamma_1^0(x_0, \gamma_2^0(x_0), \gamma_2^1(x(1)), u_1(1) = \gamma_2^0(x(1), \gamma_2^1(x(1)))$. Assumptions for γ_2^0 and γ_2^1 have to be made in order for the calculations to be carried out explicitly. Here it will be required that γ_2^0 and γ_2^1 are linear functions of x_0 and $x(1)$, respectively,

$$u_2(0) = \gamma_2^0(x(0)) = -F_{200}x(0)$$

$$u_2(1) = \gamma_2^1(x(1)) = -F_{211}x(1) \tag{3.17}$$

With respect to these assumptions the solution $u_1(0)$, $u_1(1)$ becomes:

$$u_1(1) = -\frac{2}{3}\{x(1) + \gamma_2^1(x(1))\}$$

$$u_1(0) = -\frac{\mu(6 + 2\chi^2)}{9 + 2\chi^2}x_0 \tag{3.18}$$

where $\mu \triangleq 1 - F_{200}$, $\chi \triangleq 1 - F_{211}$. Before P_2 announces equations (3.17) he will realize that P_1 will play according to equations (3.18). Hence P_2 will minimize $J_2(2,0)$ in which equations (3.1, 3.17, 3.18) are substituted with respect to F_{200} and F_{211};

$$J_2(2,0) = \left[F_{200}^2 + \left\{1 + F_{211} + \frac{1}{9}(1 - F_{211})^2\right\}\right.$$

$$\left.\cdot \left[\frac{3(1 - F_{200})^2}{9 + 2(1 - F_{211})^2}\right]^2\right]x_0^2 \tag{3.19}$$

Obviously P_2 can get the numerical value of $J_2(2,0)$ as close to zero as he wishes by playing $F_{200} = 0$ and $|F_{211}|$ sufficiently large. The solution $F_{200} = 0$, $F_{211} = \pm \infty$ does not exist in the ordinary sense. Equation (3.19) has a local minimum for

$$F_{200} = \frac{2}{29}, \quad F_{211} = -\frac{1}{2} \tag{3.20}$$

For these values $J_2(2,0)$ becomes $2x_0^2/29$.

If one allows the strategies to be continuous instead of linear functions of their argument, then the solution in (3.20) is not longer (locally) optimal, which can be seen as follows. Instead of with equation (3.17), start with the following possible γ_2^0, γ_2^1,

$$\gamma_2^0 = -\frac{2}{29} x_0; \quad \gamma_2^1 = \frac{1}{2} x(1) + \varepsilon x^3(1) \tag{3.21}$$

Going through the calculations again, one obtains, provided that ε is sufficiently small,

$$J_2(2,0) \approx \frac{2}{29} x_0^2 - \varepsilon \frac{1728}{29^4} x_0^4$$

Hence, for sufficiently small, positive ε, the nonlinear strategy in (3.21) is more advantageous than the linear strategy ($\varepsilon = 0$). For the solution of the general problem (i.e., no restrictions on the strategies) one has to solve very complicated functional equations. This has not been pursued.

Another kind of "minimization" of $J_2(2,0)$ exists. One could first minimize $J_2(2,1)$ with respect to F_{211} which yields a numerical answer for F_{211} and subsequently one could minimize $J_2(2,0)$ in order to find F_{200}. In this case the solution could be called Stackelberg closed-loop with preservation of the dynamic programming argument. It is easily seen that this solution equals the Stackelberg feedback solution of the previous subsection. From this consideration it follows that with the Stackelberg closed-loop structure player P_2 is better off than with the Stackelberg feedback structure, i.e., min J_2 is smaller in case of Stackelberg closed-loop.

3.7 Stackelberg Open- versus Closed-loop, Leader Plays Open-loop

Suppose again P_2 is the leader. In this subsection P_2 plays open-loop, i.e., P_2 announces $\gamma_2^i(x_0)$, i = 0, 1, and then P_1 faces the problem of minimizing $J_1(2,0)$ subject to $x(i+1) = x(i) + u_1(i) + \gamma_2^i(x_0)$, i = 0, 1, in closed-loop form. The optimal $u_1(i)$, i = 0, 1, are

$$u_1(0) = \gamma_1^0(x_0) = -\frac{1}{11}\left(8x_0 + 8\gamma_2^0(x_0) + 2\gamma_2^1(x_0)\right)$$

$$u_1(1) = \gamma_1^1(x_0, x(1)) = -\frac{2}{3}\left(x(1) + \gamma_2^1(x_0)\right) \tag{3.22}$$

P_2 will realize that P_1 will play according to (3.22). Hence P_2 will choose $u_2(i) = \gamma_2^i(x_0)$, i = 1, 2, in such a way as to minimize $J_1(2,0)$ subject to $x(i+1) = x(i) + u_1(i) + u_2(i)$, i = 1, 2, in which formulas (3.22) are substituted. The solution is

$$u_2(0) = \gamma_2^0(x_0) = -\frac{11}{145}x_0$$

$$u_2(1) = \gamma_2^1(x_0) = \frac{3}{145}x_0 \tag{3.23}$$

The solution of the overall game in this subsection is given by (3.22) and (3.23). If these equations are compared to equations (3.15) and (3.16), it is easily seen that, though the functional form of the equations differs, the numerical outcomes are equal. The reason, of course, is that in deterministic control problems there is no difference between open- and closed-loop solutions as far as the numerical values are concerned. In subsection 3.4, as well as in this subsection, P_1 solves a simple optimal control problem, in open- and closed-loop form, respectively. If noise should be added to the system equations (3.1), the numerical outcomes will become different.

3.8 Stackelberg Open- versus Closed-loop, Leader Plays Closed-loop

The i.s. as mentioned in the title is not yet uniquely determined. Three different structures can be distinguished, to be defined now.

I. At stage t = 0, P_2 announces $\gamma_2^0(x_0)$, after which P_1 has to choose both $u_1(0) = \gamma_1^0(x_0, \gamma_2^0)$ and $u_2(1) = \gamma_1^1(x_0, \gamma_2^0)$. At stage 1, P_2 will choose $u_2(1) = \gamma_2^1(x(1), \gamma_1^1(x_0))$. The solution is

$$u_2(1) = \gamma_2^1(x(1), \gamma_1^1(x_0)) = -\frac{1}{2}(x(1) + \gamma_1^1(x_0))$$

$$u_1(0) = \gamma_1^0(x_0, \gamma_2^0) = -\frac{7}{10}(x_0 + \gamma_2^0)$$

$$u_1(1) = \gamma_1^1(x_0, \gamma_2^0) = -\frac{1}{10}(x_0 + \gamma_2^0)$$

$$u_2(0) = \gamma_2^0(x_0) = -\frac{11}{111}x_0$$

II. At stage t = 0, P_2 announces both $\gamma_2^0(x_0)$ and $\gamma_2^1(x_0, x(1))$. In order to keep the calculations simple, we assume these functions to be linear;

$$u_2(0) = \gamma_2^0(x_0) = -F_{200}x_0;\ u_2(1) = \gamma_2^1(x_0,\ x(1))$$

$$= -F_{210}x_0 - F_{211}x(1) \qquad (3.24)$$

Once P_1 knows these functions, he can solve $u_1(0)$ and $u_1(1)$ by minimizing $J_1(2,0)$ subject to equations (3.1) in which equations (3.24) are substituted;

$$u_1(0) = \gamma_1^0(x_0,\gamma_2^0,\gamma_2^1) = -\frac{6\mu + 2\chi(\mu\chi - F_{210})}{9 + 2\chi^2}\, x_0$$

$$u_1(1) = \gamma_2^0(x_0,\gamma_2^0,\gamma_2^1) = -\frac{2\mu\chi - 6F_{210}}{9 + 2\chi^2}\, x_0 \qquad (3.25)$$

where $\mu \triangleq 1 - F_{200}$, $\chi \triangleq 1 - F_{211}$. Now, in order to determine F_{200}, F_{210}, F_{211}, P_2 wants to minimize $J_2(2,0)$ subject to equations (3.1) in which (3.24) and (3.25) have been substituted. As in subsection 3.6, two possible "minimizations" exist. First, $J_2(2,0)$ can be minimized with respect to F_{200}, F_{210}, and F_{211}, the answers to which will be given in III. Second, it is possible to minimize stagewise, i.e., first $J_2(2,1)$ is minimized with respect to the unknowns F, which give two relations in the unknowns, viz. $F_{211} = +\frac{1}{2}$, $F_{200} - 13F_{210} = 1$, and then $J_2(2,0)$ is minimized, which gives another relation, viz. $407\ F_{200} - 66\ F_{210} = 46$. The solution is

$$F_{200} = \frac{28}{275};\qquad F_{210} = -\frac{19}{275};\qquad F_{211} = \frac{1}{2}$$

III. If $J_2(2,0)$ is minimized with respect to the three unknowns F_{200}, F_{210}, and F_{211}, the result becomes:

$$F_{200} = 0;\qquad F_{210} = \frac{7}{10};\qquad F_{211} \to -\infty$$

This solution, which yields $J_2(2,0) \downarrow 0$, does not exist in the ordinary sense. It has not been investigated whether a local minimum exists as in subsection 3.6.

Treated in sub-section	P_1	P_2	$\frac{\min J_1}{x_0^2}$	$\frac{\min J_2}{x_0^2}$	Roles reversed P_1	P_2	$\frac{\min J_1}{x_0^2}$	$\frac{\min J_2}{x_0^2}$
3.1	OL	OL	0,382	0,119				
3.2	CL	CL	0,396	0,118				
3.3	OL	CL	0,367	0,113	CL	OL	0,411	0,109
3.4	FOL	LOL	0,628	0,076	LOL	FOL	0,286	0,306
3.5	FCL	LF	0,606	0,081	LF	FCL	0,321	0,254
3.6	FCL	LCL	$\to 1$	$\downarrow 0$	LCL	FCL	$\downarrow 0$	$\to 1$
3.6	FCL	LCLDP	0,606	0,081	LCLDP	FCL	0,321	0,254
3.7	FCL	LOL	0,628	0,076	LOL	FCL	0,286	0,306
3.8 I	FOL	LF	0,568	0,099	LF	FOL	0,365	0,211
3.8 II	FOL	LCLDP	0,559	0,096	LCLDP	FOL	0,353	0,216
3.8 III	FOL	LCL	$\to 1$	$\downarrow 0$	LCL	FOL	$\downarrow 0$	$\to 1$

The minimal values of J_1 and J_2 for the example in Section 3. For the abbreviations in the columns see Section 5.

§4. SOLUTIONS TO THE N-STAGE VECTOR CASE

In this section the solution will be given to the general linear quadratic problem as described in Section 2. This solution will only be given with respect to the first five i.s. mentioned in Section 3. The solution of the i.s. open-loop (subsection 3.1) and closed-loop (subsection 3.2) have been given in [1], but will be repeated for the sake of completeness. In [1] yet another solution has been given, of the i.s. treated in subsection 3.7; Stackelberg open- versus closed-loop, leader plays open-loop. That solution, in [1] called open- versus closed-loop, will not be given.*

*See footnote in subsection 3.3.

4.1 Open-loop Information Structure

Two ways along which the optimal solution can be obtained will be given. The first is given in [1].

I. Define

$$\underline{x} = [x^T(1) \mid x^T(2) \mid \ldots \mid x^T(N)]^T \qquad \text{(Nn vector)}$$

$$\underline{x}_0 = [x^T(0) \mid x^T(0) \mid \ldots \mid x^T(0)]^T \qquad \text{(Nn vector)}$$

$$\underline{u}_j = [u_j^T(0) \mid u_j^T(1) \mid \ldots \mid u_j^T(N-1)]^T, \; j = 1,2; \qquad (\text{Nr}_j \text{ vector})$$

$\underline{A}$ is an Nn × Nn matrix, diagonal in blocks of n × n matrices; the jjth block is given by

$$A_{jj} = \prod_{k=0}^{j-1} A(k)$$

$\underline{B}_k$ is an Nn × Nr_1 matrix, defined by means of blocks B_{kij}, which all have size n × r_k. For j > i, $B_{kij} = 0$, for j = i, $B_{kij} = B_{kjj} = B_k(j)$, and for j < i,

$$B_{kij} = \left[\prod_{k=j}^{i-1} A(k)\right] B_k(j-1); \; k = 1, 2$$

$\underline{Q}_i$, i = 1, 2; Nn × Nn matrix, diagonal in blocks of n × n; the kth block is given by $Q_i(k)$. $\underline{R}_{ij}$, i,j = 1, 2; $Nr_j \times Nr_j$ matrix, diagonal in blocks of $r_j \times r_j$; the kth block is given by $R_{ij}(k-1)$. (4.1)

By means of these definitions, equation (2.1) can be written as

$$\underline{x} = \underline{A}\underline{x}_0 + \underline{B}\underline{u} + \underline{C}\underline{v} \qquad (4.2)$$

and $J_i(N,0)$ can now be written as

$$J_i(N,0) = \underline{x}^T\underline{Q}_i\underline{x} + \underline{u}_1^T\underline{R}_{i1}\underline{u}_1 + \underline{u}_2^T\underline{R}_{i2}\underline{u}_2, \; i = 1, 2 \tag{4.3}$$

The unique $\underline{u}_i$ that minimizes $J_i(N,0)$ subject to (4.2) and for fixed $\underline{u}_j$ is given by

$$\underline{u}_i = - \left[\underline{R}_{ii} + \underline{B}_i^T\underline{Q}_i\underline{B}_i\right]^{-1} \underline{B}_i^T\underline{Q}_i\left[\underline{A}\underline{x}_0 + \underline{C}\underline{u}_j\right],$$

$$i,j = 1,\ 2;\ i \neq j \tag{4.4}$$

Equations (4.4) constitute two vector equations with two unknown vectors $\underline{u}_1$ and $\underline{u}_2$. The solution is

$$\underline{u}_i = - \left[\underline{R}_{ii} + K_i\underline{B}_i\right]^{-1} K_i\underline{A}\underline{x}_0 \tag{4.5}$$

where

$$K_i = \underline{B}_i^T\underline{Q}_i\left[I - \underline{B}_j\left(\underline{R}_{jj} + \underline{B}_j^T\underline{Q}_j\underline{B}_j\right)^{-1} \underline{B}_j\underline{Q}_j\right]$$

$$i,j = 1,\ 2;\ i \neq j$$

The solution, of course, only exists if the matrices $\underline{R}_{ii}$ + $K_i\underline{B}_i$, i = 1, 2, are nonsingular. If the solution exists, it is globally optimal [1].

II. Suppose for the time being that the functions $u_2(i) = \gamma_2^i(x_0)$ are fixed. P_1 wants to choose $u_1(i)$, i = 0, $\cdots$, N-1, in such a way as to minimize

$$\sum_{j=0}^{N-1}\left[x^T(j+1)Q_1(j+1)x(j+1) + u_1^T(j)R_{11}(j)u_1(j)\right]$$

subject to

$$x(i+1) = A(i)x(i) + B_1(i)u_1(i) + B_2(i)\gamma_2^i(x_0)$$

The term $u_2^T(j)R_{12}(j)u_2(j)$ has been deleted in the cost function since this part is constant as long as the $u_2(i)$ are kept fixed. The solution is

$$u_j(i) = - D_j(i+1)B_j^T(i)\Big[\big(S_j(i+1) + Q_j(i+1)\big)\big(A(i)x(i) + B_k(i)u_k(i)\big) + \frac{1}{2}\, e_j(i+1)\Big]$$

$$j = 1,\ k = 2,\ i = 0, \cdots, N-1 \qquad (4.6)$$

where $S_j(i)$, $D_j(i)$, and $e_j(i)$ are given by the following recurrence relations:

$$S_j(N) = 0;\ e_j(N) = 0;\ D_j(N) = 0$$

$$D_j(i) = \big[B_j^T(i-1)\big(S_j(i) + Q_j(i)\big) + R_{jj}(i-1)\big]^{-1}$$

$$S_j(i-1) = - A^T(i-1)\big(S_j(i) + Q_j(i)\big)B_j(i-1)D_j(i)B_j^T(i-1) \cdot \big(S_j(i) + Q_j(i)\big)A(i-1) + A^T(i-1)\big(S_j(i) + Q_j(i)\big)A(i-1)$$

$$e_j^T(i-1) = - \Big[2u_k^T(i-1)B_k^T(i-1)\big(S_j(i) + Q_j(i)\big) + S_j^T(i)\Big] B_j(i-1)D_j(i)B_j^T(i-1)\big(S_j(i) + Q_j(i)\big)A(i-1) + 2u_k^T(i-1)B_k^T(i-1)\big(S_j(i) + Q_j(i)\big)A(i-1) + e_j^T(i)A(i-1),$$

$$j,\ k = 1,\ 2;\ j \neq k \qquad (4.7)$$

The proof of (4.6) is a well known application in the dynamic programming argument and will not be given here. The solution $u_1(i)$ in (4.6) has been expressed in closed-loop form. By substituting

$$x(i) = \left[\prod_{k=0}^{i-1} A(k)\right] x_0 + \sum_{m=1}^{2} \left[\sum_{j=0}^{i-1} \left[\left[\prod_{k=j}^{i-1} A(k)\right] B_m(j) u_m(j)\right]\right]$$

$$\triangleq \Delta(i) x_0 + \sum_{m=1}^{2} \left[\sum_{j=0}^{i-1} \left[\Lambda_m(j+1, i) u_m(j)\right]\right] \tag{4.8}$$

into equations (4.6) and (4.7) and writing the terms $e_1^T(i)$ in (4.6) completely in terms of $u_2(i)$ by means of equations (4.7), equation (4.6) ultimately becomes a vector equation consisting of $N \times r_1$ linear scalar equations in the unknowns $u_1(i)$, $u_2(i)$, $i = 0, \cdots, N-1$, and the vector x_0.

Another $N \times r_2$ equation in the same unknowns can be obtained if $u_1(i)$, $i = 0, \cdots, N-1$ are kept fixed and player P_2 minimizes $J_2(N,0)$ with respect to $u_2(i)$, $i = 0, \cdots, N-1$. Together there are $N \times (r_1 + r_2)$ equations with the same number of unknowns. If these unknowns can be solved, they are expressed in x_0 and they form the open-loop solution. With this latter method, an $N(r_1 + r_2) \times N(r_1 + r_2)$ matrix has to be inverted and if method I is used, matrices of sizes $Nr_1 \times Nr_1$ and $Nr_2 \times Nr_2$ have to be inverted. Method II, however, can be converted to a slightly different method such that the maximum size of matrices to be inverted is $(r_1 + r_2) \times (r_1 + r_2)$, which may cause considerable savings in computer time.

To this end, the linear equations (4.6) are symbolically written as

$$u_i(j) = f_{ij}(u_k(j), x(j), e_i(j+1)) \tag{4.9a}$$

$$e_i(j) = g_{ij}(u_k(j), e_i(j+1)), \; j = 0, \cdots, N-1;$$
$$i,k = 1, 2; \; i \neq k \tag{4.9b}$$

For $j = N-1$, $u_1(N-1)$ and $u_2(N-1)$ can be solved from (4.9a), $i = 1, 2$, and expressed in $x(N-1)$, symbolically written as

$$u_i(N-1) = h_{i,N-1}(x(N-1)),\ i = 1,\ 2 \tag{4.10}$$

It is assumed that this solution exists; otherwise the construction breaks down. Note that $e_i(N) = 0$, $i = 1,2$. Next, substitute $x(N-1) = A(N-2)x(N-2) + B_1(N-2)u_1(N-2) + B_2(N-2) \cdot u_2(N-2)$ into equations (4.10). These relations are used to express $f_{i,N-2}$ and $g_{i,N-2}$ completely in terms of $x(N-2)$, $u_i(N-2)$. In this way again in (4.9a), $i = 1, 2$, we get for $j = N-2, r_1 + r_2$ linear equations in the unknowns $u_1(N-2)$ and $u_2(N-2)$. If these equations are solved, we get

$$u_i(N-2) = h_{i,N-2}(x(N-2)),\ i = 1,\ 2$$

In this way we continue with a result

$$u_i(j) = h_{ij}(x(j)),\ j = 0,\ \cdots,\ N-1,\ i = 1,\ 2 \tag{4.11}$$

Equations (4.11) constitute the open-loop solution, which is, however, given in the open-eye form as defined in subsection 3.2. To obtain the closed-eye form, i.e., $u_1(j)$ and $u_2(j)$ must only be expressed in x_0, we start in (4.11) with $j = 0$ after which $x(1)$ can via $x(1) = A(0)x(0) + B_1(0)u_1(0) + B_2(0)u_2(0)$ be expressed completely in terms of $x(0)$. Then the $x(0)$ dependence of $u_1(1)$ and $u_2(1)$ via (4.11) is known also, etc.

4.2 Closed-loop Information Structure

There exists a recursive closed-loop, no memory Nash solution to the problem described in Section 2, for each $n \in \theta$

given by

$$u_i(k) = \gamma_i^k(x(k)) = -F_{ikk}x(k), \; i = 1, 2 \tag{4.12}$$

$$F_{ikk} = [R_{ii}(k) + H_i(k)B_i(k)]^{-1} H_i(k)A(k)$$

$$H_i(k) = B_i^T(k)Y_i(k+1)\Big\{I - B_j(k)[R_{jj}(k) + B_j^T(k)Y_j(k+1)B_j(k)]^{-1}B_j^T(k)Y_j(k+1)\Big\}$$

$$L(k) = A(k) - B_i(k)F_{ikk} - B_j(k)F_{jkk}$$

$$Y_i(k) = L^T(k)Y_i(k+1)L(k) + F_{ikk}^T R_{ii}(k)F_{ikk} + F_{jkk}^T R_{ij}(k)F_{jkk} + Q_i(k)$$

$$Y_i(N) = Q_i(N), \; i,j = 1, 2; \; i \neq j$$

provided that $[R_{ii}(k) + H_i(k)B_i(k)]$ is nonsingular $\forall\, k \in \theta$. If the solution exists, it is unique and also globally optimal [1]. The matrices $Y_i(k)$ have the following interpretation. If both players play optimally, then $\min J_i(N,k) = x^T(k)Y_i(k)x(k)$.

4.3 Open- Versus Closed-loop Information Structure

As in subsection 3.3, it is assumed that P_1 plays open-loop and P_2 plays closed-loop.

We first assume that $u_1(i) = \gamma_1^i(x_0)$ is kept fixed and that P_2 minimizes $J_2(N,0)$ with respect to $u_2(i)$, $i = 0, \cdots, N-1$, in closed-loop form. The solution is given in equation (4.6) with $j = 2$.

We next assume that $u_2(i) = \gamma_2^i(x_0, x(i))$ is kept fixed and that P_1 minimizes $J_1(N,0)$ with respect to $u_1(i)$, $i \in \theta$, in open-loop form. For that reason $J_1(N,0)$ is written as

$$J_1(N,0) = \sum_{j=0}^{N-1} \Bigg[\Delta(j+1)x_o + \sum_{k=1}^{j+1} \Big[\Lambda_1(k,j+1)u_1(j) + \Lambda_2(k,j+1)u_2(j) \Big] \Bigg]^T Q_1(j+1)$$

$$\Bigg[\Delta(j+1)x_0 + \Bigg[\sum_{k=1}^{j+1} \Lambda_1(k,j+1)u_1(j) + \Lambda_2(k,j+1)u_2(j) \Bigg] \Bigg]$$

$$+ u_1^T(j)R_{11}(j)u_1(j) + u_2^T(j)R_{12}(j)u_2(j)$$

This expression will be differentiated with respect to $u_1(\ell)$, $\ell = 0, \cdots, N-1$. We must be careful with the dependence of $u_2(j)$ on $u_1(\ell)$. If we define

$$\tilde{L}(i) = - [D_2(i+1)C(i)\big(S_2(i+1) + Q_2(i+1)\big)A(i)]$$

then

$$\frac{\partial u_2(j)}{\partial u_1(\ell)} = \frac{\partial \mathbf{u}_2(j)}{\partial x(j)} \frac{\partial x(j)}{\partial u_1(\ell)} = \begin{cases} \tilde{L}^T(j)\Lambda_1(\ell+1,j) & \text{if } j > \ell \\ 0 & \text{if } j \le \ell \end{cases}$$

Now after some cumbersome manipulations, the derivative of $J_1(N,0)$ with respect to $u_1(\ell)$, which must be equal to zero, becomes

$$2\Bigg[\sum_{j=\ell}^{N-1}\Bigg[\sum_{k=1}^{j+1} \Lambda_1^T(\ell+1,j)\tilde{L}(j)\Lambda_2^T(k,\ell+1)\Bigg]Q_1(j+1)\Delta(j+1)x_0\Bigg]$$

$$+ 2\sum_{m=1}^{2}\Bigg\{\Bigg[\sum_{j=\ell}^{N-1}\Bigg[\sum_{k=1}^{j+1} \Lambda_1^T(\ell+1,j)\tilde{L}(j)\Lambda_2^T(k,j+1)\Bigg]Q_1(j+1) \cdot \Bigg[\sum_{k=1}^{j+1} \Lambda_m(k,j+1)\Bigg]u_m(j)\Bigg]\Bigg\}$$

$$+ 2R_{11}(\ell)u(\ell) + 2\sum_{j=\ell+1}^{N-1} R_{12}(j)u_2(j) = 0,$$

$$\ell = 0,1, \cdots, N-1 \qquad (4.13)$$

where $\Lambda_1^T(\ell+1,\ell)\tilde{L}(\ell) \triangleq I$.

In this way N-1 vector equations are obtained. In order to obtain the strategies we proceed as follows. In (4.6), j = 2, the terms x(i) are expressed in x_0, u_1's and u_2's by means of (4.8).

The resulting $u_2(i)$'s are substituted in (4.13) which then constitute Nr_i linear (!) equations in the Nr_1 unknown $u_1(i)$'s. If these equations are solvable, then the solution $\underline{u}_1$ is purely expressed as a linear function of x_0. Once it is known that $u_1(i) = \gamma_1^i(x_0)$, these functions are substituted in (4.6) with j = 2 and then these equations constitute the optimal solution $u_2(i) = \gamma_2^i(x_0, x(i))$. Again a straightforward procedure will prove that, if the solution exists, it is also globally optimal.

4.4 Stackelberg Open-loop Information Structure

As in section 3.4 it is assumed that P_2 is the leader. We start from equations (4.2-4.4). If P_2 has announced $\underline{u}_2$, then P_1 will choose $\underline{u}_1$ according to (4.4), which will here symbolically be written as

$$\underline{u}_1 = \underline{X}(\underline{A}\underline{x}_0 + \underline{B}_2\underline{u}_2) \tag{4.14}$$

P_2 of course realizes that P_1 will choose this $\underline{u}_1$ and hence P_2 wants to minimize

$$\underline{x}^T\underline{Q}_2\underline{x} + (\underline{A}\underline{x}_0 + \underline{B}_2\underline{u}_2)^T\underline{X}^T\underline{R}_{21}\underline{X}(\underline{A}\underline{x}_0 + \underline{B}_2\underline{u}_2) + \underline{u}_2^T\underline{R}_{22}\underline{u}_2$$

subject to

$$\underline{x} = \underline{A}\underline{x}_0 + \underline{B}_1\underline{X}\underline{A}\underline{x}_0 + \underline{B}_1\underline{X}\underline{B}_2\underline{u}_2 + \underline{B}_2\underline{u}_2 = (I + \underline{B}_1\underline{X})(\underline{A}\underline{x}_0 + \underline{B}_2\underline{u}_2)$$

The solution is

$$\underline{u}_2 = - \left[\underline{B}_2^T\left\{(I + \underline{B}_1\underline{X})^T\underline{Q}_2(I + \underline{B}_1\underline{X}) + \underline{X}^T\underline{R}_{21}\underline{X}\right\}\underline{B}_2 + \underline{R}_{22}\right]^{-1} \underline{B}_2^T\left[(I + \underline{B}_1\underline{X})^T\underline{Q}_2(I + \underline{B}_1\underline{X}) + \underline{X}^T\underline{R}_{21}\underline{X}\right]\underline{A}x_0 \tag{4.15}$$

4.5 Stackelberg Feedback Information Structure

Assume P_2 is the leader. It is straightforward to derive the following recursive solution:

$$u_1(i) = X(i)[A(i)x(i) + B_2(i)u_2(i)]$$

$$u_2(i) = - F(i)x(i)$$

$$X(i) = - [R_{11}(i) + B_1^T(i)W_1(i+1)B_1(i)]^{-1}B_1^T(i)W_1(i+1)$$

$$\begin{aligned} F(i) = - [&B_2^T(i)(I - B_1(i)X(i))^TW_2(i+1)(I - B_1(i)X(i))B_2(i) \\ &+ B_2^T(i)X^T(i)R_{21}(i)X(i)B_2(i) + R_{22}(i)]^{-1} \\ &\cdot [B_2^T(i)(I - B_1(i)X(i))^T \\ &\cdot W_2(i+1)(I - B(i)X(i))A(i) \\ &+ B_2^T(i)X^T(i)R_{21}(i)X(i)A(i)] \end{aligned}$$

$$\begin{aligned} W_j(i) = [&A(i) + B_j(i)X(i)(A(i) - B_k(i)F(i)) \\ &- B_k(i)F(i)]^TW_j(i+1)[A(i) + B_j(i)X(i)(A(i) \\ &- B_k(i)F(i)) - B_k(i)F(i)] \\ &+ (A(i) - B_k(i)F(i))^TX^T(i)R_{j1}(i)X(i)(A(i) \\ &- B_k(i)F(i)) + F^T(i)R_{j2}(i)F(i) + Q_j(i) \end{aligned}$$

$$W_j(N) = Q_j(N); \; j,k = 1, 2; \; j \neq k$$

4.6 Some Remarks on the Stackelberg Closed-loop Information Structure

For the general linear quadratic problem (Section 2), the leader, say P_2, will in general not be able to get his costs arbitrarily close to zero as in the example in subsection 3.6.

The following assumptions are made (only in this subsection): $Q_i > 0$, the system is autonomous, i.e., A, B_1, B_2, Q_i, R_{ij} independent of the stage, $N > n$, the pairs $\{A, B_i\}$ are controllable. If P_2 chooses and announces $u_2(i) = -F_{2ii}x(i)$ with $F_{2ii} = 0$, $i = 0, \cdots, n-1$, and $F_{2nn} \to \pm\infty$ in some norm, then P_1 will surely choose F_{1ii} in his strategy $u_1(i) = -F_{1ii}x(i)$ in such a way that $x(n) = 0$, which is possible because of the controllability assumption. [If $x(n) \neq 0$, P_1 faces enormous costs because of the terms $u_2^T(n)R_{12}u_2(n)$ and $x^T(n)Q_1x(n) + u_1^T(n)R_{22} \cdot u_1(n)$ in J_1.] The appropriate F_{1ii} will be written as F_{1ii}^*. An upper bound for the costs of P_2 can be given. With the (probably nonoptimal from the viewpoint of P_2) strategy described above we get

$$J_2 \to \sum_{i=0}^{n-2} x^T(i+1)Q_2x(i+1)$$

$$+ \sum_{i=0}^{n-1} x^T(i)F_{1ii}^{*T}R_{21}F_{1ii}^{*}x(i), \text{ as } F_{2nn} \to \pm\infty$$

with $x(i+1) = (A - B_1F_{1ii}^*)x(i)$, $i = 0, \cdots, n-1$. For $n = 1$ and $R_{21} = 0$ (subsection 3.6), this is reduced to $J_2 \to 0$.

§5. THE M-PLAYER MULTISTAGE GAME

So far only two-person games have been considered. According to the rules of the games, the following types of players can be distinguished:

1. follower open-loop players (FOL)
2. follower closed-loop players (FCL)
3. leader open-loop players (LOL)
4. leader closed-loop players with preservation of dynamic programming argument (LCLDP)
5. leader-feedback players (LF)
6. leader closed-loop players (LCL)

The distinction follower-leader only makes sense if both types of players are present in a game. If only followers or only leaders participate in a game, the distinguishing feature of such a role disappears.

In a general M-person game all types of players may be present; M_1 FOL players, M_2 FCL players, etc., such that $\Sigma M_i = M$. No attempt has been made to obtain explicit expressions for the strategies of the players in such an M-player game. In [5] an M-person multistage game has been considered with only LOL and FCL players.

Of course, this list of types of players is by no means exhaustive. In Sections 6 and 7 we will consider two other types of players, or--which is the same--two 2-person games with i.s. other than those considered so far.

§6. AN INFORMATION STRUCTURE WITH A TIME DELAY IN THE OBSERVATION

Instead of equation (2.1) the following equation will be considered in this section:

$$x(i+1) = A(i)x(i) + B_1(i)u_1(i) + B_2(i)u_2(i) + w(i); \quad x(0) = x_0 \tag{6.1}$$

where $\{w(i)\}$ is a zero-mean independent stochastic process. The cost functions will accordingly be replaced with

$$E\{J_1(N,0)\} \quad \text{and} \quad E\{J_2(N,0)\} \tag{6.2}$$

where E stands for expectation. Both players make exact measurements of the state. At stage i P_1 measures $x(i)$ and P_2 measures $x(i-k)$, where k is a positive integer. Here it is assumed $i - k \geq 0$, for $i - k < 0$ P_2 only knows the initial value x_0. Hence P_2 has a time delay of k steps in his observations. These facts about observations are known to both players. At stage i P_1 chooses $u_1(i)$ in such a way as to minimize $E\{J_1(N,i)\}$ and P_2 chooses $u_2(i)$ in such a way as to minimize $E\{J_2(N,i)\}$. The choice of $u_1(i)$ is based on the knowledge of $x(0), \cdots, x(i), u_1(0), \cdots, u_1(i-1)$, and the choice of $u_2(i)$ is based on the knowledge of $x(0), \cdots, x(i-k), u_2(0), \cdots, u_2(i-1)$.

In order to make his decision at stage i player P_2 calculates $\hat{x}(i)$, the best estimate of $x(i)$ based on $x(0), \cdots, x(i-1), u_2(0), \cdots, u_2(i-1)$. This estimate is determined by the following recurrence relations:

$$\hat{x}(i-k) = x(i-k)$$

$$\hat{x}(j+1) = A(j)\hat{x}(j) + B_1(j)\hat{u}_1(j) + B_2(j)u_2(j)$$

$$\hat{u}_1(j) = - F_{1jj}\hat{x}(j), \; j = i-k, \cdots, i-1, \text{ if } i-k \geq 0,$$

$$j = 0, \cdots, i-1, \text{ if } i-k < 0$$

where the matrices F_{ijj} have been defined in equation (4.12). Now P_2 will choose

$$u_2(i) = - F_{2ii}\hat{x}(i) \tag{6.3}$$

according to (4.12). Player P_1 realizes that equation (6.3) is the best choice for P_2 to make and hence $u_1(i)$ will be a function of both $x(i)$ and $\hat{x}(i)$, where $\hat{x}(i)$ remains P_2's best choice, which, however, can be calculated by P_1 as well because P_1 can reconstruct $u_2(0), \cdots, u_2(i-1)$. P_1 wants to minimize

$$E\{J_1(N,i)\} = E\Big[x^T(i+1)Y_1(i+1)x^T(i+1) + u_1^T(i)R_{11}(i)u_1(i) + \hat{x}^T(i)F_{2ii}^T R_{12}(i)F_{2ii}\hat{x}(i) + \sum_{j=i+2}^{N} tr[(Q_1(j) + Y_1(j))\Psi(j-1)]\Big]$$

where $\Psi(j)$ is the covariance of $w(j)$, subject to

$$x(i+1) = A(i)x(i) + B_1(i)u_1(i) - B_2(i)F_{2ii}\hat{x}(i) + w(i)$$

The matrix $Y_1(i)$ is given in subsection 4.2. The solution is

$$u_1(i) = - [B_i^T(i)Y(i+1)B_1(i) + R_{11}(i)]^{-1} \cdot B_i^T(i)Y_1(i+1)\{A(i)x(i) - B_2(i)F_{2ii}\hat{x}(i)\}$$

If we should substitute $\hat{x}(i) = x(i)$ into equation (6.4), equation (6.4) is reduced to equation (4.12).

§7. STACKELBERG INFORMATION STRUCTURE IN WHICH LEADER DECIDES K STEPS AHEAD OF TIME

We are given the deterministic game as described in Section 2. Player P_2 is the leader; at stage $i = 0$ he announces $u_2(0), \cdots, u_2(k)$, after which P_1 chooses $u_1(0)$; k is a positive integer. At an arbitrary stage i with $i \leq N-k-1$, P_2 announces $u_2(i+k)$ after which P_1 decides about $u_1(i)$. For $i >$ $N-k-1$ the whole u_2 sequence has been determined and P_1 faces an ordinary optimal control problem. Both players have perfect information and recall, i.e., at stage i they know all states $x(j)$, $j = 0, \cdots, i$, and their own previous decisions. The i.s. described so far is known to both players.

The problem has not been uniquely defined yet. It has not been defined what form $u_2(i+k)$ should have. At stage i P_2 has to decide about $u_2(i+k)$. On the one hand P_2 can assign a numerical value to $u_2(i+k)$. This case will be treated in subsection 7.1. On the other hand P_2 can assign a functional form to $u_2(i+k)$ which depends on arguments not yet known but which will be known at stage i+k, at which $u_2(i+k)$ must be applied. P_2 could, for instance, choose a function $f(x(i), \cdots, x(i+k), u_2(i), \cdots, u_2(i+k-1))$ such that $u_2(i+k) = f(\ldots)$. This will be treated in subsection 7.2.

Intuitively, it is easily argued that the i.s. to be described in subsections 7.1 and 7.2 will give rise to different solutions. In the approach in subsection 7.1 the numerical value of $u_2(i+k)$ cannot influence the numerical value of $u_2(j)$ for $j < i+k$. In the approach in subsection 7.2 this is quite well possible; the choice of $u_2(i+k)$ influences the choice of $u_1(i)$ and because $u_2(j)$ for $i < j < i+k$ has already been given in a functional form whose arguments depend on $u_1(i)$ and hence on $u_2(i+k)$.

7.1 Leader Chooses the Numerical Value k Steps Ahead of Time

In order to describe the solution, model (2.1) is replaced with model (7.1) for $i = 0, 1, \cdots, N-k-1$. The y-vectors have the same number of components as x has. Except for the beginning ($i = 0$) and the end ($i \geq N-k-1$) the optimal solution to this problem is of the Stackelberg i.s., for which recurrence relations have been given in subsection 4.5.

$$\begin{pmatrix} x(i+1)\\ y(i+1)\\ \vdots\\ \vdots\\ y(i+k)\end{pmatrix} = \begin{pmatrix} A(i) & I & 0 & \cdots & 0\\ 0 & 0 & \ddots & & 0\\ \vdots & & \ddots & \ddots & I\\ 0 & \cdots & 0 & & 0\end{pmatrix} \begin{pmatrix} x(i)\\ y(i)\\ \vdots\\ \vdots\\ y(i+k-1)\end{pmatrix} + \begin{pmatrix}B_1(i)\\ 0\\ \vdots\\ \vdots\\ 0\end{pmatrix} u_1(i) + \begin{pmatrix}0\\ \vdots\\ \vdots\\ 0\\ B_2(i+k)\end{pmatrix} u_2(i+k) \tag{7.1}$$

At stage $i = N-k-1$, P_2 must announce $u_2(N-1)$ and then P_1 faces an ordinary control problem; minimize $J_1(N,N-k-1)$ with respect to $u_1(j)$, $j = N-k-1, N-k, \cdots, N-1$ and subject to

$$x(j+1) = A(j)x(j) + B_1(j)u_1(j) + B_2(j)u_2(j) \tag{7.2}$$

$$u_2(j) = y(j), \quad j = N-k-1, \cdots, N-2 \tag{7.3}$$

$u_2(N-1)$ announced by P_2

Because this control problem is deterministic, it does not matter whether it is solved by P_1 according to an open-loop or closed-loop structure. Suppose P_1 solves the game according to the open-loop structure, then the $u_1(j)$, $j = N-k-1, \cdots, N-1$ become (linear) functions of $y(j)$, $j = N-k-1, \cdots, N-2$ and $u_2(N-1)$. Player P_2 of course realizes what P_1 will do once he knows $u_2(N-1)$ and hence P_2 chooses $u_2(N-1)$ in such a way as to minimize $J_2(N, N-k-1)$ with respect to $u_2(N-1)$ and subject to (7.2) and (7.3). Of course the optimal $u_1(j)$, $j = N-k-1, \cdots, N-1$ and $u_2(N-1)$ depend linearly on $x(N-k-1)$, $y(N-k-1), \cdots, y(N-k-2)$, which are not known explicitly. Now we work backwards according to the formulas in subsection 4.5 and with respect to the model given in (7.1), until we reach stage $i = 0$. Note that the matrices $Q_\ell(j)$, $R_{\ell 1}(j)$, in the formulas of subsection 4.5, must be adapted to the new model as given in equation (7.1). The new $Q_\ell(j)$ equals a matrix of size $(k+1)n \times (k+1)n$ and consists of only zeros except for the left upper block which corresponds to the original $Q_\ell(j)$. The new $R_{i2}(j)$ equals the old $R_{i2}(j+k)$; R_{i1} remains the same. At stage i = 0 P_2 announces $u_2(0), \cdots, u_2(k)$ after which P_1 chooses $u_1(0)$. Of course, P_1 chooses $u_1(0)$ in such a way as to minimize $J_1(N,0)$ and thus $u_1(0)$ becomes a function of $u_2(0), \cdots, u_2(k)$. Player P_2 of course realizes that P_1 will play according to this function. Hence P_2 will choose $u_2(0), \cdots, u_2(k)$ in such a way as to minimize $J_2(N,0)$. Note that with respect to the minimization of $J_2(N,0)$ as a function of $u_2(0), \cdots, u_2(k)$, two possibilities exist. First, the absolute minimum is sought; second, one can first minimize $J_2(N,k)$ with respect to $u_2(k)$ and subsequently minimize $J_2(N,j)$ with respect to $u_2(j)$, $j = k-1, \cdots, 0$.

Ultimately, N-1 vector equations in the N-1 unknown vectors $u_2(i)$, $i \in \theta$, are obtained; at stage i with $0 < i \le N-k-1$ the vector $u_2(N-1)$ was determined as a function of $y(j) = u_2(j)$, $j = N-k-1, \cdots, N-2$; and at stage i = 0 the remaining k vector equations are obtained. If these linear equations are solvable, then the solution exists.

7.2 Leader Chooses the Functional Form k Steps Ahead of Time

We restrict ourselves to the case that at stage i, $0 < i \le N-k-1$, player P_2 announces the $n \times r_2$ matrix $F_{2\ i+k\ i+k}$, which means that at the future stage i+k P_2 will apply $u_2(i+k) = -F_{2\ i+k\ i+k}x(i+k)$. At stage i = 0, P_2 announces F_{2jj}, $0 \le j \le k$. At stage i = N-k-1, P_2 announces $F_{2\ N-1,N-1}$ after which P_1 chooses $u_1(j) = -F_{1jj}x(j)$ for $j = N-k-1, N-k, \cdots, N-1$ in such a way as to minimize $J_1(N, N-k-1)$. In this way F_{1jj}, $j = N-k-1, \cdots, N-1$ become (in general nonlinear) functions of $F_{2\ell\ell}$, $\ell = N-k, \cdots, N-1$. Player P_2 realizes that P_1 will choose F_{1jj} in this way and hence P_2 wants to choose $F_{2\,N-1,\,N-1}$ in such a way as to minimize $J_2(N, N-k-1)$ subject to

$$x(i+1) = A(i)x(i) - B_1(i)F_{1ii}x(i)$$

$$- B_2(i)F_{2ii}x(i), \quad i = N-k-1, \cdots, N-1$$

where F_{2ii}, $i = N-k-1, \cdots, N-2$, are assumed to be known and F_{1ii} depend on F_{2jj} as described above.

Now go back in time as in the previous subsection to derive more equations for the matrices F_{2jj} until the stage $i = 0$ is reached at which, in the same way as in subsection 7.1, P_2 has to announce F_{2jj}, $0 \leq j \leq k$. Player P_1 gets to know these F_{2jj} and chooses F_{100}. Of course, P_2 realizes this and substitutes this into the state equations. Subsequently P_2 chooses F_{2jj}, $0 \leq j \leq k$ in such a way as to minimize $J_2(N,0)$, which again can be done with or without the dynamic programming argument.

§8. CONCLUSIONS

Several information structures (i.s.) have been illustrated by means of a simple example. For some i.s. the optimal solutions have been given if the underlying model is linear and the cost functions quadratic. For nonzero-sum games different i.s. generally give rise to different numerical answers (it is well known that for deterministic two-person zero-sum games the numerical answers are the same independent of the fact which i.s. has been chosen). It is important to realize that these i.s. of both players are fixed a priori and that each player is expected to play rationally and make utmost use of the information available in arriving at an optimal decision. A possible extension would be that players themselves can choose their own i.s. as a part of the game and that the i.s. may be time dependent.

No attempt has been made to introduce an ordering among the information structures (an i.s. is "better" than another i.s. if, for instance, the costs are lower for all players). It is easily proved [3], that in a Stackelberg game the leader is always better off than in the corresponding game in which there is no leader (e.g., the minimal value of J_2 in subsection 3.4 is less than the minimal value of J_2 in subsection 3.1).

Another possible extension is that one allows the players to have noisy measurements instead of exact measurements. Then, once a particular i.s. has been given, the procedure of obtaining the optimal controls becomes much more complicated (see, for instance, [7,8]).

REFERENCES

1. Başar, T., On the Uniqueness of the Nash Solution in Linear-Quadratic Differential Games, International J. of Game Theory, vol. 4 (1975).

2. Starr, A. W. and Y. C. Ho, Nonzero-sum Differential Games, JOTA, vol. 3, no. 3 (1969), pp. 184-219.

3. Simaan, M. and J. B. Cruz, Jr., On the Stackelberg Strategy in Nonzero-sum Games, JOTA, vol. 11, no. 5 (1973), pp. 533-555.

4. Kwakernaak, H. and R. Sivan, Linear Optimal Control Systems, Wiley, New York, 1972.

5. Başar, T., A New Class of Nash Strategies for M-person Differential Games with Mixed Information Structures, Preprints 6th World IFAC Congress, Boston (1975).

6. Lukes, D. H. and D. H. Russell, A Global Theory for Linear-Quadratic Differential Games, Journal of Math. An. and Appl. 33 (1971), pp. 96-123.

7. Ho, Y. C., On the Minimax Principle and Zero-Sum Stochastic Differential Games, JOTA, vol. 13, no. 3 (1974), pp. 343-361.

8. Rhodes, I. B. and D. G. Luenberger, Stochastic Differential Games with Constrained State Estimators, IEEE Trans. on Aut. Control, vol. AC-14, no. 5 (1969), pp. 476-481.

MARTINGALES AND OPTIMAL CONTROL

Robert J. Elliott

University of Hull
Hull, England

ABSTRACT

Using martingale methods a general dynamic programming result is described for the optimal control of a partially observed stochastic system.

§1. INTRODUCTION

We describe below a very general dynamic programming condition for the optimal control of a partially observed stochastic system. The evolution of the system is described by a functional differential equation whose solutions are interpreted by the Girsanov measure transformation method [4]. Complicated continuity and differentiability conditions for the existence of solutions of stochastic and parabolic equations are not needed, and the main tools are decomposition and representation results for martingales. A 'Hamiltonian' for the system is introduced, where the role of the co-state variables is played by the integrand that occurs in the representation of the minimum cost as a stochastic integral. The partially observable optimum control is then almost surely obtained by minimizing the conditional expectation of the Hamiltonian with respect to the observed σ-field. The results are an extension of those of Davis and Varaiya [2], because only the one Hamiltonian is involved. (In [2] there is a different Hamiltonian for each control.) Also, (although we do not consider state constraints), they are a simplification of the work of Haussmann [5] who bases his work on a result of Neustadt on extremals. The dynamic programming condition is obtained by differentiating a certain inequality, and to justify this procedure delicate L^P estimates are needed for the Radon-Nikodym derivatives that arise from Girsanov's theorem. Details are given in [3].

2. THE STOCHASTIC SYSTEM

We first describe the standard conditions on the dynamics and cost. Consider a system whose dynamics are described by an equation of the form

$$dx_t = f(t,x,u)dt + \sigma(t,x)dB_t \tag{2.1}$$

Here $t \in [0,1]$, B is an m-dimensional Brownian motion, and $x \in C$, the space of continuous functions from [0,1] to R^m. The drift term f can depend at time t on the past $\{x_s : s \leq t\}$ of the process. Write F_t for the σ-field $\{x_s : x \in C,\ s \leq t\}$.

Definition 2.1 We suppose the m × m matrix satisfies:

i) for $1 \leq i, j \leq m$, $\sigma_{ij}(t,\cdot) : C \to R$ is F_t measurable, and for each $x \in C$ $\sigma_{ij}(\cdot,x) : [0,1] \to R$ is Lebesgue measurable.

ii) $\sigma(t,x)$ is nonsingular.

iii) each σ_{ij} satisfies a uniform Lipschitz condition in x, where $x \in C$ is given the uniform norm $\|x\|_s = \sup_{0 \leq t \leq s} |x(t)|$.

iv) there is a constant $k_0 < \infty$ such that for almost every $x \in C$ $\Sigma \int_0^1 \sigma_{ij}^2(s)ds < k_0$.

Having given an m-dimensional Brownian motion B_t on a probability space (Ω,μ) the solution of the equation

$$x(t) = x_0 + \int_0^t \sigma(s,x)dB_s$$

enables us to define a probability measure P on (C,F_1) by putting

$$P(A) = \mu\{\omega : x(\omega) \in A\}, \text{ for } A \in F_1.$$

Suppose now that $x \in R^m$ is written in terms of two components

$$x = (y,z), \text{ where } y \in R^n, \; z \in R^{m-n}$$

Correspondingly f and σ will be written

$$f = (f_1,f_2), \qquad \sigma = \begin{pmatrix}\sigma_1\\ \sigma_2\end{pmatrix}$$

where f_1 (resp. f_2) is an n (resp. m - n) dimensional vector function, and σ_1 (resp. σ_2) is an $n \times m$ (resp. $(m - n) \times m$) matrix function. The variables y represent the (noisy) partial observations that are made of the whole process, and

$$dy_t = f_1(t,x,u)dt + \sigma_1(t,x)dB_t$$

The observation σ-field, Y_t, is the sub σ-field of F_t defined by

$$Y_t = \sigma\{y_s : s \le t\}$$

Of course, if n = m we say the system is completely observable, so $Y_t = F_t$.

Suppose the control space U is a compact metric space with the Borel field $\mathcal{U}$.

Definition 2.2 A partially observable (completely observable) feedback control over $[s,t] \subset [0,1]$ is a measurable function $u : [s,t] \times C \to U$ such that

i) for each $\tau \in [s,t], u(\tau,\cdot)$ is $\mathcal{Y}_\tau$ (resp. $\mathcal{F}_\tau$) measurable

ii) for each $x \in C$, $u(\cdot,x)$ is Lebesgue measurable

Write $\mathcal{N}_s^t$ (resp. $\mathcal{M}_s^t$) for the set of such controls.

Definition 2.3 The drift function $f : [0,1] \times C \times U \to R^m$ is a measurable function satisfying the following conditions:

i) $f(\cdot,x,u)$ is Lebesgue measurable

ii) $f(t,\cdot,u)$ is $\mathcal{F}_t$ measurable

iii) $f(t,x,\cdot)$ is continuous on U

iv) $|\sigma^{-1}(t,x)f(t,x,u)| \leq K(1 + \|x\|_t)$

The conditions on f ensure that for each control function $u \in \mathcal{M}_0^1$

$$E\left[\rho_s^t(u) \,|\, \mathcal{F}_s\right] = 1 \qquad \text{a.s.}$$

where

$$\rho_s^t(u) = \exp \xi_s^t(f^u)$$

$$f^u(\tau,x) = f(\tau,x,u(\tau,x))$$

and

$$\xi_s^t(f^u) = \int_s^t \{\sigma^{-1}(\tau,x)f^u(\tau,x)\}' dB_\tau - \frac{1}{2}\int_s^t |\sigma^{-1}(\tau,x)f^u(\tau,x)|^2 d\tau$$

Girsanov's theorem [4] then states the following.

Theorem 2.4 Define a new probability measure μ_u on Ω by putting

$$d\mu_u/d\mu = \exp \xi_0^1(f^u)$$

Then under μ_u w_t^u is a Brownian motion where

$$dw_t^u = \sigma^{-1}(t,x)(dx_t - f^u(t,x)dt)$$

That is,

$$dx_t = f^u(t,x)dt + \sigma(t,x)dw_t^u$$

and

$$x(0) = x_0$$

We suppose the cost associated with the process is of the form

$$g(x(1)) + \int_0^1 c(t,x,u)dt$$

where g and c are real, bounded and measurable. Corresponding to a control u, the expected total cost is

$$J(u) = E_u\left[g(x(1)) + \int_0^1 c(t,x,u(t,x))dt\right]$$

Here E_u denotes expectation with respect to μ_u. The optimal control problem for the partially observed system is to determine how $u \in N_0^1$ should be chosen so that J(u) is minimized.

§3. A PARTIALLY OBSERVED OPTIMALITY PRINCIPLE

Suppose control $u \in N_0^t$ is used to time t and control $v \in N_t^1$ is used from time t to time 1. Then a control $w \in N_0^1$ can be defined by concatenation. The expected remaining cost in this situation, given the information Y_t, is

$$\tilde{\psi}_w(t) = E_w\left[g(x(1)) + \int_t^1 c_s^v ds \,|\, \mathcal{Y}_t\right]$$

where $c_s^v = c(s,x,v(s,x))$. From 24.4 of [6] this is

$$= \frac{E\left[\rho_0^t(u)\rho_t^1(v)\left(g(x(1)) + \int_t^1 c_s^v ds\right) | \mathcal{Y}_t\right]}{E\left[\rho_0^t(u)\rho_t^1(v) \,|\, \mathcal{Y}_t\right]}$$

and the denominator is just $E[\rho_0^t(u)|\mathcal{Y}_t]$. Because $L^1(\Omega,\mu)$ is a complete lattice, for a fixed $u \in \mathcal{N}_0^t$ the infimum of the functions $\tilde{\psi}_w(t)$ exists; call it $W_u(t)$, so that $W_u(t)$ represents the minimum remaining cost that can be incurred from time t to time 1, given $\mathcal{Y}_t$.

From Theorem 3.1 of [2] we quote the following result.

Theorem 3.1

i) $u^* \in \mathcal{N}_0^1$ is optimal if and only if for each $t \in [0,1]$ and $h > 0$

$$W_{u^*}(t) = E_{u^*}\left[\int_t^{t+h} c_s^{u^*} ds \,|\, \mathcal{Y}_t\right] + E_{u^*}[W_{u^*}(t+h)\,|\,\mathcal{Y}_t] \quad \text{a.s.}$$

ii) in general, for $u \in \mathcal{N}_0^1$

$$W_u(t) \le E_u\left[\int_t^{t+h} c_s^u ds \,|\, \mathcal{Y}_t\right] + E_u[W_u(t+h)\,|\,\mathcal{Y}_t]$$

Suppose there is a partially observable optimal control $u^* \in \mathcal{N}_0^1$. Write

$$\tilde{W}(t) = E_{u^*}\left[g(x(1)) + \int_t^1 c_s^{u^*} ds \,|\, \mathcal{F}_t\right]$$

so that, because u^* is optimal,

$$W_{u^*}(t) = E_{u^*}[\tilde{W}(t)\,|\,\mathcal{Y}_t]$$

By adding the $\mathcal{Y}_t$ measurable quantities $E_{u*}\left[\int_0^t c_s^{u*} ds \mid \mathcal{Y}_t\right]$ (resp. $E_{u*}\left[\int_0^t c_s^u \, ds \mid \mathcal{Y}_t\right]$) to the above relations, we have the following 'optimality principle':

Corollary 3.2

i) $u^* \in \mathcal{N}_0^1$ is optimal if and only if $E_{u*}[N_t^* \mid \mathcal{Y}_t]$ is a $(\mathcal{C}, \mathcal{Y}_t, \mu_{u*})$ martingale, where $N_t^* = \int_0^t c_s^{u*} ds + \tilde{W}(t)$.

ii) for general $u \in \mathcal{N}_0^1$ and $h \geq 0$

$$E_{u*}[N_t^u \mid \mathcal{Y}_t] \leq E_{u*}[E_u[N_{t+h}^u \mid \mathcal{F}_t] \mathcal{Y}_t]$$

where

$$N_t^u = \int_0^t c_s^u ds + \tilde{W}(t)$$

§4. DYNAMIC PROGRAMMING CONDITIONS

Suppose u^* is an optimal partially observable control. Then by the martingale representation result (see [1]), because N_t^* is a square integrable $\mathcal{F}_t$ martingale there is a predictable process g^* such that $\int_0^1 E^*(g_s^*)^2 ds < \infty$ and

$$N_t^* = J^* + \int_0^t g^* dw^*$$

Here E^* denotes expectation with respect to $\mu^* = \mu_{u*}$, $J^* = \tilde{W}(0)$ and w^* is the Brownian motion on (Ω, μ^*) given by

$$dw_t^* = \sigma^{-1}(t,x)(dx_t - f^{u*}(t,x)dt)$$

By simple manipulation we have the following representation for N_t^u:

$$N_t^u = J^* + \int_0^t g^* dw^u + \int_0^t (g^*\sigma^{-1} f^u + c^u) - (g^*\sigma^{-1} f^{u*} + c^{u*}) ds \tag{4.1}$$

Here w^u is the Brownian motion on (Ω, μ_u) given by

$$dw_t^u = \sigma^{-1}(t,x)(dx_t - f^u(t,x)dt)$$

Write

$$\phi_t^u = (g^*\sigma^{-1}f^u + c^u) - (g^*\sigma^{-1}f^{u^*} + c^{u^*})$$

Substituting in the inequality of Corollary 3.2 (ii), we have the following fundamental result.

Proposition 4.1 Suppose $u^* \in \mathcal{N}_0^1$ is an optimal control and g^* is the process obtained in the integral representation of N_t^*. Then for any $t \in [0,1]$ and all $h \geq 0$

$$E_{u^*}\left[E_u\left[\int_t^{t+h} \phi_s^u ds \mid \mathcal{F}_t\right] \mathcal{Y}_t\right] \geq 0, \qquad \text{a.s.}$$

Remark 4.2 What we would like to do now is differentiate, that is, divide the above inequality by $h > 0$ and let $h \to 0$ to obtain

$$E_{u^*}\left[\phi_s^u | \mathcal{Y}_t\right] \geq 0$$

However, because the expectations E_{u^*} and E_u with respect to different measures occur, this is not immediately possible, and very delicate L^P bounds for the Radon-Nikodym derivatives $\rho_0^t(u^*)$, $\rho_0^t(u)$ are needed. Great care has also to be taken with the sets of measure zero on which the above inequality is valid. Detailed calculations justifying the 'differentiation' of the above expression are given in [3] and the final dynamic programming result is now given.

Theorem 4.3 Suppose $u^* \in \mathcal{N}_0^1$ is an optimal partially observable control, and $u \in \mathcal{N}_0^1$ is any other control. Then there is a set of zero measure $T \subset [0,1]$ such that if $t \notin T$

$$E^*\left[\left(g^*\sigma^{-1}f_t^u + c_t^u\right) - \left(g^*\sigma^{-1}f_t^{u^*} + c_t^{u^*}\right) | \mathcal{Y}_t\right] \geq 0 \qquad \text{a.s.}$$

That is, the partially observable optimal control minimizes the conditional expectation of the 'Hamiltonian':

$$E^*[g^*\sigma^{-1}f^u + c^u | \mathcal{Y}_t]$$

§5. CONCLUDING REMARKS

If the system is completely observable, so $\mathcal{Y}_t = \mathcal{F}_t$, then the processes N_t^*, N_t^u are $\mathcal{F}_t$ measurable and Corollary 3.2 (ii) states that

$$N_t^u \leq E_u[N_{t+h}^u | \mathcal{F}_t]$$

that is, N_t^u is a submartingale under μ_u. Therefore, N_t^u has a unique Doob-Meyer decomposition as $J^* + M_t^u + A_t^u$, where M_t^u is a martingale on (Ω, μ_u) and A_t^u is a predictable increasing process. However, N_t^u is certainly a 'semi-martingale speciale' (see Meyer [7]) so its decomposition into a martingale plus a predictable integrable variation process is unique. The Doob-Meyer decomposition of N_t^u must, therefore, be the same as the decomposition given in (4.1) above. In particular

$$A_t^u = \int_0^t \phi_s^u ds$$

is an increasing process, so $\phi_s^u \geq 0$ a.s. In the completely observable case, therefore, we obtain immediately the following dynamic programming result without incurring the differentiation problems described above.

<u>Theorem 5.1</u> If $u^* \in \mathcal{M}_0^1$ is a completely observable optimal control then:

$$g^{*\,-1}f^{u^*} + c^{u^*} = \inf_{u \in U} g^{*\,-1}f^u + c^u \text{ a.s.}$$

Finally, if there is not an optimal control we consider

a sequence of controls $u_n \in N_0^1$ such that $J(u_n)$ converges to the minimum J^*. Using compactness of a certain measure space, even though there is not an optimal control, there is an optimum measure μ^* and analogs of the above results can be obtained. For details see [3].

REFERENCES

1. Clark, J. M. C., The Representation of Functionals of Brownian Motion by Stochastic Integrals, Ann. Math. Stat. 41 (1970), 1282-1295.
2. Davis, M. H. A. and P. P. Varaiya, Dynamic Programming Conditions for Partially Observable Stochastic Systems, SIAM Jour. Control 11 (1973), 226-261.
3. Elliott, R. J., The Optimal Control of a Stochastic System, Hull University U. K., March 1976, preprint.
4. Girsanov, I. V., On Transforming a Certain Class of Stochastic Processes by Absolutely Continuous Substitution of Measures, Theory Prob. App. 5 (1960), 285-301.
5. Haussmann, U. G., General Necessary Conditions for the Optimal Control of Stochastic Systems, Symposium on Stochastic Systems, University of Kentucky, June 1975.
6. Loeve, M., Probability Theory, 3rd Ed., Van Nostrand, Princeton, New Jersey, 1963.
7. Meyer, P. A., Un cours sur les intégrales stochastiques, Sem. Prob. Univ. Strasbourg, 1974-5. Lecture Notes in Math., Springer-Verlag, vol. 511, Berlin, Heidelberg, New York, 1976.

GENERALIZED SOLUTIONS IN OPTIMAL STOCHASTIC CONTROL

Wendell H. Fleming

Brown University
Providence, Rhode Island

§1. INTRODUCTION

Generalized solutions in calculus of variations were introduced many years ago by L. C. Young [13]. By doing so, he obtained solutions in some wider sense to problems which have no ordinary solution. Similar ideas reappeared in optimal control theory under such names as relaxed controls, chattering controls, or sliding regimes [2, Chap. IV, 4, 10, 12]. Under appropriate assumptions a relaxed optimal control was shown to exist. If, in addition, a Filippov-type convexity condition holds, the methods show that there is an optimal control in the ordinary sense.

In Sections 2, 3, 4 we make a straightforward adaptation of the idea of relaxed control, to a class of problems which are stochastic perturbations of the standard Pontryagin control problem. In Section 5, we prove the existence of a relaxed optimal control, by a method of [7]. That result applies if the controller has complete information at each time t, and also if the control is open loop.

If the controller has complete information and noise enters every system component, then the methods of Davis [5] imply the existence of an ordinary (feedback) optimal control. These methods make no use of Filippov-type convexity assumptions. Hence, relaxed controls are unnecessary for this particular class of problems. See the Remark at the end of Section 5.

The general question of existence of an optimal stochastic control is open, in case the controller has partial information. In the last part of the paper, a possible approach to this problem is outlined, which considers a new kind of generalized solution. Following Benes [1] the stochastic control problem is reformulated in Section 6 as one of choosing an optimal Girsanov density ρ in a certain class $\mathcal{U}$. Under appropriate assumptions, $\mathcal{U}$ is a bounded subset of the appropriate L^2-space. Moreover, $\mathcal{U} \subset \mathcal{U}_0$ where $\mathcal{U}_0$ is the corresponding set of densities if the controller is allowed complete information. Benes [1] showed that $\mathcal{U}_0$ is convex and weakly closed, if the Filippov convexity condition holds. However, examples given in [1] and by Duncan-Varaiya [6] show that $\mathcal{U}$ need be neither convex nor weakly closed in case of partial information. In Section 7, we define a generalized solution to be a density $\rho \in \bar{\mathcal{U}}$, where $\bar{\mathcal{U}}$ is the weak closure of the convex hull of $\mathcal{U}$. Existence of an optimal generalized solution is immediate. There remains the problem of characterizing generalized solutions. A partial characterization is given in Section 8, using the idea of auxiliary randomizations (i.e., mixtures of densities).

2. STOCHASTIC VERSION OF THE PONTRYAGIN PROBLEM

We use the following notation. Let t denote time, $0 \le t \le T$. The system state at time t is a vector $x(t) = (x_1(t),$

$\cdots, x_n(t))$ in n-dimensional E^n. The control applied at time t is u(t). We require that $u(t) \in U$, where U is a given compact metric space called the control space. Both state and control are stochastic processes, defined on some probability space (Ω, F, P). The state process is modelled as the solution to a system of stochastic differential equations (Itô sense), written in vector-matrix notation as

$$dx = f(t,x(t),u(t))dt + \sigma(t,x(t))dw \tag{2.1}$$

where $w = (w_1, \cdots, w_d)$ is a Brownian motion of some dimension d.

It is assumed that f is continuous on $[0,T] \times E^n \times U$, and σ continuous on $[0,T] \times E^n$. Moreover, $f(t,\cdot,u)$, $\sigma(t,\cdot)$ satisfy Lipschitz conditions with Lipschitz constant not depending on t or u. The initial data x(0) for (2.1) have a given distribution π, such that $E|x(0)|^k < \infty$ for any $k > 0$. If M is a metric space, by M-valued random variable we mean a function v from a probability space (Ω, F, P) into M such that v is measurable $(F, \mathcal{B}(M))$, where $\mathcal{B}(M)$ is the σ-algebra of Borel subsets of M. If $\{F_t\}$ is an increasing family of σ-algebras, with $F_t \subset F$ for $0 \le t \le T$, then by nonanticipative stochastic process with respect to $\{F_t\}$ we mean a measurable process u such that u(t) is F_t-measurable for $0 \le t \le T$.

We consider triples (x(0),u,w) which satisfy, for some such family $\{F_t\}$: (1) x(0) is F_0-measurable and has distribution π; (2) u is nonanticipative with respect to $\{F_t\}$; (3) w is a P-Brownian motion adapted to $\{F_t\}$; i.e., w is nonanticipative with respect to $\{F_t\}$, and the increments of w for times $\ge t$ are independent of F_t. Given (x(0),u,w), equations (2.1) have a unique solution x, which is nonanticipative with respect to $\{F_t\}$ and has continuous sample paths. Moreover, $E\|x\|^k < \infty$ for each $k > 0$, where $\| \ \|$ is the sup norm in the space $C^n[0,T]$ of continuous functions from $[0,T]$ into E^n (see [9], p. 107).

Let Φ be a continuous real-valued function on $C^n[0,T]$ such that for suitable constants M, k

$$|\Phi(h)| \le M(1 + \|h\|)^k, \text{ all } h \in C^n[0,T] \tag{2.2}$$

We take $E\Phi(x)$ as the criterion to be minimized.

The probability space is not fixed. Let us say that the controller has *complete information* if any triple $(x(0),u,w)$ is allowed, subject only to conditions (1), (2), (3) above with respect to the probability space (Ω, F, P) on which this triple is defined and corresponding family $\{F_t\}$. If instead of (2) we require that u is a function on $[0,T]$ (not a stochastic process), then we say that the control is *open loop*.

There is no optimal control without further restrictions on the problem. However, we show in Section 5 that there is an optimal relaxed control.

§3. RELAXED CONTROL PROBLEMS

In the relaxed form of a control problem, the controller chooses at time t a probability measure μ_t on the control set U rather than an element $u(t) \in U$. The state process x is required to satisfy, instead of (2.1),

$$dx = \int_U f(t,x(t),u)\,d\mu_t(u)\,dt + \sigma(t,x(t))\,dw \tag{3.1}$$

with initial data $x(0)$. We call μ_t the *relaxed control* applied at time t. If μ_t is an atomic measure, concentrated at a single point u(t) for each t, then we get an ordinary control as a special case of a relaxed control.

In order to describe the relaxed problem more precisely, let $\mathcal{M}(U)$ denote the set of all probability measures on $\mathcal{B}(U)$. We give $\mathcal{M}(U)$ a metric compatible with weak* convergence of sequences of measures. Let

$$\tilde{f}(t,x,\mu) = \int_U f(t,x,u)\,d\mu(u) \tag{3.2}$$

Then $\tilde{f}$ is continuous on $[0,T] \times E^n \times \mathcal{M}(U)$; and $\tilde{f}$ is linear in μ. Equations (3.1) can be rewritten as

$$dx = \tilde{f}(t,x(t),\mu_t)dt + \sigma(t,x(t))dw \tag{3.3}$$

We require that the relaxed control process μ be $\{\mathcal{F}_t\}$ nonanticipative. Now $\tilde{f}(t,\cdot,\mu)$ satisfies a Lipschitz condition, with the same Lipschitz constant as f. Therefore, (3.3) with the initial data x(0) has a unique solution x, which is nonanticipative, with continuous sample paths, and with $E\|x\|^k < \infty$ for any $k > 0$.

By introducing relaxed controls, we have merely replaced the control space U by a larger metric space $\mathcal{M}(U)$, and f by $\tilde{f}$. We have gained the advantage that $\mathcal{M}(U)$ is both compact and convex, and $\tilde{f}$ is linear in the relaxed control variable μ.

§4. CHATTERING LEMMA

In order for the relaxed control problem to be truly an extension of the original problem, the infimum of the criterion $E\Phi(x)$ among relaxed controls must be the same as the infimum of $E\Phi(x)$ taken among ordinary controls. For this purpose we adapt a chattering lemma for the Pontryagin program [2,IV.4].

In Lemmas 1 and 2, (Ω, F, P) is a probability space and $\{F_t\}, \{G_t\}$ increasing families of σ-algebras with $G_t \subset F_t \subset F$ for $0 \le t \le T$. We consider a relaxed control process μ, which is $\{G_t\}$-nonanticipative. Thus μ is a measurable process with values in $M(U)$, and μ_t is G_t-measurable for $0 \le t \le T$.

Lemma 1 For $r = 1, 2, \cdots$, $0 \le t \le T$, there exists μ_t^r such that:

(a) μ_t^r is an atomic measure, concentrated at $u^r(t) \in U$.

(b) u^r is a $\{G_t\}$-nonanticipative process.

(c) For any g continuous on $[0,T] \times M(U)$ such that $g(t,\cdot)$ is linear, and for each $\omega \in \Omega$,

$$\lim_{r\to\infty} \int_0^t g(s,\mu_s^r)\,ds = \int_0^t g(s,\mu_s)\,ds$$

uniformly for $0 \le t \le T$.

To prove Lemma 1 one can slightly modify a standard construction, taking care to arrange that (b) holds. Let us outline the four steps of this construction.

Step 1 Given $\varepsilon > 0$ there is a finite set $V \subset U$ such that each $u \in U$ is distant less than ε from V. Let ψ be Borel measurable from U onto V, with $|\psi(u) - u| < \varepsilon$; and let $\hat{\mu}_t = \psi(\mu_t)$.

Step 2 Let $u_1, \cdots, u_m$ be the points of V, and $\alpha_{it} = \hat{\mu}_t(\{u_i\})$, $i = 1, \cdots, m$. Given $\delta > 0$ let

$$\beta_{it} = \delta^{-1} \int_{t-\delta}^t \alpha_{is}\,ds$$

where $\alpha_{is} = \alpha_{i0}$ if $s < 0$. Note that β_{it} is continuous in t, and G_t-measurable.

Step 3 Given a subdivision of $[0,T]$, namely $0 = t_0 < t_1 < \cdots < t_M = T$, let $\gamma_{it} = \beta_{it_k}$ for $t_k \le t < t_{k+1}$. Then γ_{it} is a step process, and γ_{it} is $\mathcal{G}_t$-measurable. Moreover, $\sum_{i=1}^m \gamma_{it} = 1$.

Step 4 Subdivide $[t_k, t_{k+1})$ into m subintervals I_{ik}, where I_{ik} has length $\gamma_{it_k}(t_{k+1} - t_k)$. Let $\hat{\hat{\mu}}_t$ be atomic, concentrated at u_i for $t \in I_{ik}$.

In Lemma 1 we take $\mu_t^r = \hat{\hat{\mu}}_t$, for $\varepsilon = \varepsilon_r$, $\delta = \delta_r$ sufficiently small, and sufficiently fine subdivision of $[0,T]$ depending on r.

Lemma 2 Let x satisfy (3.3), and let x^r satisfy the corresponding equation $dx^r = f(t,x^r(t),\mu_t^r)dt + \sigma(t,x^r(t))dw$ with $x^r(0) = x(0)$ and μ_t^r as in Lemma 1. Then $E\|x^r - x\|^2 \to 0$ and $E\Phi(x^r) \to E\Phi(x)$ as $r \to \infty$.

Proof We write the stochastic differential equations for x^r and x in integrated form and subtract, obtaining

$$x^r(t) - x(t) = \int_0^t [\tilde{f}(s,x^r(s),\mu_s^r) - \tilde{f}(s,x(s),\mu_s^r)]ds$$

$$+ \int_0^t [\sigma(s,x^r(s)) - \sigma(s,x(s))]ds + \eta^r(t) \qquad (*)$$

$$\eta^r(t) = \int_0^t [\tilde{f}(s,x(s),\mu_s^r) - \tilde{f}(s,x(s),\mu_s)]ds$$

By Lemma 1, with $g(t,\mu) = \tilde{f}(t,x(t),\mu)$, $\|\eta_r\| \to 0$ as $n \to \infty$, for each $\omega \in \Omega$. Moreover, there is a constant B such that $|\tilde{f}(t,x,\mu)| \le B(1 + \|x\|)$ for all (t,x,μ). Thus, $\|\eta_r\| \le 2BT(1 + \|x\|)$. Since $E\|x\|^2 < \infty$, the dominated convergence theorem implies that $E\|\eta_r\|^2 \to 0$ as $r \to \infty$. Let

$$h_r(t) = E \max_{0 \le s \le t} |x^r(s) - x(s)|^2$$

We square both sides of (*), and use the Lipschitz condition on $\tilde{f}(t,\cdot,\mu)$, $\sigma(t,\cdot)$ together with $(a + b + c)^2 \le 3(a^2 + b^2 + c^2)$ to obtain

$$h_r(t) \le K\left[\int_0^t h_r(s)ds + E\|\eta_r\|^2\right]$$

for suitable constant K. Gronwall's inequality then implies that $h_r(T) = E\|x^r - x\|^2$ tends to 0 as $n \to \infty$.

Finally, $E\|x^r\|^\ell$ is bounded by a constant depending on ℓ, but not on r [9, p. 107]. We take $\ell > k$, with k as in (2.2). Then $\Phi(x^r)$ is uniformly integrable. Since $\|x^r - x\| \to 0$ in mean square (hence also in probability) this implies that $E\Phi(x^r) \to E\Phi(x)$ as $n \to \infty$. This proves Lemma 2.

Note that in Lemma 2, $\tilde{f}(t,x^r(t),\mu_t^r) = f(t,x^r(t),u^r(t))$. Thus x^r is the solution of (2.1) with $x^r(0) = x(0)$ corresponding to the ordinary control u^r. In Lemma 2 we may take in particular $\mathcal{G}_t = \mathcal{F}_t$ (the complete information case), or $\mathcal{G}_t$ the trivial σ-algebra (the open loop case). In either case, it is immediate from Lemma 2 that <u>the infimum</u> of $E\Phi(x)$ <u>is the same among relaxed controls</u> μ <u>as among ordinary controls</u> u.

§5. EXISTENCE OF AN OPTIMAL RELAXED CONTROL

Let us consider the relaxed problem, either with complete information or open loop.

<u>Theorem 1</u> An optimal relaxed control exists.

This theorem will be proved by the method of [7]. In order to adapt the proof there we use Lemma 3.

Let C(U) denote the space of continuous real-valued functions on U. We consider the time integral M_t of μ_t:

$$M_t(\phi) = \int_0^t \mu_s(\phi)ds, \text{ for all } \phi \in C(U)$$

The process M_t is measure-valued, with values in the dual $C^*(U)$. Moreover:

(a) $\|M_t - M_s\| \leq |t - s|$ where $\| \ \|$ is the norm in $C^*(U)$

(b) $M_t(\phi)$ is increasing in t if $\phi \geq 0$

(c) $M_t(1) = t$

We also have $dM_t/dt = \mu_t$ in the w* sense. Finally, M_t has values in $N(U) = \{\nu \in C^*(U) : \nu \geq 0, \nu(1) \leq T\}$. We give $N(U)$ a metric d compatible with w* convergence of sequences of measures.

In Lemma 3, x^r, x, M^r, M are defined, for r = 1, 2, $\cdots$, on the same probability space. Convergence of M^r to M refers to the metric d. Moreover, $\mu_t^r = dM^r/dt$, $\mu_t = dM/dt$.

<u>Lemma 3</u> Suppose that $x^r \to x$ and $M^r \to M$ uniformly on [0,T], with probability 1. Then

$$\lim_{r\to\infty} \int_0^t \tilde{f}(s,x^r(s),\mu_s^r)ds = \int_0^t \tilde{f}(s,x(s),\mu_s)ds$$

with probability 1.

<u>Proof</u> First of all, $\tilde{f}(s,x^r(s),\mu) - \tilde{f}(s,x(s),\mu)$ tends to 0 with probability 1, uniformly with respect to s and μ. In particular, this holds for $\mu = \mu_s^r$. Therefore, it suffices to show that with probability 1

$$\lim_{r\to\infty} \int_0^t g(s,\mu_s^r)ds = \int_0^t g(s,\mu_s)ds$$

where $g(s,\mu) = \tilde{f}(s,x(s),\mu)$. Consider a subdivision $0 = t_0 < t_1 < \cdots < t_M = t$ of $[0,t]$. Let $\phi_k(u) = f(t_k,x(t_k),u)$ for $t_k \le s < t_{k+1}$. Since f is continuous

$$\int_0^t g(s,\mu_s^r)ds = \sum_{k=0}^{M-1}\left[M_{t_{k+1}}^r(\phi_k) - M_{t_k}^r(\phi_k)\right] + \theta^r \qquad (*)$$

$$\int_0^t g(s,\mu_s)ds = \sum_{k=0}^{M-1}\left[M_{t_{k+1}}(\phi_k) - M_{t_k}(\phi_k)\right] + \theta \qquad (**)$$

where with probability 1, $\theta^r \to 0$ uniformly with respect to r and $\theta \to 0$, as the mesh of the subdivision tends to 0. Since $M_{t_k}^r(\phi) \to M_{t_k}(\phi)$ as $r \to \infty$ for each $\phi \in C(U)$, the sum on the right side of (*) tends to the corresponding sum on the right side of (**) as $r \to \infty$. This gives Lemma 3.

Let us now outline a proof of Theorem 1 in the complete information case, following [7]. The open loop case is similar. In the relaxed problem we consider all triples $(x(0), \mu, w)$, defined on some probability space (Ω, F, P) which is provided with an increasing family $\{F_t\}$ of σ-algebras ($F_t \subset F$), such that

(1) $x(0)$ is F_0-measurable and has given distribution π;

(2) μ is a process with values in $M(U)$ nonanticipative with respect to $\{F_t\}$; and

(3) w is a P-Brownian motion adapted to $\{F_t\}$.

Such triples $(x(0), \mu, w)$ are called *admissible*. Let x be the solution of (3.3) with initial data $x(0)$ corresponding to an admissible triple.

Let a denote the infimum of $E\Phi(x)$ taken among all admissible triples. Consider any sequence $(x^r(0), \mu^r, w^r)$, r =

1, 2, $\cdots$, with corresponding solution x^r of (3.3) such that $E\Phi(x^r) \to a$ as $r \to \infty$. Let M_t^r be the time integral of μ_t^r as above. By theorems of Skorokhod and Prohorov [7, p. 781] there exist processes $\bar{x}^r$, $\bar{M}^r$, $\bar{w}^r$ such that

(i) $\bar{x}^r$, $\bar{M}^r$, $\bar{w}^r$ are defined on the same probability space $(\bar{\Omega}, \bar{F}, \bar{P})$ for all $r = 1, 2, \cdots$;

(ii) the triples (x^r, M^r, w^r), $(\bar{x}^r, \bar{M}^r, \bar{w}^r)$ are identical in probability law; and

(iii) $\bar{x}^r$, $\bar{M}^r$, $\bar{w}^r$ tend to respective limits $\bar{x}$, $\bar{M}$, $\bar{w}$ as $r \to \infty$, uniformly with probability 1.

In applying these theorems we use the fact that $N(U)$ is a compact metric space, in the metric d. Since $E\|\bar{x}^r\|^k$ is bounded independent of r for each $k > 0$, and Φ satisfies (2.2), $E\Phi(\bar{x}^r) = E\Phi(x^r)$ tends to $E\Phi(x)$ as $r \to \infty$. Hence $E\Phi(x) = a$.

We take $\bar{F}_t$ as the σ-algebra generated by $\bar{x}(s)$, $\bar{M}(s)$, $\bar{w}(s)$ for $s \le t$. Let $\bar{\mu}_t = d\bar{M}_t/dt$. Then the triple $(\bar{x}(0), \bar{\mu}, \bar{w})$ is admissible. It remains to show that $\bar{x}$ is the corresponding solution of (3.3). Now

$$\bar{x}^r(t) = \bar{x}^r(0) + \int_0^t \tilde{f}(s,\bar{x}^r(s),\bar{\mu}_s^r)ds + \int_0^t \sigma(s,\bar{x}^r(s))d\bar{w}^r$$

We apply Lemma 3 to the first integral on the right side (with $\bar{x}^r$, $\bar{M}^r$ in place of x^r, M^r). The second integral on the right side tends in probability to $\int_0^t \sigma(s,\bar{x}(s))d\bar{w}(s)$. (See [7, Lemma 6]. Hence

$$\bar{x}(t) = \bar{x}(0) + \int_0^t f(s,\bar{x}(s),\bar{\mu}_s)ds + \int_0^t \sigma(s,\bar{x}(s))d\bar{w}$$

This is the integrated form of (3.3).

We have shown that the triple $(\bar{x}(0), \bar{\mu}, \bar{w})$ is admissible, with $E\Phi(\bar{x}) = a$. This proves Theorem 1.

Remarks Theorem 1 can be applied to the case when the criterion to be minimized has the form

$$E\int_0^T L(t,x(t),u(t))dt$$

This is done by the standard device of introducing an additional system component x_{n+1} such that $dx_{n+1} = L\,dt$. If $\sigma(t,x)$ is an $n \times n$ matrix with bounded inverse $\sigma^{-1}(t,x)$, then Davis [5] showed by other methods that in the complete information case an ordinary optimal control exists. Similar existence results can be obtained using dynamic programming arguments and results from partial differential equations [8, VI.6, VI.8]. However, these methods depend on the complete information assumption. The following example shows that there is no corresponding result in the open loop case.

Example The problem is to minimize

$$E\int_0^T \{x(t)^2 + [1 - u(t)^2]^2\}dt$$

where equation (2.1) has the form $dx = u\,dt + dw$ with $x(0) = 0$. Here x, u, w are 1-dimensional, and $U = [-1, 1]$. The control u (open loop) is a measurable function from $[0,T]$ into U. The separation principle [8, VI.11] applies to this example. The optimal control minimizes

$$\int_0^T \{\hat{x}(t)^2 + [1 - u(t)^2]^2\}dt$$

where $\hat{x}(t) = Ex(t)$ satisfies $d\hat{x} = u(t)dt$ with $\hat{x}(0) = 0$. This problem has no ordinary solution. A relaxed solution is to let μ_t be the measure assigning probability 1/2 to each of the points $u = \pm 1$.

6. PROBLEMS WITH PARTIAL INFORMATION

In a model for stochastic control problems with partial information which has often been considered, it is supposed that the controller can observe a process y of some dimension m, satisfying a stochastic differential equation

$$dy = f_1(t,x(t),y(t))dt + \sigma_1(t,x(t),y(t))dw_1 \qquad (6.1)$$

with $y(0) = 0$. Here x is a solution of (2.1) and w_1 a Brownian motion independent of w. See [8, VI.10] and references cited there. Consider the process $z = (x,y)$ of dimension $p = n + m$. Then (2.1) and (6.1) can be rewritten in the form

$$dz = F(t,z(t),u(t))dt + \Sigma(t,z(t))dW \qquad (6.2)$$

At time t, the controller knows the components y(s) of the vector z(s) for all $s \le t$.

In the remainder of this paper, we consider the following problem. As in [1], [5], [6] we fix Ω and an increasing family of σ-algebras. In fact, $\Omega = C^p[0,T]$ and $\mathcal{F}_t$ is generated by the past up to t of functions in Ω. Let $\mathcal{F} = \mathcal{F}_T$. The probability measure P on $\mathcal{F}$ depends on the control. We also fix another family of σ-algebras $\mathcal{G}_t$, with $\mathcal{G}_t \subset \mathcal{F}_t$. The σ-algebra $\mathcal{G}_t$ contains the information available to the controller in time t. In (6.2), we require that u(t) be $\mathcal{G}_t$-measurable. As in Section 2, $u(t) \in U$, where U is a compact metric space.

For simplicity we take $z(0) = 0$. The problem is to choose a control process u minimizing a criterion of the form $E\Psi(z)$.

We assume that F and Σ are bounded, continuous functions; and that $F(t,\cdot,u)$, $\sigma(t,\cdot)$ satisfy a uniform Lipschitz condition. Moreover, $\Sigma(t,z)$ is a $p \times p$ matrix with bounded inverse $\Sigma^{-1}(t,z)$.

In the particular case above, $z = (x,y)$, $F = (f, f_1)$, $\Sigma = (\sigma, \sigma_1)$, $\Psi(z) = \Phi(x)$, $W = (w, w_1)$, and G_t is the σ-algebra generated by $y(s)$ for $s \le t$.

We use the Girsanov theorem to reformulate this problem. Let P_0 be the unique probability measure on (Ω, F) such that W_0 is a P_0-Brownian motion, where W_0 satisfies $dW_0 = \Sigma^{-1}(t,z)dz$ with $W_0(0) = 0$. Given a control process u, let

$$\rho = \exp\left\{\int_0^T F_t A_t^{-1} dz - \frac{1}{2}\int_0^T F_t A_t^{-1} F_t dt\right\} \tag{6.3}$$

where $F_t = F(t,z(t),u(t))$, $A_t = A(t,z(t))$ and $A = \Sigma\Sigma^{-1}$. Let P be the measure on F whose Radon-Nikodym density with respect to P_0 is ρ ($dP/dP_0 = \rho$). Since F is bounded, $P(\Omega) = 1$. The Girsanov theorem [1], [9] implies that W is a P-Brownian motion, where $W(0) = 0$ and

$$dW = \Sigma^{-1}(dz - Fdt)$$

See also [6, pp. 588-9]. This equation is equivalent to (6.2). Let

$$J(u) = \int_\Omega \Psi(z)\rho(z)dP_0(z) \tag{6.4}$$

Let $\mathcal{U}_0$ denote the set of all densities ρ, such that in (6.3) u is any $\{F_t\}$-nonanticipative process with values in U.

Let $\mathcal{U}$ be the set of such ρ such that u(t) is G_t-measurable for $0 \leq t \leq T$. Clearly, $\mathcal{U} \subset \mathcal{U}_0$. Now

$$J(u) = \int_\Omega \Psi(z)dP(z) = E\Psi(z)$$

The problem is whether there exists a density $\rho \in \mathcal{U}$ minimizing J(u).

Since F is bounded, $\mathcal{U}_0$ is a bounded subset of $L^2 = L^2(\Omega, F, P_0)$. See [9, p. 157]. Benes [1] showed that $\mathcal{U}_0$ is also weakly closed and convex if the Filippov convexity condition holds. He used that fact to prove existence of an optimal density, if all densities $\rho \in \mathcal{U}_0$ are admitted (this corresponds to complete information). As shown by examples [1], [6] $\mathcal{U}$ may be neither convex nor weakly closed, even if the Filippov condition holds. In the next section, we consider generalized solutions, which correspond to points in the smallest weakly closed convex set $\bar{\mathcal{U}}$ which contains $\mathcal{U}$.

§7. GENERALIZED SOLUTIONS TO PROBLEMS WITH PARTIAL INFORMATION

In Section 3 relaxed controls were introduced, in order to obtain system equations linear in a new control variable μ which belongs to a convex control space $M(U)$. We now consider a new concept of generalized solution, which is introduced for quite different reasons.

Let co $\mathcal{U}$ denote the convex hull of $\mathcal{U}$, and $\bar{\mathcal{U}}$ the closure of co $\mathcal{U}$ in the space $L^2 = L^2(\Omega, F, P_0)$. Since co $\mathcal{U}$ is convex, $\bar{\mathcal{U}}$ is also the weak closure of co $\mathcal{U}$. Since $\mathcal{U}$ is bounded, $\bar{\mathcal{U}}$ is a weakly compact subset of L^2.

Definition Any density $\rho \in \bar{\mathcal{U}}$ is a generalized solution to the control problem formulated in Section 6.

Let us assume in (6.4) that $\Psi \in L^2$. From the form (6.4) of the criterion J(u) to be minimized we obtain immediately:

Theorem 2 A generalized solution exists.

It is also immediate that

$$\inf_{\rho \in \mathcal{U}} J(u) = \min_{\rho \in \bar{\mathcal{U}}} J(u) \tag{7.1}$$

There remains the unsolved problem of characterizing the densities $\rho \in \bar{\mathcal{U}}$. A partial result in this direction is given in the next section. By (6.4) the minimum of J(u) is attained at an extreme point of $\bar{\mathcal{U}}$. Thus, it would be of interest to characterize such extreme points. Finally, there is the question of giving conditions under which J(u) has an ordinary minimum, i.e., under which a minimizing $\rho \in \mathcal{U}$ exists.

§8. MIXTURES OF DENSITIES

Let Θ be some set and $\mathcal{B}$ a σ-algebra of subsets of Θ. Let $\hat{\Omega} = \Omega \times \Theta$, $\hat{F} = F \times \mathcal{B}$. We write $\langle\psi,\rho\rangle = \int_\Omega \psi dP_0$ for the usual inner product in L^2.

Definition The density ρ is a mixture over Θ if there exists a probability measure ν on $\mathcal{B}$ and, for ν-almost all $\theta \in \Theta$, a density $\rho_\theta \in \bar{\mathcal{U}}$ such that $\rho_\theta(z)$ is $\hat{F}$-measurable and

$$\langle\psi,\rho\rangle = \int_\Theta \langle\psi,\rho_\theta\rangle d\nu(\theta) \text{ for all } \psi \in L^2 \tag{8.1}$$

Since $\bar{\mathcal{U}}$ is weakly closed and convex, $\rho \in \bar{\mathcal{U}}$. Note that $J(u) = \langle\Psi,\rho\rangle$. Hence, if the mixture ρ minimizes J(u) on $\bar{\mathcal{U}}$, then ρ_θ minimizes $J(u)$ for ν-almost all θ. In particular, if it should happen that $\rho_\theta \in \mathcal{U}$, then there is a minimum in $\mathcal{U}$. It is interesting, though apparently difficult, open problem to find general conditions insuring that $\rho_\theta \in \mathcal{U}$ if ρ minimizes $J(u)$ on $\bar{\mathcal{U}}$.

Let us describe some particular ways in which mixtures arise.

1. For each $\theta \in \Theta$ let u_θ be a control process, corresponding to density $\rho_\theta \in \mathcal{U}$. Consider the process v defined on $\hat{\Omega}$ by $v(t,\omega,\theta) = u_\theta(t,\omega)$. We require that v be a measurable process. The corresponding density $\hat{\rho}(z,\theta)$ obtained from the Girsanov theorem (applied on $(\hat{\Omega}, \hat{F})$) is as follows. Let

$$\hat{\rho}(z,\theta) = \exp\left\{\int_0^T F_t A_t^{-1} dz - \frac{1}{2}\int_0^T F_t A_t^{-1} F_t dt\right\} \tag{8.2}$$

with $F_t = F(t,z(t),u_\theta(t))$. Let $\hat{P}_0 = P_0 \times \nu$ and $\hat{P}$ the measure on $\hat{F}$ with $d\hat{P}/d\hat{P}_0 = \hat{\rho}$. By the Girsanov theorem there is a $\hat{P}$-Brownian motion W such that $dz = Fdt + \Sigma dW$ and $z(0) = 0$, $\hat{P}$-almost surely. The distribution P of z has density $dP/dP_0 = \rho$, where

$$\rho(z) = \int_\Theta \hat{\rho}(z,\theta)d\nu(\theta)$$

For each θ, $\hat{\rho}(z,\theta) = \rho_\theta(z)$. Hence ρ is a mixture of densities $\rho_\theta \in \mathcal{U}$.

Note that the control u_θ appears in (8.2) in a highly nonlinear way. Even when $F(t,z,u)$ is linear in u, one does not get ρ by simply mixing u_θ with respect to θ.

Intuitively, θ is an auxiliary parameter randomly chosen from the distribution ν independently of everything else. A more general procedure, which we shall not describe here, would be to allow auxiliary randomizations depending on time t and available data.

2. Finite mixtures of $\mathcal{U}$. Let $\rho = \sum_{i=1}^{N} \lambda_i \rho_i$, where $\lambda_i \geq 0$, $\sum_{i=1}^{N} \lambda_i = 1$ and each $\rho_i \in \mathcal{U}$. In 1, we take $\Theta = \{\theta_1, \cdots, \theta_N\}$ a finite set. Let ν assign θ_i measure λ_i, and let u_i be a control corresponding to density ρ_i.

3. Let Θ be the set of extreme points of $\mathcal{U}$. By Choquet's theorem every $\rho \in \bar{\mathcal{U}}$ is a mixture of Θ [11]. It would be interesting to find conditions under which extreme points of $\bar{\mathcal{U}}$ which minimize J(u) must be in $\mathcal{U}$.

4. If $\mathcal{U}$ is weakly closed, then every $\rho \in \bar{\mathcal{U}}$ is a mixture of $\mathcal{U}$ [11, Chap. 1]. (This result is less sophisticated than Choquet's theorem.) Of course, if $\mathcal{U}$ is weakly closed, then J(u) clearly has a minimum on $\mathcal{U}$. There is less interest in considering mixtures.

By a result of Bismut [3, Theorem IV.3] $\mathcal{U}$ is weakly closed in the following Markov (or zero memory) case. Let the control space U be finite dimensional, compact and convex. Let z = (x,y), and F linear in u. Admit control processes $u(t) = \underline{u}(t,y(t))$, where $\underline{u}$ is any Borel measurable function from $[0,T] \times E^m$ into U.

ACKNOWLEDGMENT

This research was supported in part by the Air Force Office of Scientific Research, under Grant No. AF-AFOSR 71-2078D and in part by the National Science Foundation, under Grant No. GP-38428X.

REFERENCES

1. Benes, V. E., Existence of Optimal Stochastic Control Laws, SIAM J. Control, 9 (1971), pp. 446-472.
2. Berkovitz, L. D., Optimal Control Theory, Applied Math. Sciences, vol. 12, Springer-Verlag, 1974.
3. Bismut, J. M., Théorie probabiliste du contrôle des diffusions, Memoirs Amer. Math. Soc., No. 167, 1975.
4. Cesari, L., Existence Theorems for Weak and Usual Optimal Solutions in Lagrange Problems with Unilateral Constraints, Trans. Amer. Math. Soc., 124 (1966), pp. 413-430.
5. Davis, M. H. A., On the Existence of Optimal Policies in Stochastic Control, SIAM J. Control, 11 (1973), pp. 587-594.
6. Duncan, T. and P. P. Varaiya, On the Solutions of a Stochastic Control System, SIAM J. Control, 9 (1971), pp. 354-371.
7. Fleming, W. H. and M. Nisio, On the Existence of Optimal Stochastic Controls, J. Math. Mech., 15 (1966), pp. 777-794.
8. Fleming, W. H. and R. W. Rishel, Deterministic and Stochastic Optimal Control. Applications of Math No. 1, Springer-Verlag, 1975.
9. A. Friedman, Stochastic Differential Equations and Applications, vol. 1, Academic Press, 1975.
10. Gamkrelidze, R. V., On Sliding Optimal Regimes, Soviet Math. Doklady, 3 (1962), pp. 390-395.
11. Phelps, R. P., Lectures on Choquet's Theorem, Van Nostrand, Princeton, 1966.
12. Warga, J., Relaxed Variational Problems, J. Math. Anal. Appl., 4 (1962), pp. 111-128.
13. Young, L. C., Generalized Curves and the Existence of an Attained Absolute Minimum in the Calculus of Variations, Compt. Rend. Soc. Sci. et Lettres Varsovie Cl III, 30 (1937), pp. 212-234.

TWO-PLAYER CONTROL PROBLEMS WITH SUBSPACE TARGETS

M. Heymann, M. Pachter, and R. J. Stern

Technion - Israel Institute of Technology
Haifa, Israel

Council for Scientific and Industrial Research
Pretoria, South Africa

and

Concordia University
Montreal, Canada

ABSTRACT

A two-input linear autonomous control system is considered. For subspace target sets, controllability and holdability properties are investigated. Capture sets under various rules of play are given. A certain related linear-quadratic max-min control problem is discussed.

§1. PRELIMINARIES

Our major object of consideration is the two-controller system

$$\dot{x} = Ax + Bu + Cv; \; t \geq 0 \tag{1.1}$$

where $A \in R^{n\times n}$, $B \in R^{n\times p}$, and $C \in R^{n\times e}$ ($R^{i\times j}$ denoting the space of $i \times j$ real matrices). The space of locally bounded measurable mappings $[0,\infty) \to R^p(R^e)$ is denoted $U(V)$, the space of *pursuer controls* (*evader controls*). We are given a subspace $S \subset R^n$ called the *target*.

Consider the (single-input or "pursuer") control system

$$\dot{x} = Ax + Bu; \qquad t \geq 0 \tag{1.2}$$

As in [1], a subspace $\mathcal{W} \subset R^n$ is called (A, B)-*invariant* if $A\mathcal{W} \subset \mathcal{W} + \mathcal{R}(B)$ (where $\mathcal{R}$ denotes range). By core $(\mathcal{W})$ we denote the set of initial states $x_0 \in \mathcal{W}$ such that for some $u \in U$ the associated solution of (1.2) satisfies $x(t) \in \mathcal{W}\ \forall\ t \geq 0$. It is readily noted that core $(\mathcal{W})$ is a subspace of $\mathcal{W}$. We shall require the following (see refs. [1-3]):

<u>Lemma 1.1</u> $\mathcal{W}$ is (A, B)-invariant iff for some $Q \in R^{p\times n}$ we have

$$(A + BQ)\mathcal{W} \subset \mathcal{W} \tag{1.3}$$

and

$$\text{core } (\mathcal{W}) \text{ is the maximal (A, B)-invariant subspace of } \mathcal{W}. \tag{1.4}$$

Other properties of cores may be found in [2-6].

We let $[A|B]$ denote the $n \times np$ matrix $[B, AB, A^2B, \cdots, A^{n-1}B]$ and we write $\{A|B\} = \mathcal{R}([A|B])$. It is well known that

$$\{A|B\} = \bigcup_{u\in U} e^{AT} \int_0^T e^{-At} Bu(t)dt \tag{1.5}$$

where $T > 0$ is arbitrary, and it is readily verified that for each $Q \in R^{p\times n}$

$$\{A + BQ|B\} = \{A|B\} \tag{1.6}$$

Given a subspace $\mathcal{W} \subset R^n$, we denote by $\mathcal{W}_A$ the largest A-invariant subspace of $\mathcal{W}$. The empty set is denoted ϕ.

In the next section we focus attention on the problems of controllability to and holdability within S under an "open loop" information scheme. In Section 3 more general differential games are considered from the controllability point of view. In the last section we return to the open loop rules of play and solve a linear-quadratic game with subspace terminal state restraint.

§2. MAX-MIN CONTROLLABILITY AND MAX-MIN CORES

Let $t > 0$ be given and for a subspace $S \subset R^n$ denote by $X_t(S)$ the set of initial states $x_0 \in R^n$ such that given any announced $v \in V$, there exists $u \in U$ for which the associated solution of (1.1) satisfies $x(t) \in S$. If $x_o \in X_t(S)$ we shall say that x_0 is *max-min controllable to* S *in time* t. If $x_0 \in \bigcup_{t>0} X_t(S)$ then we call x_0 *max-min controllable to* S; if $x_0 \in \bigcap_{t>0} X_t(S)$ then x_0 is *perfectly max-min controllable* to S. It is a consequence of (1.5) that if $C = 0$ [i.e., if system (1.1) reduces to system (1.2)], then the three sets defined above are equal for $S = 0$. That this is not the case for more general subspace targets is evidenced by the following example (see [2]):

$$A = \begin{pmatrix} 1 & 0 & 0 \\ 0 & 2 & 0 \\ 0 & 0 & 0 \end{pmatrix}, \quad B = \begin{pmatrix} 0 \\ 0 \\ 1 \end{pmatrix}, \quad S = R\left(\begin{pmatrix} 1 \\ 1 \\ 0 \end{pmatrix}\right)$$

Easy calculations show that here the perfectly controllable set is given by

$$R\left(\begin{pmatrix} 0 \\ 0 \\ 1 \end{pmatrix}\right)$$

while the controllable set is $0 \cup \{x : x_1 > x_2 \geq 0\} \cup \{x : x_1$

$< x_2 \le 0\}$.

The following result appears in [7] (see also [8], where the special case $S = 0$ was dealt with).

Theorem 2.1

$$\bigcap_{t>0} X_t(S) = \begin{cases} (S + \{A|B\})_A & \text{iff } \{A|C\} \subset S + \{A|B\} \\ \phi, & \text{otherwise} \end{cases}$$

We denote by $\tilde{\text{core}}\ (S)$ the *max-min core* of S. This is the set of points $x_0 \in S$ such that for any announced $v \in V$ there exists $u \in U$ such that the associated solution of (1.1) satisfies $x(t) \in S\ \forall\ t \ge 0$. We note that $\tilde{\text{core}}\ (S)$ may possibly be empty, but whenever nonempty it is a subspace of S. The set $\tilde{\text{core}}\ (S)$ is of interest because the capability of "hitting and then holding in S" (in our two-player framework) is then equivalent to the capability of "hitting $\tilde{\text{core}}\ (S)$." We have the following results concerning $\tilde{\text{core}}\ (S)$:

Proposition 2.2 Let $\tilde{\text{core}}\ (S) \ne \phi$. Then

$$A\ \tilde{\text{core}}\ (S) \subset R(B) + S \tag{2.1}$$

and

$$R(C) \subset R(B) + S \tag{2.2}$$

Outline of proof Equation (2.1) follows from the obvious fact that $\tilde{\text{core}}\ (S) \subset \text{core}\ (S)$, and (1.4). If (2.2) did not hold, then for some $\tilde{v} \in R^e$ we have $C\tilde{v} \notin R(B) + S$. A straightforward argument shows that if the evader uses this (constant) control, then for any $u \in U$, the solution of (1.1) emanating from $x_0 = 0$ initially exits S.

A slightly more stringent condition than (2.2) is required

to insure that $\tilde{\text{core}}$ (S) equals core (S); this equality is desirable in view of the characterization which is then afforded by (1.4).

Proposition 2.3

$$\tilde{\text{core}}\ (S) = \text{core}\ (S) \text{ iff } R(C) \subset R(B) + \text{core}\ (S)$$

Proof Assume $R(C) \subset R(B) + \text{core}\ (S)$, and let $x_0 \in$ core (S). Let $v \in V$ be announced by the evader. We shall exhibit $u \in U$ such that $x(t) \in \text{core}\ (S)\ \forall\ t \geq 0$. To this end, let $\tilde{u} \in U$ and w (measurable, locally bounded R^n-valued) be such that $Cv(t) = -B\tilde{u}(t) + w(t)$. Consider the pursuer law $u = Qx + \tilde{u}$, where Q is as in (1.3). The result of applying this law is that x(t) holds in core (S), as required. The converse of the proof follows from (2.2).

Theorem 2.4 Let $\tilde{\text{core}}$ (S) = core (S). Then

$$\bigcup_{t>0} X_t(\tilde{\text{core}}\ (S)) = \bigcap_{t>0} X_t(\tilde{\text{core}}\ (S)).$$

Proof We only need to show that the union is contained in the intersection. If $x_0 \in X_t(\tilde{\text{core}}\ (S))$ for $t > 0$, then $e^{At}x_0 \in \tilde{\text{core}}\ (S) + \{A|B\}$, since the evader could employ $v = 0$. The (A, B)-invariance of $\tilde{\text{core}}$ (S) implies A-invariance of $\tilde{\text{core}}\ (S) + \{A|B\}$, and so $x_0 \in (\tilde{\text{core}}\ (S) + \{A|B\})_A$. Thus, in view of Theorem 2.1, the proof is complete if $\{A|C\} \subset \tilde{\text{core}}\ (S) + \{A|B\}$. But this follows readily from the (A, B)-invariance of $\tilde{\text{core}}$ (S) and the fact that $R(C) \subset \tilde{\text{core}}\ (S) + R(B)$.

3. CAPTURE SETS FOR LINEAR GAMES

A class E of *evader strategies* is a set of pairs (T, e) where $T > 0$ and e is a causal map $e : U \to V$. For convenience we regard the evader as the announcer of the duration T. Three types of evader strategies are considered:

(E_1) e is constant. This is the "blind" or "open loop" case, in which there is $v \in V$ such that $e(u) = v$ $\forall\ u \in U$.

(E_2) e is given via a feedback law $v = Kx$ where $K \in R^{e\times n}$.

(E_3) e is given by a combination of E_1 and E_2 strategies; that is, $v = Kx + \tilde{v}$ where $K \in R^{e\times n}$ and $\tilde{v} \in V$ is fixed.

The pursuer's goal is to steer x to S and the evader seeks to prevent this. By G_i (i = 1, 2, 3) we denote the differential game in which the evader selects a strategy $(T, e) \in E_i$ and announces it to the pursuer. The pursuer then announces $u \in U$; hence pursuer strategies are maps $p : E_i \to U$.

An initial state $x_0 \in R^n$ is called *capturable* in game G_i (i = 1, 2, 3) if for every announced $(T, e) \in E_i$ there exists $u \in U$ for which the associated solution of (1.1) satisfies $x(T) \in S$. The set of capturable states in game G_i is denoted X_i. The following result gives a complete characterization of the X_i. Note that the set X_1 is already known due to Theorem 2.1; for the elementary details involved in obtaining the other two characterizations from the first, the reader is referred to [7].

Theorem 3.1

$$X_1 = \begin{cases} (\mathcal{S} + \{A|B\})_A & \text{iff } \{A|C\} \subset \mathcal{S} + \{A|B\} \\ \phi & \text{otherwise} \end{cases}$$

$$X_2 = \bigcap_{K \in R^{e\times n}} (\mathcal{S} + \{A + CK|B\})_{A+CK}$$

$$X_3 = \begin{cases} X_2 & \text{iff } \{A|C\} \subset X_2 \\ \phi & \text{otherwise} \end{cases}$$

Remarks

1. Consider the target $\mathcal{S} = 0$. Then X_2 equals the intersection of all the (A,C)-invariant subspaces which contain $\mathcal{R}(B)$. Hence it is always true that $\mathcal{R}(B) \subset X_2$. Thus, from the evader's viewpoint, the situation in G_2 is "optimal" if $\mathcal{R}(B) = X_2$. It does not follow, however, that then there is a corresponding "optimal" $\tilde{K}$ such that $X_2 = \mathcal{R}(B) = \{A + C\tilde{K}|B\}$. Consider the case

$$A = \begin{pmatrix} 0 & 1 & 0 \\ 0 & 0 & 0 \\ 1 & 0 & 0 \end{pmatrix}; \quad B = \begin{pmatrix} 1 \\ 0 \\ 0 \end{pmatrix}; \quad C = \begin{pmatrix} 0 \\ 1 \\ 0 \end{pmatrix}$$

One can readily check that $\mathcal{R}(B)$ is not (A,C)-invariant, while the spaces

$$\mathcal{S}_1 = \mathcal{R}\left(\begin{pmatrix} 1 & 0 \\ 0 & 0 \\ 0 & 1 \end{pmatrix}\right) \quad \text{and} \quad \mathcal{S}_2 = \mathcal{R}\left(\begin{pmatrix} 1 & 0 \\ 0 & 1 \\ 0 & 1 \end{pmatrix}\right)$$

are (A,C)-invariant. Now note $\mathcal{S}_1 \cap \mathcal{S}_2 = \mathcal{R}(B)$.

2. If $X_1 = \phi$, then the evader prefers to play in G_1

rather than in G_2. If $X_1 \neq \phi$, then the preference is reversed.

3. Consider game G_3. It is possible that the evader has available a $\tilde{K} \in R^{e\times n}$ such that the resulting G_1-type game has an empty set of capturable states. Such "controllability destroyers" $\tilde{K}$ are precisely those for which

$$\{A|C\} \not\subset S + \{A + C\tilde{K}|B\} \tag{3.1}$$

In [7] the following was proven: If $S = S_A$ then a necessary and sufficient condition for the existence of K satisfying (3.1) is

$$R(C) \not\subset S_A + R(B) \tag{3.2}$$

§4. A MAX-MIN CONTROL PROBLEM WITH SUBSPACE TARGET

The general problem considered is the following: Consider system (1.1) with target S. Assume $x_0 \in X_1$; i.e., x_0 is max-min controllable to S. $T > 0$ is the fixed duration of the game. The evader announces $v \in V$. The pursuer responds with $u \in U$ such that $x(T) \in S$. Both players make their choices in accordance with the optimization (evader maximizing, pursuer minimizing) of the payoff functional

$$P(u,v) = \int_0^T [\|u(t)\|^2 - \|v(t)\|^2]dt \tag{4.1}$$

(Here $\|\cdot\|$ denotes Euclidean norm.)

It will be convenient to decompose (1.1) as follows:

$$\dot{x}_p = Ax_p + Bu; \quad x_p(0) = 0 \tag*{(1.1)$_p$}$$

$$\dot{x}_e = Ax_e + Cv; \quad x_e(0) = x_0 \tag*{(1.1)$_e$}$$

Hence $x(t) = x_p(t) + x_e(t)$, where $x(t)$ is the solution of (1.1) under inputs u, v and initial condition $x(0) = x_0$.

We introduce the evader "controllability Grammian"

$$W_e = \int_0^T e^{-AT} C C' e^{-A't} dt$$

the prime denoting transpose. It is well known that $\mathcal{R}(W_e) = \{A|C\}$. One similarly defines the pursuer controllability Grammian W_p.

Let the evader specify $v \in V$ (on the interval [0, T]). Then for some $y \in R^n$ we have

$$[A|C]y = x_0 - e^{-AT} x_e(T) \tag{4.2}$$

The pursuer responds with $u = u_v \in U$ such that

$$x_e(T) + x_p(T) \in S \tag{4.3}$$

Note that there must exist $z \in R^n$ such that

$$-e^{-AT} x_p(T) = W_p z \tag{4.4}$$

Subject to (4.3) and (4.4), the pursuer is to minimize $\int_0^T \|u_v(t)\|^2 dt$. According to a standard least-squares result (see, e.g., [6]) the u_v^* accomplishing this is given by

$$u_v^*(t) = - B' e^{-A'T} z \tag{4.5}$$

Then

$$P(u_v, v) = z' W_p z - \int_0^T \|v(t)\|^2 dt \tag{4.6}$$

We now turn attention to finding the optimal evader control. The minimizer of $\int_0^T \|v(t)\|^2 dt$ subject to the state transfer $x_0 \to x_e(T)$ is given by

$$v^*(t) = - C' e^{-A'T} y \tag{4.7}$$

Then $\int_0^T \|v^*(t)\|^2 dt = y'W_e y$. Also,

$$P(u_{v^*}, v^*) = z'W_p z - y'W_e y \tag{4.8}$$

and so our problem reduces to

$$\max_y \min_z [z'W_p z - y'W_e y] \tag{4.9}$$

subject to (4.3) or equivalently

$$[A|B]z + [A|C]y \in x_0 \; e^{-AT} S \tag{4.10}$$

We note here that the problem is consistent (that is, (4.10) is solvable in y for any given z) iff x_0 is a max-min controllable initial state. The optimization problem (4.9) decomposes to the following pair of quadratic programming problems:

(P_1) $\min z'W_p z = P_1(y)$
subject to $[A|B]z \in x_0 - [A|C]y + e^{-AT} S$

(P_2) $\max[P_1(y) - y'W_e y]$
subject to $y \in R^n$

Define a new variable $\xi = [A|B]z$. Then in terms of the Moore-Penrose inverse, $z = [A|B]^+\xi + N([A|B])$ (see [9]; here N denotes null space. We note that the positive semi-definite symmetric matrix W_p has a square root $W_p^{1/2}$. Also, $W_p q = 0$ for any $q \in N([A|B])$. We will now make the simplifying assumption that the "pursuer-system" (1.2) is completely controllable, i.e.,

$$R(W_p) = \{A|B\} = R^n \tag{4.11}$$

Then, upon defining $\eta = W_p^{1/2}[A|B]^+\xi$ and noting that $N([A|B]^*)$

= 0, we can write problem (P_1) as

$$(P_1)' \quad \min\|\eta\|^2 = P_1(y)$$

subject to

$$\eta \in W_p^{1/2}[A|B]^+[x_0 - [A|C]y + e^{-AT}S] + N\left[[A|B]W_p^{-1/2}\right]$$

This problem consists of finding the minimum norm element in a linear variety, and as such is a problem which can be solved explicitly in terms of projections or generalized inverses. In fact, a result in [10] yields that the solution to $(P_1)'$ is given by

$$\eta^*(y) = N^+Na \tag{4.12}$$

where $a = W_p^{1/2}[A|B]^+[x_0 - [A|C]y]$ and where N is any full rank matrix whose rows span the orthogonal complement of

$$W_p^{1/2}[A|B]^+e^{-AT}S + N[[A|B]W_p^{-1/2}]$$

and

$$P_1(y) = a'N^+Na \tag{4.13}$$

The solutions of (P_1) are all vectors satisfying

$$z^* \in [A|B]^+[A\ B]W_p^{-1/2}\ \eta^*(y) + N[[A|B]]. \tag{4.14}$$

Now, in view of (4.5), an optimal pursuer control, for evader specified $v \in V$, is given by

$$u_v^*(t) = -B'e^{-A'T}[A|B]^+[A|B]W_p^{-1/2}\eta^*(y) \tag{4.15}$$

It remains to determine an optimal y; that is, problem (P_2) must still be solved. Since this problem is constraint free, simply setting the gradient of the objective function equal to zero yields the required vector y^*, provided the

Hessian of $[P_1(y) - y'W_e y]$ is negative definite. In [10] it was proven that in case controllability conditions $\{A|C\} = \{A|B\} = R^n$ hold, negative definiteness holds iff $[W_p - W_e]$ is negative definite, and in this case, (P) is solvable for any initial state $x_0 \in R^n$. For further detail on this definiteness problem without the above controllability conditions (in case $S = 0$ and (1.1) is not necessarily autonomous) the reader is referred to [8].

REFERENCES

1. Wonham, W. M. and A. S. Morse, Decoupling and Pole Assignment in Linear Multivariable Systems: A Geometric Approach, SIAM J. Control, 8 (1970), pp. 1-18.

2. Pachter, M. and R. J. Stern, The Controllability Problem for Affine Targets, to appear.

3. Pachter, M. and R. J. Stern, On Cores of Subspaces in Linear Autonomous Control Systems, to appear.

4. Heymann, M., Weak Invariance, Cores and Feedback, to appear.

5. Hájek, Cores of Targets in Linear Control Systems, Math. Systems Theory, 8 (1974), pp. 203-6.

6. Lee, E. B. and L. Markus, Foundations of Optimal Control Theory, Wiley, 1976.

7. Heymann, M., M. Pachter, and R. J. Stern, On Linear Games with Subspace Target, to appear.

8. Heymann, M., M. Pachter, and R. J. Stern, Max-min Control Problems: A System Theoretic Approach, IEEE Trans. on Automatic Control, AC21, 4 (1976).

9. Ben-Israel, A., and T. N. E. Greville, Generalized Inverses, Wiley, 1974.

10. Pachter, M., System Theoretic Approach to Max-min Control Problems and Generalized Targets, Doctoral dissertation, Technion - Israel Institute of Technology (1975).

CONTROL OF A STRUCTURED POPULATION MODELLED BY A MULTIVARIATE BIRTH-AND-DEATH PROCESS

W. M. Getz

National Research Institute of Mathematical Sciences
Pretoria, South Africa

ABSTRACT

A continuous-time matrix model is formulated for a structured population governed by a simple linear multivariate birth-and-death process. In particular, using the matrix Kronecker product, the derivation is discussed of a system of differential equations for the variance and covariance parameters of the joint multivariate probability distribution of the population. A bilinear scalar control term is introduced into the model and a cost-performance index suitable for measuring the profitability of various harvesting strategies is considered. The optimal harvesting strategy is shown to be bang-bang.

§0. INTRODUCTION

Systems of linear equations, usually difference equations rather than differential equations, have been used extensively to model the dynamic behavior of populations structured into a number of distinct age classes. The first models of this type were introduced by Lewis [1] and Leslie [2,3]. More recently, Lefkovitch [4], Doubleday [5], and Rorres and Fair [6] have used these models to formulate harvesting policies for the exploitation of various animal populations. These models are all deterministic in the sense that the state-space variables account only for the size of each class of the population. Using the Kronecker product, Pollard [7] has proposed a method for extending the state-space of such models to include variance and covariance terms.

In this paper we formulate a system of ordinary differential equations as a model of a structured population, under the assumption that the behavior of the population is governed by a simple birth-and-death process. The homogeneous system of equations is linear and the state-space variables are the mean, variance, and covariance parameters of the joint probability distribution of the population.

Although our treatment in deriving a system of differential equations for the variance and covariance parameters differs completely from that of Pollard, we show that the two systems are closely related. Our approach is, however, more general in that it embraces all populations with ordered class structures, such as those classed according to size, weight, and age. In addition, our method is applicable to a wider class of processes, since Pollard's results were derived for certain multitype Galton-Watson processes, while our results hold for all linear multivariate birth-and-death processes [8].

The effect of scalar control, applied at a given intensity, on the birth-and-death process governing the populations, is considered. This results in the control variable appearing bilinearly in the system equations. The harvesting of populations modelled by these systems of equations is discussed with reference to suitable performance criteria. In particular, a cost is associated with nonzero variance so that the uncertainty in the mean size of the population classes is penalized costwise. Using Pontryagin's maximum principle, it is shown in the case where the cost of control is linear that the optimal strategy is a sequence of open and closed harvesting seasons, i.e., that the optimal strategy is bang-bang. Computation of the optimal strategy is briefly discussed.

§1. HOMOGENEOUS POPULATION MODEL

In this section we derive a system of homogeneous linear differential equations in the state variables (i.e., mean, variance, and covariance parameters). A number of steps in the derivation, not of direct applicability to the issues central to this paper, have been glossed over. A detailed account of the general methodology used in the derivation is given in [8] and [9].

Consider a population divided into m classes. Let n_i, $i = 1, \cdots, m$, the ith element of the m-vector $\underset{\sim}{n}$, denote the number of individuals in the ith class. Let $P(\underset{\sim}{n}; t)$ denote the probability that the population is in state $\underset{\sim}{n}$ at time t. Suppose that the population is governed by the following birth-and-death process.

(i) An individual in the ith class may give birth to an individual in the first class in the interval $[t, t + \Delta t]$ with probability $f_i(t)\Delta t + o(\Delta t)$, $i = 1, \cdots, m$.

(ii) An individual in the ith class may die in the interval $[t, t + \Delta t]$ with probability $\ell_i(t)\Delta t + o(\Delta t)$, $i = 1, \cdots, m$.

(iii) An individual in the ith class may enter the (i + 1)th class (by virtue of increasing age or size, etc.) in the interval [t, t + Δt] with probability $p_i(t)\Delta t + o(\Delta t)$, $i = 1, \cdots, m - 1$.

(iv) The probability of more than one event occurring in the interval [t, t + Δt] is o(Δt).

Under these assumptions the following difference-differential equation for P(n; t) can be derived in the usual manner [8,9].

$$\frac{dP(\underset{\sim}{n};t)}{dt} = \left[f_1(t)(n_1 - 1) + \sum_{i=2}^{m} f_i(t)n_i\right]P(\underset{\sim}{n} - \underset{\sim}{e}_1;\ t)$$
$$+ \sum_{i=1}^{m} \ell_i(t)(n_i + 1)P(\underset{\sim}{n} + \underset{\sim}{e}_i;\ t)$$
$$+ \sum_{i=1}^{m-1} p_i(t)(n_i + 1)P(\underset{\sim}{n} + \underset{\sim}{e}_i - \underset{\sim}{e}_{i+1};\ t)$$
$$- \left[\sum_{i=1}^{m} \bigl(f_i(t) + \ell_i(t)\bigr)n_i + \sum_{i=1}^{m-1} p_i(t)n_i\right]$$
$$\cdot\ P(\underset{\sim}{n};\ t) \qquad (1.1)$$

where $\underset{\sim}{e}_i$ is the m-vector whose ith entry is unity and all its other entries are zero. Since it is a physical requirement of the process that the elements of $\underset{\sim}{n}$ can assume nonnegative values only, i.e., $\underset{\sim}{n} \geq \underset{\sim}{0}$, we define $P(\underset{\sim}{n};\ t) \equiv 0$ whenever $\underset{\sim}{n} < \underset{\sim}{0}$.

Define the probability-generating function

$$G(s;t) = \sum_{n_1=0}^{\infty} \cdots \sum_{n_m=0}^{\infty} P(\underset{\sim}{n};t)s_1^{n_1}s_2^{n_2}\cdots s_m^{n_m}$$
$$\triangleq \sum_{\underset{\sim}{n}\geq\underset{\sim}{0}} P(\underset{\sim}{n};t)\underset{\sim}{s}^{\underset{\sim}{n}} \qquad (1.2)$$

Then it is easily seen that

$$s_i^{1-r} \frac{\partial G(\underset{\sim}{s};t)}{\partial s_i} = \sum_{\underset{\sim}{n} \geq 0} (n_i + r) P(\underset{\sim}{n} + r\underset{\sim}{e}_i;t) \underset{\sim}{s}^{\underset{\sim}{n}} \qquad (1.3)$$

$$r = -1, 0, 1$$

$$s_i s_j \frac{\partial G(\underset{\sim}{s};t)}{\partial s_j} = \sum_{\underset{\sim}{n} \geq 0} n_j P(\underset{\sim}{n} - \underset{\sim}{e}_i;t) \underset{\sim}{s}^{\underset{\sim}{n}} \qquad (1.4)$$

$$s_j \frac{\partial G(\underset{\sim}{s};t)}{\partial s_i} = \sum_{\underset{\sim}{n} \geq 0} (n_i + 1) P(\underset{\sim}{n} + \underset{\sim}{e}_i - \underset{\sim}{e}_j;t) \underset{\sim}{s}^{\underset{\sim}{n}} \qquad (1.5)$$

$$\frac{\partial G(\underset{\sim}{s};t)}{\partial t} = \sum_{\underset{\sim}{n} \geq 0} \frac{dP(\underset{\sim}{n};t)}{dt} \underset{\sim}{s}^{\underset{\sim}{n}} \qquad (1.6)$$

Multiplying (1.1) by $\underset{\sim}{s}^{\underset{\sim}{n}}$, summing over $\underset{\sim}{n} \geq 0$, and using (1.3-1.6) we obtain the following partial differential equation in $G(\underset{\sim}{s},t)$:

$$\frac{\partial G(\underset{\sim}{s};t)}{\partial t} = \sum_{i=1}^{m} \left[f_i(t) s_i (s_1 - 1) - \ell_i(t)(s_i - 1) \right] \frac{\partial G(\underset{\sim}{s};t)}{\partial s_i}$$

$$+ \sum_{i=1}^{m-1} p_i(t)(s_{i+1} - s_i) \frac{\partial G(\underset{\sim}{s};t)}{\partial s_i} \qquad (1.7)$$

From (1.7) we can extract a system of ordinary differential equations in the mean, variance, and covariance parameters of the distribution $P(\underset{\sim}{n}; t)$.

Let $\bar{n}_i$ denote the mean of the ith population, i.e.,

$$\bar{n}_i = \sum_{\underset{\sim}{n} \geq 0} n_i P(\underset{\sim}{n};t) \qquad i = 1, \cdots, m \qquad (1.8)$$

Similarly, the variance parameters σ_{ii} and covariance parameters σ_{ij} by definition satisfy:

$$\sigma_{ij} = \sum_{\underset{\sim}{n} \geq 0} n_i n_j P(\underset{\sim}{n};t) - \bar{n}_i \bar{n}_j \qquad i,j = 1, \cdots, m \qquad (1.9)$$

By suitably differentiating (1.2) we see that

$$s_i^2 \frac{\partial^2 G(\underset{\sim}{s};t)}{\partial s_i^2} + s_i \frac{\partial G(\underset{\sim}{s};t)}{\partial s_i} = \sum_{\underset{\sim}{n} \geq \underset{\sim}{0}} n_i^2 P(\underset{\sim}{n};t) \underset{\sim}{s}^{\underset{\sim}{n}} \qquad (1.10)$$

$$i = 1, \cdots, m$$

and

$$s_i s_i \frac{\partial^2 G(\underset{\sim}{s};t)}{\partial s_i \partial s_j} = \sum_{\underset{\sim}{n} \geq \underset{\sim}{0}} n_i n_j P(\underset{\sim}{n};t) \underset{\sim}{s}^{\underset{\sim}{n}} \qquad (1.11)$$

$$i \neq j, \quad i,j = 1, \cdots, m$$

Using (1.2-1.5) and (1.8-1.11) and setting $\underset{\sim}{s} = \underset{\sim}{1}$ (m-vector with entries all unity), it follows that

$$\bar{n}_i = \frac{\partial G(\underset{\sim}{1};t)}{\partial s_i} \qquad i = 1, \cdots, m \qquad (1.12)$$

$$\sigma_{ii} = \frac{\partial^2 G(\underset{\sim}{1};t)}{\partial s_i^2} + \frac{\partial G(\underset{\sim}{1};t)}{\partial s_i} \left(1 - \frac{\partial G(\underset{\sim}{1};t)}{\partial s_i}\right) \qquad (1.13)$$

$$i = 1, \cdots, m$$

$$\sigma_{ij} = \frac{\partial^2 G(\underset{\sim}{1};t)}{\partial s_i \partial s_j} - \frac{\partial G(\underset{\sim}{1};t)}{\partial s_i} \frac{\partial G(\underset{\sim}{1};t)}{\partial s_j} \qquad (1.14)$$

$$i \neq j \quad i,j = 1, \cdots, m$$

Thus, differentiating (1.7) with respect to s_i, then setting $\underset{\sim}{s} = \underset{\sim}{1}$ and using (1.6) and (1.8), we derive the following system of ordinary differential equations in $\bar{\underset{\sim}{n}}^T = (\bar{n}_1, \cdots, \bar{n}_m)$:

$$\frac{d\bar{\underset{\sim}{n}}}{dt} = \big(L(t) - D(t)\big)\, \bar{\underset{\sim}{n}} \qquad \bar{\underset{\sim}{n}}(t_0) = \bar{\underset{\sim}{n}}_o \qquad (1.15)$$

where $\bar{\underset{\sim}{n}}_o$ is an estimate of the mean size of the initial population, and the matrices L(t) and D(t) have the form

$$L(t) = \begin{vmatrix} f_1(t) & f_2(t) & \cdots & f_{m-1}(t) & f_m(t) \\ p_1(t) & 0 & \cdots & \cdots & 0 \\ 0 & p_2(t) & \ddots & & \vdots \\ \vdots & \ddots & \ddots & \ddots & \vdots \\ 0 & \cdots & 0 & p_{m-1}(t) & 0 \end{vmatrix} \tag{1.16}$$

and

$$D(t) = \begin{vmatrix} \ell_1(t)+p_1(t) & 0 & \cdots & \cdots & 0 \\ 0 & \ell_2(t)+p_2(t) & \ddots & & \vdots \\ \vdots & \ddots & \ddots & \ddots & \vdots \\ \vdots & & \ddots & \ell_{m-1}(t)+p_{m-1}(t) & 0 \\ 0 & \cdots & \cdots & 0 & \ell_m(t) \end{vmatrix} \tag{1.17}$$

Similarly, after differentiating (1.7) with respect to s_i and then s_j, $i,j = 1, \cdots, m$, and setting $\underset{\sim}{s} = \underset{\sim}{1}$, we can derive a system of differential equations for the parameters σ_{ij}. This system of equations can be compactly written in two equivalent forms.

Define the m^2 vector $\underset{\sim}{\sigma}$ and $m \times m$ matrix S as

$$\underset{\sim}{\sigma}^T = (\sigma_{11}, \cdots, \sigma_{1n}, \sigma_{21}, \cdots, \sigma_{2n}, \sigma_{31}, \cdots, \sigma_{mm})$$

and

$$(S)_{ij} = \sigma_{ij}$$

respectively.

Let $[K]_i$ denote the ith row of any arbitrary matrix K. Then since the Kronecker product of two matrices [10] $A \in R^{m\times n}$ and $B \in R^{p\times q}$ is defined as the $mp \times nq$ matrix

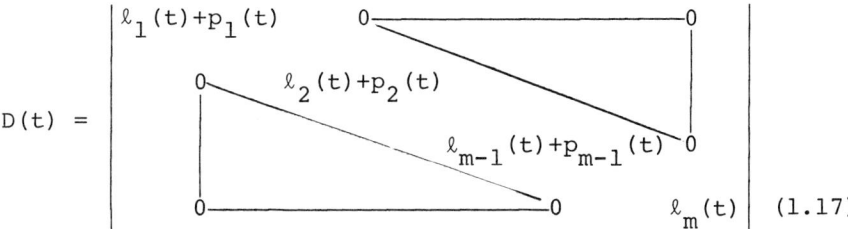

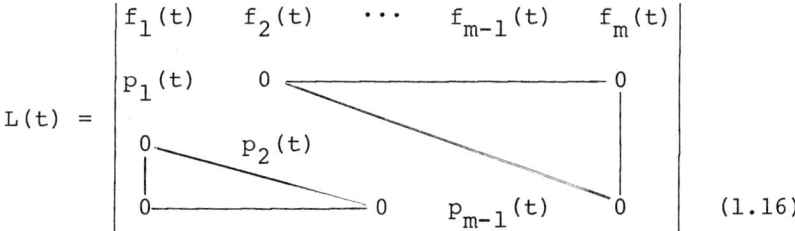

$$A \otimes B = \begin{bmatrix} a_{11}B & a_{12}B & \cdots & a_{1n}B \\ a_{21}B & & & \vdots \\ \vdots & & & \vdots \\ a_{m1}B & \cdots & \cdots & a_{mn}B \end{bmatrix}$$

where $a_{ij}B$ is the $p \times q$ block consisting of the elements of B each multiplied by the ijth element of A, the system can be written in the form

$$\frac{d\underset{\sim}{\sigma}}{dt} = \begin{bmatrix} [L(t) + D(t)]_1 \\ 0 \quad 0 \quad \cdots \quad 0 \\ \vdots \quad \vdots \quad\quad \vdots \\ 0 \quad 0 \quad \cdots \quad 0 \\ [L(t) + D(t)]_2 \\ 0 \quad 0 \quad \cdots \quad 0 \\ \vdots \quad \vdots \quad\quad \vdots \\ \vdots \quad \vdots \quad\quad \vdots \\ 0 \quad 0 \quad \cdots \quad 0 \\ [L(t) + D(t)]_m \end{bmatrix} \bar{\underset{\sim}{n}} + [(L(t) - D(t)) \otimes I + I \otimes (L(t) - D(t))]\underset{\sim}{\sigma}$$

$$\underset{\sim}{\sigma}(t_o) = \underset{\sim}{\sigma}_o \tag{1.18}$$

We note the following: the first matrix is $m^2 \times m$ and contains m rows of zeros between each pair of nonzero rows $[L(t) + D(t)]_i$ and $[L(t) + D(t)]_{i+1}$, $i = 1, \cdots, m-1$; the second matrix is $m^2 \times m^2$ (since I is taken as the $m \times m$ identity matrix); and the initial variance $\underset{\sim}{\sigma}_o$ relates to our initial estimate for $\bar{\underset{\sim}{n}}_o$.

Alternatively, if we define $\mathrm{diag}[(L(t) + D(t))\bar{\underset{\sim}{n}}]$ to be the $m \times m$ diagonal matrix with diagonal elements given by the vector $(L(t) + D(t))\bar{\underset{\sim}{n}}$, (1.18) is equivalent to the following matrix equation in the covariance matrix S:

$$\frac{dS}{dt} = \text{diag}[(L(t) + D(t))\bar{\underset{\sim}{n}}] + (L(t) - D(t))^T S + S(L(t) - D(t)) \qquad S(t_o) = S_o \tag{1.19}$$

The usefulness of the form (1.18) is evident if we observe that (1.15) and (1.18) form an extended linear system that can be regarded as a state-space model of the essential parameters (mean, variance, and covariance) describing the population. Although (1.19) is not as easily adjoined to (1.15), its form allows us to directly verify that the matrix $S(t)$ is indeed a covariance matrix (i.e., nonnegative definite) for all $t > t_o$. This property of $S(t)$ is verified as follows.

For a real population the initial estimates of the state-space variables must satisfy $\bar{\underset{\sim}{n}}_o \geq 0$ (i.e., $\underset{\sim}{n}_i \geq \underset{\sim}{0}$, $i = 1, \cdots, m$) and S_o nonnegative definite. Further, the elements of L(t) and D(t) are all positive, thus since D(t) is diagonal, it is easily seen that the nonnegative quadrant in R^n is an invariant set associated with (1.15), i.e., $\bar{\underset{\sim}{n}}(t) \geq 0$ for all $t > t_o$, whenever $\bar{\underset{\sim}{n}}(t_o) \geq \underset{\sim}{0}$. Hence, for $t \geq t_o$ the vector $(L(t) + D(t)]\bar{\underset{\sim}{n}}(t)$ contains nonnegative elements only. If $\Phi(t,\tau)$ denotes the matrix solution of

$$\frac{d\Phi(t,\tau)}{dt} = [L(t) - D(t)]\Phi(t,\tau) \qquad \Phi(\tau,\tau) = I$$

then it is well known [11] that the solution to (1.19) can be written as:

$$S(t) = \Phi^T(t,t_o)S_o\Phi(t,t_o) + \int_{t_o}^{t} \Phi^T(t,\tau) \text{diag}[(L(t) + D(t))\bar{\underset{\sim}{n}}(t)]\Phi(t,\tau)d\tau \tag{1.20}$$

Thus the nonnegative definiteness of $S(t)$, $t \geq t_o$, follows directly from the nonnegative definiteness of the matrices S_o and diag $[(L(t) + D(t))\bar{\underset{\sim}{n}}(t)]$.

Finally, we note that $m(m - 1)/2$ equations in (1.18) and (1.19) are redundant, since $\sigma_{ij} = \sigma_{ji}$, $i,j = 1, \cdots, m$, but they have been included to enable the system of equations in the variance and covariance parameters to be written in a compact form.

§2. RELATION TO THE LESLIE-POLLARD MODEL

The discrete model proposed by Leslie [2] has the form

$$\underset{\sim}{n}(t + 1) = L(t)\ \underset{\sim}{n}(t) \tag{2.1}$$

with $L(t)$ defined as in (1.16). The stochastic extension to this model proposed by Pollard [7] has the form

$$\underset{\sim}{\sigma}(t + 1) = K(t) + L(t) \otimes L(t)\underset{\sim}{\sigma}(t) \tag{2.2}$$

where the $m^2 \times m$ matrix $K(t)$ [having the same sparse structure as the $m^2 \times m$ matrix defined in (1.18)] is defined in [7].

Discretizing (1.15), we have

$$\underset{\sim}{n}(t + \Delta t) = [(L(t) - D(t))\Delta t + I]\underset{\sim}{n}(t) \tag{2.3}$$

Putting $\Delta t = 1$, (2.3) is equivalent to (2.1) when $D(t) = I$, i.e., from (1.17) when

$$\ell_i(t) + p_i(t) = 1 \qquad i = 1, \cdots, m-1$$

and

$$\ell_m(t) = 1$$

Recalling the birth-and-death process assumptions at the beginning of Section 1, we see that the above constraints on

$\ell_i(t)$ and $p_i(t)$ imply that during each interval of time, each member of the ith class must either die or pass into the (i + 1)th class. This is obviously true, except for the mth class, for populations structured into age classes. In general, however, the class structure may not be directly linked to the unit time interval, e.g., for a population divided into weight classes. Hence, it is a natural extension of the basic Leslie matrix model to allow a variable rate for the progress of each individual through the class structure. Apart from this simple generalization, system (1.15) is the continuous time analog of (2.1).

The stochastic extensions of the discrete system (2.2) and the continuous system (1.18) are closely related. The homogeneous term in (1.18) is the continuous-time analog of the homogeneous term in (2.2). The elements of the $m^2 \times m$ matrix in (1.18) are, however, not the continuous time analogs of the entries in K(t) in (2.2), although the two matrices have the same sparse structure.

By considering the difference equation

$$\underset{\sim}{\sigma}(t + \Delta t) = [(L(t) - D(t))\Delta t + I] \otimes [(L(t) - D(t))\Delta t + I]\underset{\sim}{\sigma}(t) \qquad (2.4)$$

it is easily seen that the $m^2 \times m^2$ coefficient matrices in (1.18) and (2.2) are the continuous and discrete analogs of each other. Putting $\Delta t = 1$ and $D(t) = I$, we have the homogeneous part of (2.2) while considering the limit in (2.4) of $(\underset{\sim}{\sigma}(t + \Delta t) - \underset{\sim}{\sigma}(t))/\Delta t$ as $\Delta t \to 0$ we derive the homogeneous part of (1.18).

Finally, it is worth noting that Pollard's extension was

derived for the multi-type Galton-Watson process, while our extension has been shown [8] to hold for the general linear multivariate birth-death-and-migration process.

§3. A BILINEAR SCALAR CONTROL HARVESTING PROBLEM

Suppose the dynamics of a population that we wish to harvest can be modelled by systems (1.15) and (1.18). Let $u(t)$ be the intensity at which we apply control to the population, i.e., $u(t)$ is a measure of the harvesting effort. Let $b_i(t)$, $i = 1, \cdots, m$, be the 'catchability coefficient' for the ith class of the population. In this problem we will constrain $u(t)$ to be bounded above by a maximum harvesting effort $\hat{u}$, say, to be nonnegative. In problems where we allow negative control to be interpreted as stocking the population, the nonnegativity constraint can be replaced by a maximum stocking intensity. Then, under assumptions (i)-(iv) in Section 1 and the assumption

(v) the probability that an individual in the ith class is removed in an interval $[t, t + \Delta t]$ due to the application of $u(t)$ is $b_i(t)\, u(t)\, \Delta t + o(\Delta t)$, $i = 1, \cdots, m$,

the following system of differential equations can be derived, following the method sketched in Section 1:

$$\frac{d\bar{\underset{\sim}{n}}}{dt} = (L(t) - D(t) - u(t)B(t))\bar{\underset{\sim}{n}} \qquad \bar{\underset{\sim}{n}}(t_o) = \bar{\underset{\sim}{n}}_o \tag{3.1}$$

and

$$\frac{d\underset{\sim}{\sigma}}{dt} = \begin{bmatrix} [L(t)+D(t)+u(t)B(t)]_1 \\ \underset{\sim}{0}^T \\ \vdots \\ \underset{\sim}{0}^T \\ [L(t)+D(t)+u(t)B(t)]_2 \\ \underset{\sim}{0}^T \\ \vdots \\ \vdots \\ \underset{\sim}{0}^T \\ [L(t)+D(t)+u(t)B(t)]_m \end{bmatrix} \bar{\underset{\sim}{n}} + [(L(t) - D(t) - u(t)B(t)) \otimes I + I \otimes (L(t) - D(t) - u(t)B(t))]\underset{\sim}{\sigma}$$

$$\underset{\sim}{\sigma}(t_o) = \underset{\sim}{\sigma}_o \tag{3.2}$$

The matrices L(t) and D(t) are those defined in (1.16) and (1.17), B(t) is the diagonal matrix, $\mathrm{diag}(b_1(t), \cdots, b_m(t))$, and the $m^2 \times m$ matrix on the right-hand side of (3.2) is defined in the manner discussed after (1.18). In many practical problems some of the $b_i(t)$ will be zero for all t. For example, a fish population may be harvested with nets which have a mesh size chosen to capture only fish in the kth class and above.

In order to find a harvesting strategy that is 'optimal' in some sense, we now formulate a criterion that will allow us to identify an 'optimal' strategy. This criterion should account for: the value of the harvest; the cost of the harvesting strategy; under- and over-exploited population levels; and the uncertainty in our knowledge of the actual number of individuals in each class of the population. This last consideration is in our case easily incorporated into the framework of deterministic optimal control theory, since (3.1) and (3.2) are deterministic systems in the parameters of a stochastic process.

(a) Value of harvest (yield). Suppose that the average weight of an individual in the ith class is $w_i(t)$ (where w_i may be a function of time due to seasonal fluctuations in weight). Then, if over $[t_o, t_f]$ the harvesting effort is u(t), we have

$$\text{Expected Yield} = \int_{t_o}^{t_f} \underset{\sim}{w}^T(t)B(t)\bar{\underset{\sim}{n}}(t)u(t)dt \qquad (3.3)$$

where

$$\underset{\sim}{w}^T(t) = \left(w_1(t), \cdots, w_m(t)\right)$$

(b) Cost of harvest. The cost of implementing a given strategy u(t), $t \in [t_o, t_f]$ will be a function of u. In general, this function will be nonlinear, but a linear relation might be a reasonable assumption to make in many real cases, especially when u(t) is constrained to a finite interval $[0, \hat{u}]$. Thus in general, the cost incurred over $[t_o, t_f]$ is

$$\text{Cost of Harvest} = \int_{t_o}^{t_f} k(u(t), t)dt \qquad (3.4)$$

where k(u(t),t) is some suitable function of u(t) and t.

(c) Under- and over-exploited populations. At the end of the harvesting program, i.e., when $t = t_f$, say, the population should be at a level where harvesting can recommence immediately, or at least soon afterwards. Hence a population which is over-exploited in this sense is undesirable. On the other hand, a large population near its saturation level has a low growth potential [this phenomenon is not reflected in the idealized linear model (1.15)] and can be regarded as under-exploited. If we can determine a priori a population level $\hat{\underset{\sim}{n}}(t)$ that can in some sense be regarded as the ideal level for exploitation (perhaps the 'intrinsic growth' rate of the population is a maximum at this level), then the following criterion would be a suitable measure:

$$\text{Cost of population level} = \frac{1}{2}\int_{t_o}^{t_f} (\bar{\underset{\sim}{n}}(t) - \hat{\underset{\sim}{n}}(t))^T Q(t)(\bar{\underset{\sim}{n}}(t) - \hat{\underset{\sim}{n}}(t))dt + \frac{1}{2}(\bar{\underset{\sim}{n}}(t_f) - \hat{\underset{\sim}{n}}(t_f))^T Q_f(\bar{\underset{\sim}{n}}(t_f) - \hat{\underset{\sim}{n}}(t_f)) \tag{3.5}$$

The weighting matrices $Q(t)$ and Q_f are included for generality and will usually be nonnegative definite. It may be sufficient to take $Q(t) \equiv 0$ if our major concern is the population level at the end of the harvesting period. A nonzero $Q(t)$, however, is useful if we do not want the population during $[t_o, t_f]$ to fall below a critical level from which the population will have difficulty in recovering [we note again that this phenomenon is not reflected in the idealized linear model (1.15)].

(d) Cost of uncertainty. A deterministic process is an ideal, in the sense that we can predict the exact effect of any harvesting policy. In a stochastic process we can only estimate the effect of any harvesting policy with a confidence that decreases as a function of increasing variance and covariance. The discrepancy between expected and actual yield and expected and actual population sizes can result in a harvesting policy that is optimal for a deterministic process, actually being (a posteriori) suboptimal for the particular state trajectory realized by the stochastic process. It is thus desirable to keep the uncertainty in the system as low as possible. Clearly, if the trace of the covariance matrix $S(t)$ or its square $S(t)S(t)$ is zero, we have a deterministic process. If for some classes the uncertainty in the mean level is more critical than for others we can include a 'weighting' matrix $P(t)$ (nonnegative definite). Then trace $S(t)P(t)S(t)$ would be a suitable measure. Since it is easily shown that

$$\underset{\sim}{\sigma}(t)^T I \otimes P(t)\underset{\sim}{\sigma}(t) = \text{trace } S(t)P(t)S(t)$$

we have

$$\begin{aligned}\text{Cost of uncertainty} &= \frac{1}{2}\int_{t_o}^{t_f} \underset{\sim}{\sigma}(t)^T I \otimes P(t)\underset{\sim}{\sigma}(t)\,dt \\ &= \frac{1}{2}\,\underset{\sim}{\sigma}(t_f)^T I \otimes P_f\underset{\sim}{\sigma}(t_f) \qquad (3.6)\end{aligned}$$

In addition to considering the cost associated with nonzero variance and covariance, we can associate a cost in estimating the initial population parameters $\underset{\sim}{n}_o$ and $\underset{\sim}{\sigma}_o$. In fact, $\underset{\sim}{\sigma}_o$ can be considered a control parameter (as is done for the scalar case in [12]). In this paper, however, we shall consider $\underset{\sim}{\sigma}_o$ as given.

Since we should want to maximize (3.3) and minimize (3.4), (3.5), and (3.6) over all possible harvesting programs u(t), $t \in [t_o, t_f]$, i.e., over all piecewise continuous $u(t) \in [0, u]$, our problem can be stated as follows.

Minimize over admissible u the functional

$$J(u) = \int_{t_o}^{t_f} [-\underset{\sim}{w}(t)^T B(t)\bar{\underset{\sim}{n}}(t)u(t) + k(u(t),t)$$
$$+ \frac{1}{2}(\bar{\underset{\sim}{n}}(t) - \hat{\underset{\sim}{n}}(t))^T Q(t)(\bar{\underset{\sim}{n}}(t) - \hat{\underset{\sim}{n}}(t))$$
$$+ \frac{1}{2}\underset{\sim}{\sigma}(t)^T I \otimes P(t)\underset{\sim}{\sigma}(t)]dt$$
$$+ \frac{1}{2}(\bar{\underset{\sim}{n}}(t_f) - \hat{\underset{\sim}{n}}(t_f))^T Q_f(\bar{\underset{\sim}{n}}(t_f) - \hat{\underset{\sim}{n}}(t_f))$$
$$+ \frac{1}{2}\underset{\sim}{\sigma}(t_f)^T I \otimes P_f\underset{\sim}{\sigma}(t_f) \qquad (3.7)$$

subject to the dynamic constraint equations (3.1) and (3.2).

Finally, as was done in Section 1, we can readily verify that the positive quadrant of R^m is an invariant set associated with (3.1) and that the solution to (3.2) is a covariance matrix.

§4. OPEN AND CLOSED SEASON HARVESTING STRATEGY

If k(u(t),t) in (3.4) is a linear function of u(t) say, i.e.,

$$k(u(t),t) = c(t)u(t) \qquad (4.1)$$

then u(t) appears only linearly in (3.7). Since (3.1) and (3.2) are also linear in u(t), if the control problem is nonsingular (defined below), it is well known that the optimal control will be bang-bang [13]. This corresponds to an alternating sequence of closed seasons ($u(t) = 0$) and seasons of maximum harvesting activity ($u(t) = \hat{u}$).

To simplify our discussion we omit the function argument t and rewrite the problem formulated in Section 3 as follows. Minimize

$$J(u) = \int_{t_o}^{t_f} [(-\underset{\sim}{w}^T B\underset{\sim}{\bar{n}} + c)u + \frac{1}{2}(\underset{\sim}{\bar{n}} - \underset{\sim}{\hat{n}})^T Q(\underset{\sim}{\bar{n}} - \underset{\sim}{\hat{n}})$$

$$+ \frac{1}{2}\underset{\sim}{\sigma}^T I \otimes P\underset{\sim}{\sigma}]dt + \frac{1}{2}[(\underset{\sim}{\bar{n}} - \underset{\sim}{\hat{n}})^T Q_f(\underset{\sim}{\bar{n}} - \underset{\sim}{\hat{n}})$$

$$+ \underset{\sim}{\sigma}^T I \otimes P_f\underset{\sim}{\sigma}]_{t=t_f} \qquad (4.2)$$

over piecewise continuous $u \in [0,\hat{u}]$, subject to

$$\frac{d\underset{\sim}{\bar{n}}}{dt} = (A - uB)\underset{\sim}{\bar{n}} \qquad \underset{\sim}{\bar{n}}(t_o) = \underset{\sim}{n}_o \qquad (4.3)$$

$$\frac{d\underset{\sim}{\sigma}}{dt} = (A_e^+ + uB_e)\underset{\sim}{\bar{n}} + [(A - uB) \otimes I$$

$$+ I \otimes (A - uB)]\underset{\sim}{\sigma} \qquad \underset{\sim}{\sigma}(t_o) = \underset{\sim}{\sigma}_o \qquad (4.4)$$

with the obvious definitions of A, A_e^+, and B_e following from (3.1) and (3.2).

For any given u let $\underset{\sim}{p}$ and $\underset{\sim}{q}$ satisfy the system adjoint to (4.3) and (4.4), i.e.,

$$\frac{d\underset{\sim}{q}}{dt} = -(A - uB)^T\underset{\sim}{p} - (A_e^+ + uB_e)^T\underset{\sim}{q} \qquad (4.5a)$$

$$\frac{d\underset{\sim}{q}}{dt} = -[(A - uB)^T \otimes I + I \otimes (A - uB)^T]\underset{\sim}{q} \qquad (4.6a)$$

with final value conditions

$$\underset{\sim}{p}(t_f) = Q_f(\underset{\sim}{\bar{n}}(t_f) - \underset{\sim}{\hat{n}}(t_f)) \tag{4.5b}$$

$$\underset{\sim}{q}(t_f) = I \otimes P_f\, \underset{\sim}{\sigma}(t_f) \tag{4.6b}$$

Define

$$H(\underset{\sim}{\bar{n}},\underset{\sim}{\sigma},\underset{\sim}{p},\underset{\sim}{q},u,t) = \text{Integrand of } J(u) + \underset{\sim}{p}^T \frac{d\underset{\sim}{\bar{n}}}{dt} + \underset{\sim}{q}^T \frac{d\underset{\sim}{\sigma}}{dt}$$

and define $\underset{\sim}{\bar{n}}^*$, $\underset{\sim}{\sigma}^*$, $\underset{\sim}{p}^*$, and $\underset{\sim}{q}^*$ to be the solutions to (4.3-4.6) when $u = u^*$. Then if u^* minimize $J(u)$ subject to the control constraints and dynamic constraints, Pontryagin's minimum principle [13] states that

$$\min_{\text{admissible } u} H(\underset{\sim}{\bar{n}}^*,\underset{\sim}{\sigma}^*,\underset{\sim}{p}^*,\underset{\sim}{q}^*,u,t)$$
$$= H(\underset{\sim}{\bar{n}}^*,\underset{\sim}{\sigma}^*,\underset{\sim}{p}^*,\underset{\sim}{q}^*,u^*,t) \tag{4.7}$$

Since the coefficient of u in H is

$$H_u(\underset{\sim}{\bar{n}},\underset{\sim}{\sigma},\underset{\sim}{p},\underset{\sim}{q},t) = -\underset{\sim}{w}^T B\underset{\sim}{\bar{n}} + c - \underset{\sim}{p}^T B\underset{\sim}{\bar{n}} + q^T B_e\underset{\sim}{\bar{n}}$$
$$- q^T[B \otimes I + I \otimes B]\underset{\sim}{\sigma} \tag{4.8}$$

we minimize $H(\underset{\sim}{\bar{n}}^*,\underset{\sim}{\sigma}^*,\underset{\sim}{p}^*,\underset{\sim}{q}^*,u,t)$ over piecewise continuous $u \in [0,\hat{u}]$ by choosing

$$u^* = 0, \text{ whenever } H_u^*(t) \equiv H_u(\underset{\sim}{\bar{n}}^*,\underset{\sim}{\sigma}^*,\underset{\sim}{p}^*,\underset{\sim}{q}^*,t) > 0 \tag{4.9}$$

$$u^* = \hat{u}, \text{ whenever } H_u^*(t) < 0 \tag{4.10}$$

If $H_u^*(t)$ is zero only at isolated points on the t axis, u^*, although it is undetermined on this set, can be given arbitrary values. If, however, $H_u^*(t)$ is zero on some open subinterval $(t_1,t_2) \subset [t_o,t_f]$, then the problem is singular and u^* can only be determined by examining the time derivatives of H_u. In this case u* will not, in general, be bang-bang on (t_1,t_2).

Even in the non-singular case we are still faced with the problem of finding u*. The dependence of the solutions $\bar{\underset{\sim}{n}}^*$, $\underset{\sim}{\sigma}^*$, $\underset{\sim}{p}^*$, and $\underset{\sim}{q}^*$ on u* and the fact that the systems for $\bar{\underset{\sim}{n}}$, $\underset{\sim}{\sigma}$, $\underset{\sim}{p}$, and $\underset{\sim}{q}$ are coupled as a two-point boundary value problem, rules out the possibility of directly generating u* from (4.3-4.10), and an iterative algorithm must be used. Jacobson's first-order differential dynamic programming algorithm for solving bang-bang control [14,15] is suitable for the above problem. However, since the maximal principle is only a necessary condition for optimality, we are still required to verify that u* satisfying (4.7), (4.9), and (4.10) does indeed, minimize (4.2).

§5. CONCLUSION

From the deterministic population model (1.15), via (1.18) we provide a means for directly generating the covariance matrix resulting from the probabilistic nature of the birth-death-and-migration parameters of a structured population. Moreover, in our formulation of an optimal harvesting procedure we can not only determine the uncertainty associated with the expected mean level of the population, but we can also account for the costs associated with this uncertainty.

ACKNOWLEDGMENT

This paper is based on material in chapter four of the author's Ph.D. thesis, submitted to the University of the Witwatersrand, Johannesburg, February 1976 [8].

REFERENCES

1. Lewis, E. G., On the Generation and Growth of a Population, Sankya, vol. 6, 1942, pp. 93-96.

2. Leslie, P. H., On the Use of Matrices in Certain Population Mathematics, Biometrika, vol. 33, 1945, pp. 183-212.

3. Leslie, P. H., Some Further Notes on the Use of Matrices in Population Mathematics, Biometrika, vol. 35, 1948, pp. 213-245.

4. Lefkovitch, L. P., A Theoretical Evaluation of Population Growth After Removing Individuals from the Same Age Group, Bull. Ent. Res., vol. 57, 1967, pp. 437-445.

5. Doubleday, W. G., Harvesting in Matrix Populations, Biometrics, vol. 31, 1975, pp. 189-200.

6. Rorres, C. and W. Fair, Optimal Harvesting Policy for an Age-specific Population, Math. Biosci., vol. 24, 1975, pp. 31-47.

7. Pollard, J. H., Mathematical Models for the Growth of Human Populations, Cambridge University Press, 1973, pp. 112-135.

8. Getz, W. M., Modelling and Control of Birth-and-Death Processes, CSIR Special Report - WISK 196, National Research Institute for Mathematical Sciences, Pretoria, 1976, pp. 1-153 (Ph.D. Thesis, University of the Witwatersrand, Johannesburg, 1976).

9. Getz, W. M., Stochastic Equivalents of the Linear and Lotka-Volterra System of Equations - A General Birth-and-Death Process, Math. Biosci., vol. 29, 1976, pp. 235-257.

10. Bellman, R., Introduction to Matrix Analysis (second edition), McGraw-Hill, Inc., New York, 1970.

11. Brockett, R. W., Finite-dimensional Linear Systems, John Wiley and Sons, Inc., New York, 1970.

12. Getz, W. M., Optimal Control of a Birth-and-Death Population Model, Math. Biosci., vol. 23, 1975, pp. 87-111.

13. Leitmann, G., An Introduction to Optimal Control Theory, McGraw-Hill, Inc., New York, 1966.

14. Jacobson, D. H., Differential Dynamic Programming Methods for Solving Bang-bang Control Problems, IEEE Trans. Au. Contr., AC-13, no. 6, 1968, pp. 661-675.

15. Jacobson, D. H. and D. Q. Mayne, Differential Dynamic Programming, American Elsevier Publishing Co., Inc., New York, 1970.

EXISTENCE OF UNIQUE NASH EQUILIBRIUM SOLUTIONS IN NONZERO-SUM STOCHASTIC DIFFERENTIAL GAMES

Tamer Başar

Marmara Research Institute
Gebze, Kocaeli, Turkey

ABSTRACT

This paper is concerned with a fixed duration stochastic two-person nonzero-sum differential game (NZSDG) in which one player has access to closed-loop nonanticipatory state information while the other player makes no observation. The state of the game evolves according to a linear Ito stochastic differential equation and the cost function of each player is assumed to be quadratic in both the state and the control variables. For this class of problems it is shown in the paper that the Nash equilibrium strategy of the first player can be realized by affine control laws and furthermore that it will be unique when the additive independent increment process in the state dynamics has positive definite incremental covariance function and when the duration of the game is taken to be sufficiently small. This result is established by identifying the equilibrium solution as the fixed-point of an appropriately structured Banach space, and it constitutes the first proof of existence of a unique Nash equilibrium solution in stochastic NZSDG, without making any explicit assumptions on the structure of the control laws. The paper also contains exact expressions and discussion concerning the unique saddle-point solution of similarly structured zero-sum differential games.

§1. INTRODUCTION

The prime objective of this paper is to establish existence of unique Nash equilibrium solutions in two-person linear-quadratic (LQ) stochastic nonzero-sum differential games (NZSDG) and to further investigate the properties of this unique solution in the special case of zero-sum differential games. The stochastic nature of the problem arises solely from the additive independent-increment process in the state dynamics. The incremental covariance function of this stochastic process is taken to be positive definite in order to prevent the informational nonuniqueness arising from different representations as reported in Başar [1,2,3]. The information structure of the problem endows player 1 with closed-loop nonanticipative perfect state information and player 2 with only the a priori static information. A similar type of a stochastic differential game has previously been considered by Friedman [6] for the case when both players have access to closed-loop nonanticipative perfect state information. However, Friedman further restricts the strategies of every player to be only functions of the current value of the state vector. In this paper we attempt to remove this restriction and prove existence of a unique affine Nash equilibrium solution under the above information pattern and with no a priori constraint on the strategy spaces. We obtain explicit expressions for the optimal strategies of each player and establish existence and uniqueness whenever the time interval on which the game is defined is taken to be sufficiently small. This result is established by identifying the strategy of player 2 as the fixed-point of an appropriately structured Banach space, and it constitutes the first proof of its kind in the literature.

We then attempt, in the paper, to specialize the unique Nash solution to LQ zero-sum differential games and determine a relatively simple differential equation that the saddle-point solution should satisfy. We further investigate the relation between the existence conditions of that solution and the closed-loop solution for both players in such games.

Section 2 is devoted to a description of the problem treated here. Section 3 contains complete expressions for the unique equilibrium solution and a proof of existence and uniqueness. Section 4 is on the special case of zero-sum differential games, followed by a conclusion section and an appendix.

§2. FORMULATION OF THE PROBLEM

The differential game problem to be treated is a two-person fixed duration nonzero-sum differential game (NZSDG) described by the Ito stochastic differential equation

$$dx_t = F(t)x_t dt + G_1(t)u_1^t dt + G_2(t)u_2(t)dt + dw_t \qquad (1)$$

$$t_o \le t \le t_f, \qquad x_{t_o} = x_o$$

where $\dim(x) = n$, $\dim(u_i) = r_i$, and the matrices F, G_1, and G_2 have appropriate dimensions and are continuous on $[t_o, t_f]$. x_o is the value of the initial state vector and it is known to both players. The functions u_1^t and $u_2(t)$ represent the control policies of players 1 and 2, respectively, and assume values in R^{r_1} and R^{r_2}, respectively. Here $u_2(t)$ is a deterministic vector while u_1^t is allowed to be a stochastic process as to be explained in the sequel. x_t, $t \geq t_o$, is an n-dimensional vector stochastic process with continuous sample paths and w_t, $t \geq t_o$, is a zero-mean independent-increment Gaussian process, the covariance matrix of each increment being positive definite. That is, $w_{t_o} = 0$, $E[dw_t dw_t^T] = \Lambda(t) > 0$.

To delineate the information structure of the problem we let $C_n = C_n[t_o, t_f]$ denote the space of continuous functions on $[t_o, t_f]$ with values in R^n. We further let F_t be the sigma-field in C_n generated by the cylinder sets $\{x \in C_n,\ x(s) \in B\}$ where B is a Borel set in R^n and $t_o \leq s \leq t$. Then, the information gained by player 1 during the course of the game is completely determined by the information field F_t for all $t \geq t_o$, i.e., player 1 has access to perfect *nonanticipative closed-loop* information concerning the state of the game. Player 2, on the other hand, gains no information during the course of the game (i.e., he plays open-loop).

Permissible strategy for player 1 will be a mapping $\gamma_1(\cdot,\cdot)$ of $[t_o, t_f] X C_n$ into R^{r_1} with the following properties:

(i) $\gamma_1(t,\eta)$ is continuous in t for each $\eta \in C_n$.

(ii) It is uniformly Lipschitz in η, i.e.,

$$|\gamma_1(t,\eta) - \gamma_1(t,\xi)| \leq k\|\eta - \xi\|,$$

$$t \in [t_o, t_f); \quad \eta, \xi \in C_n$$

where $\|\cdot\|$ is the standard sup norm in C_n.

(iii) $u_1(t) = \gamma_1(t,x)$ is adapted to the information field F_t.

We denote the class of strategies described above by U_1, to be referred to as the permissible strategy set for player 1. Since player 2 has access to open-loop information, we let the permissible strategy set U_2 for player 2 be C_{r_2}.

Since the strategy set U_1 does not only contain Markovian controls, equation (1) is actually a functional differential equation rather than an ordinary differential equation, which should better be written as

$$dx_t = F(t)x_t dt + G_1(t)\gamma_1[t,x_s, s \le t]dt + G_2(t)\gamma_2(t)dt + dw_t, \quad t \ge t_o, \; x_{t_o} = x_o \tag{2}$$

It is known that corresponding to any pair of strategies $\{\gamma_1 \in U_1, \gamma_2 \in U_2\}$ this stochastic differential equation admits a unique solution that is a sample-path-continuous second-order process [4,5,10]. Furthermore, the control process $u_1^t = \gamma_1[t,x_s; s \le t]$ is a second order stochastic process with continuous sample paths and adapted to $\mathcal{B}_t$ which is the sigma-field generated by x_s and w_s, $t_o \le s \le t$.

For any pair of strategies $\{\gamma_i \in U_i, k = 1, 2\}$ the expected loss (or minus the expected payoff) incurred to player i is given by the quadratic cost function

$$J_i(\gamma_1,\gamma_2) = E\left\{x_{t_f}^T C_{i_f} x_{t_f} + \int_{t_o}^{t_f}\left[x_t^T C_i(t) x_t + u_1^{t^T} D_{i1}(t) u_1^t + u_2^T D_{i2}(t) u_2\right]dt\right\} \tag{3}$$

where $u_1^t = \gamma_1[\cdot,\cdot]$, $u_2 = \gamma_2(\cdot)$; the weighting matrices C_{i_f}, $C_i(t)$ are nonnegative definite for all $t \in [t_o,t_f]$ and with each entry of $C_i(\cdot)$ being continuous on $[t_o,t_f]$. The matrices

$D_{i1}(t) > 0$, $D_{i2}(t) > 0$ are also defined and continuous on $[t_o, t_f]$, $i = 1, 2$. Furthermore, $E\{\cdot\}$ denotes the expectation operation with respect to the statistics of $\{w_t\}$.

The objective of player i is to pick the permissible strategy that will yield the minimum value of the cost function J_i against some rationally selected strategy of player j, $j \neq i$. Allowing no direct cooperation between the players, this reasoning leads to what is known as the noncooperative Nash equilibrium solution.

Definition 1 A pair $\{\gamma_1^* \in U_1, \gamma_2^* \in U_2\}$ is said to be in *(Nash) equilibrium* if the following inequalities hold for all $\gamma_1 \in U_1$, $\gamma_2 \in U_2$:

$$J_1(\gamma_1^*, \gamma_2^*) \leq J_1(\gamma_1, \gamma_2^*) \tag{4a}$$

$$J_2(\gamma_1^*, \gamma_2^*) \leq J_2(\gamma_1^*, \gamma_2) \tag{4b}$$

Hence the prime objective of the rest of the paper is to establish existence of a unique noncooperative equilibrium solution that satisfies (4a) and (4b).

§3. MAIN RESULTS

We first note that for every fixed $\gamma_2 \in U_2$ player 1 is faced with the following stochastic control problem:

$$\min_{U_1} J_1(u_1^t, \gamma_2) \tag{5a}$$

with

$$J_1 = E\left\{x_{t_f}^T C_{1_f} x_{t_f} + \int_{t_o}^{t_f} \left(x_t^T C_1 x_t + (u_1^t)^T D_{11} u_1^t + \gamma_2^T D_{12} \gamma_2\right) dt\right\} \tag{5b}$$

$$dx_t = \left(Fx_t + G_1u_1^t + G_2\gamma_2(t)\right)dt + dw_t \tag{5c}$$

This is known as a generalized linear regulator problem in the literature and is well-known to admit the following unique solution (uniqueness is a result of positive definiteness of the incremental covariance matrix of the process w_t for each t and can be established by generalizing the proof given in Kwakernaak [7]; see Başar [3] for further details of the proof).

Lemma 1 For every fixed $\gamma_2 \in U_2$ there exists a *unique* element of U_1 that solves the stochastic control problem (5a-5c). This unique solution is given by

$$\gamma_1^*(t,x_t) = -D_{11}^{-1}G_1^T[P(t)x_t + k(t)] \tag{6a}$$

$$\dot{P} + F^TP + PF - PG_1D_{11}^{-1}G_1^TP + C_1 = 0 \tag{6b}$$

$$P(t_f) = C_{1_f}, \quad t_o \le t \le t_f$$

$$\dot{k} + \left(F^T - PG_1D_{11}^{-1}G_1^T\right)k + PG_2\gamma_2(t) = 0 \tag{6c}$$

$$k(t_f) = 0$$

It now follows from Lemma 1 that in characterizing all solutions to (4), we can restrict ourselves (without any loss of generality) to a proper subset of U_1 consisting of all measurable affine mappings γ_1 of the form

$$\gamma_1(t,x_t) = -D_{11}^{-1}G_1^T\,[P(t)x_t + \ell(t)] \tag{7}$$

where P is given by (6b) and ℓ is any element of C_n. Hence every Nash strategy for player 1 will be of the form (7) for some ℓ in C_n. Now, replacing u_1^t by the expression given by

(7) in both (1) and (3) with i = 2, we observe that every Nash policy for player 2 will be an optimizing solution to the following stochastic control problem for some $\ell \in C_n$:

$$\min_{U_2} \quad L(u_2) \tag{8a}$$

with

$$L(u_2) \triangleq E\Big\{x_{t_f}^T C_{2_f} x_{t_f} + \int_{t_o}^{t_f} \Big[x_t^T \tilde{C}_2 x_t + 2x_t^T \tilde{\ell}(t)$$

$$+ u_2^T D_{22} u_2 + \ell^T G_1 D_{11}^{-1} D_{21} D_{11}^{-1} G_1^T \ell\Big] dt\Big\} \tag{8b}$$

$$\tilde{\ell}(t) \triangleq PG_1 D_{11}^{-1} D_{21} D_{11}^{-1} G_1^T \ell(t) \tag{8c}$$

$$\tilde{C}_2(t) \triangleq C_2(t) + PG_1 D_{11}^{-1} D_{21} D_{11}^{-1} G_1^T P \tag{8d}$$

and subject to

$$dx_t = (\tilde{F}x_t + G_2 u_2 + s)dt + dw_t, \quad x_{t_o} = x_o \tag{8e}$$

$$\tilde{F}(t) \triangleq F - G_1 D_{11}^{-1} G_1^T P \tag{8f}$$

$$s(t) \triangleq - G_1 D_{11}^{-1} G_1^T \ell(t) \tag{8g}$$

where P(t) is given by (6b), and ℓ is any a priori picked element of C_n that is functionally independent of u_2. (In the above description we have intentionally suppressed the time dependence in order to avoid unnecessary repetition.)

This stochastic control problem admits a unique globally optimal solution which is given below in Lemma 2. This result can again be established by extending the proof given in [7] and the reader is referred to Basar [3] for further details:

Lemma 2 Corresponding to every a priori fixed $\ell \in C_n$, there exists a *unique* element γ_2^* of U_2 that solves the stochastic control problem (8a-8g). This optimal strategy can explicitly be written as a function of $\ell(t)$ as follows (where time dependence is again suppressed):

$$u_2^*(t) = \gamma_2^*(t) = - D_{22}^{-1}G_2^T[S(t)y(t) + b(t)] \tag{9a}$$

$$\dot{S} + \tilde{F}^T S + S\tilde{F} - SG_2D_{22}^{-1}G_2^T S + \tilde{C}_2 = 0 \tag{9b}$$

$$S(t_f) = C_{2_f}, \quad t_o \le t \le t_f$$

$$\dot{b} + \tilde{F}^T b - SG_2D_{22}^{-1}G_2^T b + Ss + \tilde{\ell} = 0 \tag{9c}$$

$$b(t_f) = 0$$

$$\dot{y} + \left(G_2D_{22}^{-1}G_2^T S - \tilde{F}\right)y + G_2D_{22}^{-1}G_2^T b - s = 0 \tag{9d}$$

$$y(t_o) = x_o$$

Lemmas 1 and 2 provide us with a characterization of all possible Nash strategies of players 1 and 2, respectively, in terms of a continuous function $\ell(\cdot)$ yet to be determined. Now, for (7) and (9a) to be mutually consistent as a permissible Nash strategy pair, they have to be optimal against each other, which implies that $k(t)$ should be equivalent to $\ell(t)$ when $\gamma_2(t)$ is replaced in (6c) by its optimal value from (9a). It is now not difficult to see that the NZSDG under investigation will admit a noncooperative equilibrium solution iff the above mentioned compatibility condition is satisfied. This brings us to the following important result which directly follows from Lemmas 1 and 2 and the discussion given above:

Theorem 1 Every noncooperative equilibrium solution to the LQNZSDG introduced in this paper is given by

$$u_1^t = \gamma_1^*(t,x_t) = -D_{11}^{-1}G_1^T(Px_t + \bar{k}) \quad (10a)$$

$$u_2^*(t) = \gamma_2^*(t) = -D_{22}^{-1}G_2^T(S\bar{y}(t) + \bar{b}) \quad (10b)$$

where P(t) and S(t) are as defined by (6b) and (9b), respectively, and $\bar{k}$, $\bar{y}$, $\bar{b}$ satisfy the coupled differential equations

$$\dot{\bar{k}} + \tilde{F}^T\bar{k} - PG_2D_{22}^{-1}G_2^T(S\bar{y} + \bar{b}) = 0, \quad \bar{k}(t_f) = 0 \quad (11a)$$

$$\dot{\bar{b}} + \left(\tilde{F}^T - SG_2D_{22}^{-1}G_2^T\right)\bar{b} + \left(PG_1D_{11}^{-1}D_{21} - SG_1\right)D_{11}^{-1}G_1^T\bar{k} = 0$$

$$\bar{b}(t_f) = 0 \quad (11b)$$

$$\dot{\bar{y}} - \left(\tilde{F} - G_2D_{22}^{-1}G_2^TS\right)\bar{y} + G_2D_{22}^{-1}G_2^T\bar{b} + G_1D_{11}^{-1}G_1^T\bar{k} = 0$$

$$\bar{y}(t_o) = x_o \quad (11c)$$

Furthermore, a necessary and sufficient condition for existence of a Nash equilibrium point is existence of a solution to the two-point boundary value problem (11a-11c).

Remark 1 A proof of the last part of the statement of Theorem 1 follows from the fact that since $C_1(t) \geq 0$, $\tilde{C}_2(t) \geq 0$, $D_{22}(t) > 0$, $D_{11}(t) > 0$, both of the Riccati differential equations (6b) and (9b) admit unique bounded nonnegative definite matrix solutions (see e.g., Reid [8], p. 121).

Via Theorem 1, we have now converted the original problem of investigating existence of a unique Nash equilibrium point to existence of a unique solution to the two-point boundary value problem (11a-11c). This, in turn, is equivalent to existence of a unique

$$z \triangleq - D_{22}^{-1}G_2^T(S\bar{y} + \bar{b}) \tag{12}$$

which also constitutes the optimal unique Nash strategy for player 2, whenever it exists. Now, adopting a Lebesgue interpretation for the integral appearing in the cost functions (3), we seek existence of a unique element $z \in L^{r_2}$ such that (11a-11c) are satisfied. (Here, L^{r_2} denotes the Banach space of all r_2-dimensional real-valued Lebesgue square-integrable functions on $[t_0,t_f]$, i.e., if $z \in L^{r_2}$, then $\int_{t_0}^{t_f} z^T z dt < \infty$.) As it will be clear from the sequel any such element (if it exists) will also be in C_{r_2} because of the nature of the differential equation involved (or nature of operator L defined by (16b) below). Furthermore, it will be a unique element of C_{r_2} since any two elements of C_{r_2} are equivalent under the sup norm iff they are equivalent under the norm of L^{r_2}.

Denoting the state-transition matrices associated with the differential equations (11a-11c) by Φ_k, Φ_b, and Φ_y, respectively, we write these equations in the equivalent forms

$$\bar{k} = L_1\bar{z} \tag{13a}$$

$$\bar{b} = L_2\bar{k} \tag{13b}$$

$$\bar{y} = \Phi_y(t,t_0)x_o + L_3\bar{k} + L_4\bar{b} \tag{13c}$$

where $\bar{z}$ is defined by

$$\bar{z} = (S\bar{y} + \bar{b}) \tag{14}$$

and L_i, $i = 1, \cdots, 4$, are linear operators mapping L^n into L^n, and are defined by

$$L_1\bar{z} = \int_{t_f}^{t} \Phi_k(t,\sigma)\left[PG_2D_{22}^{-1}G_2^T\bar{z}\right](\sigma)\,d\sigma \tag{15a}$$

$$L_2\bar{k} = \int_{t_f}^{t} \Phi_b(t,\sigma)\left[SG_1D_{11}^{-1}G_1^T\bar{k} - PG_1D_{11}^{-1}D_{21}D_{11}^{-1}G_1^T\bar{k}\right](\sigma)\,d\sigma \tag{15b}$$

$$L_3\bar{k} = -\int_{t_o}^{t} \Phi_y(t,\sigma)\left[G_1D_{11}^{-1}G_1^T\bar{k}\right](\sigma)\,d\sigma \tag{15c}$$

$$L_4\bar{b} = -\int_{t_o}^{t} \Phi_y(t,\sigma)\left[G_2D_{22}^{-1}G_2^T\bar{b}\right](\sigma)\,d\sigma \tag{15d}$$

Compatibility condition now requires solvability of the operator equation

$$\bar{z} = S(t)\Phi_y(t,t_o)x_o + L\bar{z} \tag{16a}$$

$$L \triangleq (SL_3 + SL_4L_2 + L_2)L_1 \tag{16b}$$

We now find a bound on the norm $\|\cdot\|_o$ of L, where $\|\cdot\|_o$ is the standard norm on the Banach space $\mathcal{B}(L^n)$ of continuous linear transformations of L^n onto itself, i.e.,

$$\|L\|_o \triangleq \sup_{\bar{z}\in L^n} \|L\bar{z}\|_{t_f}, \qquad \|\bar{z}\|_{t_f} \le 1 \tag{17a}$$

where

$$\|\bar{z}\|_{t_f} \triangleq \int_{t_o}^{t_f} \bar{z}^T\bar{z}\,dt \tag{17b}$$

<u>Preliminary remarks and notation</u> For every $t' \in [t_o,t_f]$, let $P(t',t)$ and $S(t',t)$ denote the unique matrix solutions to the Riccati differential equations (6b) and (9b), respectively, for $t_o \le t \le t'$ and with the original boundary conditions replaced by $P(t') = C_{1_f}$, $S(t') = C_{2_f}$. We know that a unique non-negative definite solution exists for every $t' \in [t_o,t_f]$ and,

furthermore, that the solution is a continuous function of t, $t_o \le t \le t'$ (see e.g., Reid [8]).

Since Φ_k, Φ_b, and Φ_y are continuous functions of entries of S and P, they will also be continuous functions of t, $t \le t'$, for every fixed $t' \in (t_o, t_f]$. To denote the explicit dependence on t', we adjoin the variable t' to their argument; that is, we write

$$\Phi_k = \Phi_k(t', t, \sigma) \tag{18a}$$

$$\Phi_b = \Phi_b(t', t, \sigma) \tag{18b}$$

$$\Phi_y = \Phi_y(t', t, \sigma) \tag{18c}$$

We now define bounded scalar numbers α_i, i = 1, ···, 4 by

$$\alpha_1 = \max_{i,j,\sigma,t,t'} \left| \Phi_k(t',t,\sigma) \left[P(t',\sigma) G_2 D_{22}^{-1} G_2^T\right](\sigma) \right|_{ij} \tag{19a}$$

$$\alpha_2 = \max_{i,j,\sigma,t,t'} \left| \Phi_b(t',t,\sigma) \left[S(t',\sigma) G_1 D_{11}^{-1} G_1^T - P(t',\sigma) G_1 D_{11}^{-1} D_{21} D_{11}^{-1} G_1^T\right](\sigma) \right|_{ij} \tag{19b}$$

$$\alpha_3 = \max_{i,j,\sigma,t,t'} \left| \Phi_y(t',t,\sigma) G_1(\sigma) D_{11}^{-1}(\sigma) G_1^T(\sigma) \right|_{ij} \tag{19c}$$

$$\alpha_4 = \max_{i,j,\sigma,t,t'} \left| \Phi_y(t',t,\sigma) G_2(\sigma) D_{22}^{-1}(\sigma) G_2^T(\sigma) \right|_{ij} \tag{19d}$$

where $\|\cdot\|_{ij}$ denotes, in each case, the absolute value of the ijth element of the n x n matrix (·), and the maxima are taken over i,j = 1, ···, n; $t_o \le \sigma, t \le t' \le t_f$. The maximum exists in each case, since the matrices involved have continuous and bounded elements for every i,j = 1, ···, n, and every $t' \in [t_o, t_f]$, and the interval $[t_o, t_f]$ is compact. We further let

$\bar{\lambda}_s^2$ denote the maximum value of trace of the nonnegative definite matrix S(t',t)S(t',t), i.e.,

$$\bar{\lambda}_s^2 = \max_{t',t} [\mathrm{Tr}\ S(t',t)S(t',t)] \tag{19e}$$

where $t_o \le t \le t' \le t_f$.

<u>Lemma 3</u> If L is defined by (16b) and α_i, i = 1, ···, 4, $\bar{\lambda}_s$ by (19), then we have the bound

$$\|L\|_o \le \left[\alpha_2 + \bar{\lambda}_s\alpha_3 + \bar{\lambda}_s(t_f-t_o)\frac{n}{\sqrt{2}}\alpha_2\alpha_4\right]\frac{n^2\alpha_1}{2}(t_f-t_o)^2 \tag{20}$$

<u>Proof</u> Since each L_i, i = 1, ···, 4 is in $B(L^n)$ and since $B(L^n)$ is an algebra, we have (see Simmons [9], p. 222):

$$\begin{aligned}\|L\|_o &= \|SL_3 + SL_4L_2 + L_2)L_1\|_o \\ &\le \|SL_3 + SL_4L_2 + L_2\|_o \cdot \|L_1\|_o \\ &\le \left\{\|SL_3\|_o + \|SL_4\|_o \cdot \|L_2\|_o + \|L_2\|_o\right\}\|L_1\|_o\end{aligned} \tag{21a}$$

where the last relation follows from the Minkowski inequality applied to L^n. We now note that for any $v \in L^n$,

$$\begin{aligned}\|SL_3v\|_{t_f}^2 &= \int_{t_o}^{t_f} (L_3v)^T S(t_f,t)S(t_f,t)L_3v\ dt \\ &\le \max_{t_o \le t \le t_f} \lambda(S^2(t_f,t))\|L_3v\|_{t_f}^2 \\ &\le \max_{t',t} \lambda(S^2(t',t))\|L_3v\|_{t_f}^2, \quad t' \ge t \\ &= \bar{\lambda}_s^2\|L_3v\|_{t_f}^2\end{aligned}$$

which implies the inequality

$$\|SL_3\|_o \le \bar{\lambda}_s\|L_3\|_o \tag{21b}$$

Similarly,

$$\|SL_4\|_o \le \bar{\lambda}_s \|L_4\|_o \tag{21c}$$

We now refer the reader to Lemma 4, Appendix 1, for a proof of the bound

$$\|L_i\|_o \le n(t_f - t_o)\alpha_i/\sqrt{2} \qquad i = 1, \cdots, 4 \tag{21d}$$

Using relations (21b-d) in (21a), we obtain the desired result.

Lemma 3 can now be used to prove existence of a unique solution to (16a) for sufficiently small time intervals $[t_o, t_f]$. We first make the following crucial observation:

Observation 1 If the original differential game is instead defined on a shorter time interval $[t_o, t_o + \delta]$, $0 < \delta < t_f - t_o$, everything else remaining the same, then the statement of Lemma 3 will still be valid with $t_f - t_o$ replaced by δ, since $\bar{\lambda}_s$ and α_i are independent of the length δ of the time interval as long as $\delta \le t_f - t_o$. Therefore we have

$$\|L\|_o \le \left[\alpha_2 + \bar{\lambda}_s\alpha_3 + \bar{\lambda}_s\delta \frac{n}{\sqrt{2}} \alpha_2\alpha_4\right]\frac{n^2}{2} \alpha_1\delta^2 \tag{22}$$

for a differential game defined on a time interval of length $\delta \le t_f - t_o$. This implies that $\|L\|_o$ can be made arbitrarily small by a sufficiently small choice of $\delta > 0$.

Theorem 2 The nonzero-sum differential game under investigation admits a unique Nash equilibrium solution given by (10a-b), if the time interval on which the game is defined is taken to be sufficiently small.

Proof We have previously shown that there exists a unique Nash equilibrium solution if there exists a unique $\bar{z} \in L^n$ that solves (16a). Since $\|L\|_o$ can be made less than one by a proper choice of δ [this follows from (22)], L is a contraction mapping for sufficiently small δ. This consequently implies existence of a unique solution to (16a) by Banach's classical fixed point theorem (Simmons [15], p. 338), which further implies existence of a unique $z \in L^{r_2}$ through (14).

§4. SIMILARLY STRUCTURED ZERO-SUM DIFFERENTIAL GAMES

We now investigate as to how the results obtained heretofore for NZSDG can be specialized to the saddle-point solutions of similarly structured zero-sum differential games (ZSDG). To this end, we assume the state dynamics to be described by the Ito stochastic differential equation (1) and the cost function to be given by

$$J(\gamma_1,\gamma_2)=E\left\{x_{t_f}^T C_f x_{t_f} + \int_{t_o}^{t_f} \left[x_t^T C(t) x_t + u_1^{t^T} D_{11}(t) u_1^t - u_2^T D_{22}(t) u_2\right] dt\right\} \tag{23}$$

where $u_1^t = \gamma_1[\cdot,\cdot]$, $u_2 = \gamma_2(\cdot)$, and $\gamma_1 \in U_1$, $\gamma_2 \in U_2$. Furthermore, $C_f \geq 0$, $C(t) \geq 0$, $D_{11}(t) > 0$, $D_{22}(t) > 0$, $\forall\, t \in [t_o,t_f]$, and matrix entries are continuous on $[t_o,t_f]$. When the objective of player 1 is to minimize J as defined by (23) and that of player 2 to maximize the same cost function, the (Nash) equilibrium inequalities (4a) and (4b) can now be written as a *saddle-point inequality*

$$J(\gamma_1^*,\gamma_2) \leq J(\gamma_1^*,\gamma_2^*) \leq J(\gamma_1,\gamma_2^*) \tag{24}$$

for all $\gamma_1 \in U_1$, $\gamma_2 \in U_2$. We now note that the cost function (23) and the saddle-point inequality (24) can actually be seen to be special cases of (3) and (4) through the parametric relations:

$$C_{1f} = -C_{2f} = C_f \geq 0,\ C_1(t) = -C_2(t) = C(t) \geq 0,$$

$$D_{11}(t) = -D_{21}(t) > 0,\ D_{22}(t) = -D_{12}(t) > 0,$$

$$t \in [t_0, t_f] \tag{25}$$

The results of Section 3 cannot directly be applied to this special case since there we had assumed parametric restrictions $C_{i_f} \geq 0$, $C_i(t) \geq 0$, $D_{ij}(t) > 0$, $i,j = 1, 2$, $t \in [t_0, t_f]$. However, in the proof of the results given in Section 3, the only place where we made use of the assumptions $C_{2_f} \geq 0$, $C_2(t) \geq 0$, $D_{21}(t) > 0$ was in the verification of existence of a unique matrix solution to the Riccati equation (9b) (i.e., those restrictions on the eigenvalues of C_{2_f}, $C_2(t)$, and $D_{21}(t)$ were sufficient for existence of a unique solution to the said equation). Hence if we assume existence of a unique solution to (9b) a priori then the results of Section 3 are still valid and can directly be applied to the stochastic zero-sum differential game posed in this section. We can therefore quote the following result (Theorem 3) which is a special case of Theorem 1.

Preliminary notation and assumptions Let P(t) be the unique nonnegative definite matrix solution of the Riccati equation

$$\dot{P} + F^T P + PF - PG_1 D_{11}^{-1} G_1^T P + C = 0 \tag{26}$$

$$P(t_f) = C_f, \; t_o \le t \le t_f$$

Furthermore, let the parameters of the game be such that the matrix Riccati equation

$$\dot{S} + \tilde{F}^T S + S\tilde{F} + SG_2 D_{22}^{-1} G_2^T S + \tilde{C} = 0 \tag{27}$$

$$S(t_f) = C_f, \quad t_o \le t \le t_f$$

admits a unique nonnegative definite solution on the interval $[t_o, t_f]$. Here $\tilde{F}$ and $\tilde{C}$ are defined by

$$\tilde{F} \triangleq F - G_1 D_{11}^{-1} G_1^T P \tag{28a}$$

$$\tilde{C} \triangleq C(t) + PG_1 D_{11}^{-1} G_1^T P \tag{28b}$$

Remark 2 It is well known that equation (27) will in general not admit a well-defined solution for sufficiently long time intervals $[t_o, t_f]$, i.e., it is possible to find a sufficiently large t_f so that the Riccati equation (27) will have a conjugate point on $[t_o, t_f]$. However, it should also be clear that it is possible to find a sufficiently small time interval $[t_o, t_f]$ so that equation (27) will have no conjugate point.

Let us now further assume that there exists a solution to the coupled differential equations

$$\dot{k} + \tilde{F}^T k + PG_2 D_{22}^{-1} G_2^T (Sy - b) = 0, \; k(t_f) = 0 \tag{29a}$$

$$\dot{b} + \left(\tilde{F}^T + SG_2 D_{22}^{-1} G_2^T\right) b + (S - P) G_1 D_{11}^{-1} G_1^T k = 0, \; b(t_f) = 0 \tag{29b}$$

$$\dot{y} - \left(\tilde{F} + G_2 D_{22}^{-1} G_2^T S\right) y + G_2 D_{22}^{-1} G_2^T b + G_1 D_{11}^{-1} G_1^T k = 0, \; y(t_o) = x_o \tag{29c}$$

Theorem 3 Under the conditions stated above the stochastic zero-sum differential game of this section admits a saddle-point solution given by

$$u_1^{*t} = \gamma_1^*(t,x_t) = -D_{11}^{-1}G_1^T(Px_t + k) \tag{30a}$$

$$u_2^*(t) = \gamma_2^*(t) = D_{22}^{-1}G_2^T(Sy(t) - b) \tag{30b}$$

If the Ricatti equation (27) admits a unique solution on $[t_o,t_f]$, then we can directly employ Theorem 2 to establish existence of a unique solution to the coupled differential equations (29) for sufficiently small time intervals $[t_o,t_f]$. Furthermore, we also know from Remark 2 that (27) admits a unique solution when the time interval is taken to be sufficiently small. Hence it turns out that the assumptions that led to Theorem 3 are satisfied if the time interval is taken to be sufficiently small. We can therefore state the following important result which follows directly from Theorem 2 and Remark 2.

Theorem 4 The zero-sum stochastic differential game of this section admits a *unique saddle-point solution* given by (30a-b), if the time interval on which the game is defined is taken to be sufficiently small.

Investigation of a solution to (29a-c) Since the differential equations (29a-c) are linear in k,b and y it might be possible that they admit a solution in the form $k = Ay$, $b = By$ where $A(t)$ and $B(t)$ are square matrices of appropriate dimensions and y satisfies the linear differential equation

$$\dot{y} = \left(\tilde{F} + G_2D_{22}^{-1}G_2^TS - G_2D_{22}^{-1}G_2^TB - G_1D_{11}^{-1}G_1^TA\right)y,\quad y(t_o) = x_o \tag{31}$$

In order to obtain the expressions for A and B, we substitute $k = Ay$ and $b = By$ into (29a) and (29b) and obtain the following matrix differential equations for A and B:

$$\dot{A} + A\tilde{F} + \tilde{F}^T A - AG_1D_{11}^{-1}G_1^T A - AG_2D_{22}^{-1}G_2^T B$$
$$+ AG_2D_{22}^{-1}G_2^T S + PG_2D_{22}^{-1}G_2^T(S - B) = 0, \quad A(t_f) = 0 \tag{32a}$$

$$\dot{B} + B\tilde{F} + \tilde{F}^T B - BG_2D_{22}^{-1}G_2^T B - BG_1D_{11}^{-1}G_1^T A + BG_2D_{22}^{-1}G_2^T S$$
$$+ SG_2D_{22}^{-1}G_2^T B + (S - P)G_1D_{11}^{-1}G_1^T A = 0, \quad B(t_f) = 0 \tag{32b}$$

Hence we are now ready to make the following proposition.

Proposition 1 If there exists a unique nonnegative definite solution to the matrix Riccati equation (27) on the time interval $[t_o, t_f]$ and if the coupled linear matrix differential equations (32) admit a unique solution, then the stochastic zero-sum differential game of this section admits a unique saddle-point solution given by

$$\overset{*}{u}{}_1^t = \overset{*}{\gamma}_1(t, x_t) = -D_{11}^{-1}G_1^T(Px_t + Ay(t)) \tag{33a}$$

$$\overset{*}{u}_2(t) = \overset{*}{\gamma}_2(t) = D_{22}^{-1}G_2^T(S - B)y(t) \tag{33b}$$

where $y(t)$ is the unique solution of the differential equation (31).

We now seek to obtain differential equations for the matrix sums $\bar{P} \triangleq P + A$ and $\bar{S} \triangleq S - B$. The reason why we are interested in these matrices will become apparent in the sequel. If P, S, A, and B are defined as unique solutions of (26), (27), (32a), and (32b), respectively, it can be shown without much difficulty that $\bar{P}$ and $\bar{S}$ should then satisfy the following coupled Riccati equations:

$$\dot{\bar{P}} + \bar{P}F + F^T\bar{P} - \bar{P}G_1D_{11}^{-1}G_1^T\bar{P} + \bar{P}G_2D_{22}^{-1}G_2^T\bar{S} + C = 0,$$
$$\bar{P}(t_f) = C_f \tag{34a}$$

$$\dot{\bar{S}} + \bar{S}F + F^T\bar{S} + \bar{S}G_2D_{22}^{-1}G_2^T\bar{S} - \bar{S}G_1D_{11}^{-1}G_1^T\bar{P}$$
$$+ PG_1D_{11}^{-1}G_1^T(\bar{P} - \bar{S}) = 0, \ \bar{S}(t_f) = C_f \tag{34b}$$

A well-defined solution to these equations is given by

$$\bar{P}(t) = \bar{S}(t) = K(t) \qquad t \in [t_o, t_f] \tag{35a}$$

where K(t) satisfies

$$\dot{K} + KF + F^TK - K\left(G_1D_{11}^{-1}G_1^T - G_2D_{22}^{-1}G_2^T\right)K + C = 0,$$
$$K(t_f) = C_f \tag{35b}$$

We have thus shown that if there exists a unique P, S, A, and B that satisfy (26), (27), (32a), and (32b), then the matrix Riccati equation (35b) admits a unique symmetric solution. We note however that the reverse is not true, i.e., existence of a unique solution to (35b) on $[t_o, t_f]$ does not necessarily imply existence of a unique solution to the former. To see this we note that if the matrix $G_1D_{11}^{-1}G_1^T - G_2D_{22}^{-1}G_2^T$ is nonnegative definite, then (35b) admits a unique nonnegative definite solution regardless of the length of the time interval $[t_o, t_f]$ on which the game is defined (Reid [8], p. 121). However, we also know that (see Remark 2) the Riccati equation (27) has a conjugate point if the time interval is taken to be sufficiently large. Hence, we have shown that the reverse implication is not true.

We now investigate as to what kind of a saddle-point solution the matrix K is related with. To this end we first endow the second player also with closed-loop nonanticipative information and denote the class of permissible strategies for player 2 (defined in a similar manner to U_1 with only r_1 replaced by r_2) by $\bar{U}_2$. The objective now is to obtain a saddle-point strategy pair $\gamma_1^* \in U_1$, $\gamma_2^* \in \bar{U}_2$ that satisfies the saddle-point inequality (24) for all $\gamma_1 \in U_1$, $\gamma_2 \in \bar{U}_2$. The following relation between the solution of this problem and the matrix K defined earlier now follows readily.

Proposition 2 If (35b) admits a unique nonnegative definite matrix solution K(t), then the strategies

$$ {}^*u_1^t = \gamma_1^*(x_t) = -D_{11}^{-1}G_1^T K(t) x_t \tag{36a} $$

$$ {}^*u_2^t = \gamma_2^*(x_t) = D_{22}^{-1}G_2^T K(t) x_t \tag{36b} $$

constitute a saddle-point pair for the zero-sum differential game of this section with both players having closed-loop nonanticipative information.

Proof If we substitute (36a) into the state equation (1) and the cost function (23), then the left-hand-side of the inequality (24) defines a standard LQ optimum control problem with additive noise in the state dynamics, which is well-known to admit a unique solution in $\bar{U}_2$, which is given by (36b). Similarly, substitution of (36b) into (1) and (23) yields, together with the right-hand-side of inequality (24), an LQ optimum control problem which again admits a unique solution in U_1, to be given by (36a). It is important to note that in each case unicity is a consequence of not only the LQ nature of the optimum control problem but also the assumption that the additive independent-increment process has a positive definite incremental covariance function.

We conclude this section with the following important property of the saddle-point solution in LQ zero-sum differential games, which follows directly from Proposition 2 and the discussion given prior to it.

Property 1 For the linear-quadratic zero-sum stochastic differential game of this section, existence of a unique saddle-point under the original information structure (i.e., closed-loop information for player 1 and open-loop information for player 2) implies existence of a unique saddle-point under closed-loop information structure for both players. The reverse implication, however, does not hold true.

§5. CONCLUSION

In this paper we have investigated Nash equilibrium solutions of a class of two-person nonzero-sum stochastic differential games which are characterized by linear state dynamics (described by an Ito differential equation) and quadratic cost functionals. The state dynamics include additively a zero-mean independent-increment Gaussian process. The information structure of the problem endows player 1 with perfect closed-loop nonanticipative state information and player 2 with only a priori static information. It has been shown in the paper that if the incremental covariance function of the additive Gaussian process is positive definite, then the NZS differential game admits an affine Nash equilibrium solution for the first player (whenever it exists). Furthermore, it has been shown that the affine solution exists and is unique whenever the time interval on which the game is defined is taken to be sufficiently small. The paper contains exact expressions for this unique solution and a full investigation of the special case of a zero-sum differential game. In particular, it is proven that this zero-sum differential game admits a unique pure-feedback solution under closed-loop information for both players whenever it admits a unique saddle-point solution under the original information structure.

It is worth reemphasizing that the assumption concerning positive definiteness of the incremental covariance function of the additive stochastic process is crucially essential in establishing uniqueness of the solution in each case. Otherwise, the problem will admit nonunique equilibrium solutions as reported in [3].

APPENDIX 1

In this appendix we provide a proof for Lemma 4 which was used in the proof of Lemma 3.

Lemma 4 If $\mathcal{L}_i \in \mathcal{B}(L^n)$, $i = 1, \cdots, 4$, are defined by (15a-d) and α_i by (19a-d), we have the bound

$$\|\mathcal{L}_i\|_o \leq n(t_f - t_o)\alpha_i/\sqrt{2} \qquad i = 1, \cdots, 4 \tag{1-1}$$

Proof We first note that because of the structurally similar forms of (15a-d), and taking into account the modifications (18a-c), the proof will be completed if we can show that

$$\|\mathcal{S}\|_o \leq n(t_f - t_o)\ \alpha/\sqrt{2} \tag{1-2a}$$

where $S \in B(L^n)$ is defined by

$$Sv = \int_{t_o}^{t} A(t_f,t,\sigma)v(\sigma)d\sigma, \quad t \le t_f \tag{1-2b}$$

for some $A(\cdot,\cdot,\cdot)$ which has square Lebesgue-integrable elements, and α is given by

$$\alpha = \max_{i,j,\sigma,t,t'} |A(t',t,\sigma)|_{ij}, \quad t_o \le t,\sigma \le t' \le t_f,$$
$$i,j = 1, \cdots, n \tag{1-2c}$$

Since $\|S\|_o \triangleq \sup\|Sv\|_{t_f}$, $\|v\|_{t_f} \le 1$, we start with

$$\int_{t_o}^{t}\int_{t_o}^{t} v^T(\sigma)K_t(\sigma,s)v(s)dsd\sigma \tag{1-3a}$$

where

$$K_t(\sigma,s) \triangleq A^T(t_f,t,\sigma)A(t_f,t,s) \tag{1-3b}$$

Denoting the ijth component of K_t by K_t^{ij} and the ith component of v by v_i, we can write (1-3a) as follows:

$$\sum_{i,j} \int_{t_o}^{t} v_i(\sigma)d\sigma \int_{t_o}^{t} K_t^{ij}(\sigma,s)v_j(s)ds$$
$$\le \sum_{i,j} \int_{t_o}^{t} v_i(\sigma)d\sigma \left[\int_{t_o}^{t} |K_t^{ij}(\sigma,s)|^2 ds\right]^{1/2} \left[\int_{t_o}^{t} v_j^2(s)ds\right]^{1/2}$$
$$\le \sum_{i,j} \left[\int_{t_o}^{t}\int_{t_o}^{t} |K_t^{ij}(\sigma,s)|^2 ds\, d\sigma\right]^{1/2} \left[\int_{t_o}^{t} v_i^2(\sigma)d\sigma\right]^{1/2} \cdot \left[\int_{t_o}^{t} v_j^2(\sigma)d\sigma\right]^{1/2}$$
$$\triangleq \sum_{i,j} \|K_t^{ij}\|_t \|v_i\|_t \|v_j\|_t \tag{1-4}$$

where we have made repeated use of Buniakowski's inequality

since $L^n(t_o,t)$ is a Hilbert space. Expression (1-4) can be bounded from above by

$$\sum_{i,j} \|K_t^{ij}\|_t \left(\|v_i\|_t^2 + \|v_j\|_t^2\right)/2$$

$$= \sum_i \|v_i\|_t^2 \sum_j \left(\|K_t^{ij}\|_t + \|K_t^{ji}\|_t\right)/2$$

$$= \sum_i \|v_i\|_t^2 \sum_j \|K_t^{ij}\|_t \qquad (1\text{-}5)$$

where in arriving at the last equality we have made use of the symmetry property of K_t as defined by (1-3b).

It now follows from (1-5) that

$$\|Sv\|_{t_f}^2 \le \sum_{i,j} \int_{t_o}^{t_f} \|v_i\|_t^2 \|K_t^{ij}\|_t \, dt$$

Since $\|v_i\|_t$ is a monotonically nondecreasing function of t for any $v \in L^n$, we can bound the last expression from above by

$$\sum_{i,j} \|v_i\|_{t_f}^2 \int_{t_o}^{t_f} \|K_t^{ij}\|_t \, dt$$

which can further be bounded from above by

$$\sum_i \bar{h}\|v_i\|_{t_f}^2 = \bar{h}\|v\|_{t_f}^2$$

where

$$\bar{h} = \max_i \sum_j \int_{t_o}^{t_f} \|K_t^{ij}\|_t \, dt \qquad (1\text{-}6a)$$

This implies that

$$\|Sv\|_{t_f}^2 \le \bar{h}\|v\|_{t_f}^2$$

and hence

$$\|S\|_o \leq \sqrt{\bar{h}} \tag{1-6b}$$

To find an explicit bound on $\bar{h}$, we first note that

$$\|K_t^{ij}\|_t = \left[\int_{t_o}^{t}\int_{t_o}^{t} |K_t^{ij}(\sigma,s)|^2 ds\, d\sigma\right]^{1/2}$$

$$= \left[\int_{t_o}^{t}\int_{t_o}^{t} |A^T(t_f,t,\sigma)A(t_f,t,s)|^2 ds\, d\sigma\right]^{1/2}$$

$$\leq \max_{t,\sigma,s} |A^T(t_f,t,\sigma)A(t_f,t,s)|_{ij}(t-t_o); \quad \sigma,s \leq t$$

$$\leq n \max_{i,j} \max_{t,\sigma} |A(t_f,t,\sigma)|_{ij}^2 (t-t_o)$$

$$\leq n\left\{\max_{i,j,t',t,\sigma} |A(t',t,\sigma)|_{ij}\right\}^2 (t-t_o),$$

$$t_o \leq t,\sigma \leq t' \leq t_f$$

and substitution of this bound in (1-6a) yields

$$\bar{h} \leq n^2\left\{\max_{i,j,t',t,\sigma} |A(t',t,\sigma)|_{ij}\right\}^2 (t_f-t_o)^2/2$$

$$\triangleq n^2\alpha^2(t_f-t_o)^2/2 \tag{1-7}$$

which is the desired bound.

This completes verification of the bounds (1-1) for all $i = 1, \cdots, 4$, since if (1-2b) is instead defined with t_o replaced by t_f, the procedure described above will still yield the bound given by (1-7).

REFERENCES

1. Başar, T., A Counter-example in Linear-Quadratic Games: Existence of Nonlinear Nash Solutions, Journal of Optimization Theory and Applications, vol. 14, no. 4, 1974.

2. Başar, T., On the Uniqueness of the Nash Solution in Linear-Quadratic Differential Games, International Journal

of Game Theory, vol. 5, no. 4, 1976.

3. Başar, T., Informationally Nonunique Equilibrium Solutions in Differential Games, Marmara Research Institute, Applied Mathematics Division, Technical Report No. 24, 1975. To appear in SIAM Journal on Control and Optimization, 1977.

4. Fleming, W. H., Optimal Continuous Parameter Stochastic Control, SIAM Review, vol. 11, no. 4, 1969.

5. Fleming, W. H. and M. Nisio, On the Existence of Optimal Stochastic Controls, Journal of Math. Mech., vol. 15, no. 5, 1966.

6. Friedman, A., Stochastic Differential Games, Journal of Differential Equations, vol. 11, no. 1, 1972.

7. Kwakernaak, H., An Extension of the Stochastic Linear Regulator Problem, IEEE Trans. on Automatic Control, vol. AC-19, no. 2, 1974.

8. Reid, W. T., Riccati Differential Equations, Academic Press, New York, 1972.

9. Simmons, G. F., Introduction to Topology and Modern Analysis, McGraw-Hill, New York, 1963.

10. Wonham, W. M., Random Differential Equations in Control Theory, in A. T. Bharuca-Reid, ed., Probabilistic Methods in Applied Mathematics, vol. II, Academic Press, New York, 1970.

SINGULAR MANIFOLDS IN PARTIAL DIFFERENTIAL GAMES

Emilio O. Roxin

University of Rhode Island
Kingston, Rhode Island

ABSTRACT

This is a kind of progress report on the work on specific examples on differential games with partial differential equations. The goal is to find modes of behavior of solutions to such games, which are similar to the ones known for ordinary differential equations, as given in the fundamental book of Isaacs [1]. This work was begun considering some differential games with the wave and the heat equation [2,3]. Here examples are given of switching manifolds, barriers, dispersal, and universal manifolds, as well as a case of a terminal manifold showing a usable and a nonusable part. The differential games of these examples concern the wave equation.

§1. DIFFERENTIAL GAME WITH THE WAVE EQUATION

Consider the wave equation

$$u_{tt}(x,t) = u_{xx}(x,t) \tag{1}$$

for the real function $u(x,t)$, defined on $0 \leq x \leq 1$, $0 \leq t$. We will admit the concept of "solution" in a generalized sense, accepting discontinuities of the derivatives and of the function itself, originating in discontinuities of the boundary conditions.

Let the initial conditions be

$$u(x,0) = \phi(x),\ u_t(x,0) = \psi(x), \qquad 0 \leqq x \leqq 1 \tag{2}$$

and the boundary conditions

$$u(0,t) = f(t),\ u(1,t) = g(t), \qquad 0 \leqq t \tag{3}$$

Here f(t) and g(t) will be the control variables governed by the two players. Player "f" will be the minimizing player, player "g" the maximizer. The controls will be subjected to the constraints

$$|f(t)| \leqq A,\ |g(t)| \leqq B \tag{4}$$

Admissible control functions will be all piecewise C^2 functions satisfying (4). Differential games of this type were considered in [2].

The solution of (1), (2), and (3) is well known and can be expressed as the superposition

$$u(x,t) = u_1(x,t) + u_2(x,t) \tag{5}$$

where

$$u_1(x,t) = \frac{1}{2}\left[\phi(x-t) + \phi(x+t) + \int_{x-t}^{x+t} \psi(s)\,ds\right] \tag{6}$$

is the solution corresponding to boundary conditions zero (in this formula the functions ϕ and ψ should be extended as odd periodic functions of period 2), and

$$u_2(x,t) = f(t-x) + g(t-1+x) - g(t-1-x) - f(t-2+x)$$
$$+ f(t-2-x) + g(t-3+x) - \cdots \tag{7}$$

is the solution corresponding to initial conditions zero.

Let the terminal condition of the game be

$$\int_0^1 u(x,T)dx = K \tag{8}$$

where K is a given real number and T is the termination time. This termination time is therefore the smallest positive time where condition (8) is satisfied.

Let the cost be

$$J[f,g] = T \tag{9}$$

so that this is a "time optimal" ("survival") game.

§2. SOLUTION OF THE GAME

In spite of the linearity of the governing equations (1), (8), the cost (9) is really nonlinear and the solution is not trivial.

Let

$$A > B > 0 \tag{10}$$

and let "f" be the minimizing player (who then wishes to terminate the game as soon as possible).

Let T^* be the termination time corresponding to optimal play, and let $f^*(t)$, $g^*(t)$ be the control functions arising from optimal play.

From formula (7) it follows that the boundary values $f(t)$, $g(t)$ propagate to the interior of the unit x-interval along the characteristic lines, and are reflected at the boundaries $x = 0$, $x = 1$ with change of sign. Once we know T^* we can easily obtain the "best" boundary values in the reversed time-order.

We just have to take into account that, if "f" wants to increase $u(x,T^*)$ and "g" wants to decrease it (and this happens if the condition (8) is reached from the $\int u\,dx < K$ side), then the last boundary values reaching the T*-line should have been $f(t) = +A$, while those boundary values reaching the T* line after one reflection should have been $f(t) = -A$, and so on, and similarly for player "g." In Figure 1 we have indicated such optimal boundary values for $5/2 < T^* < 3$.

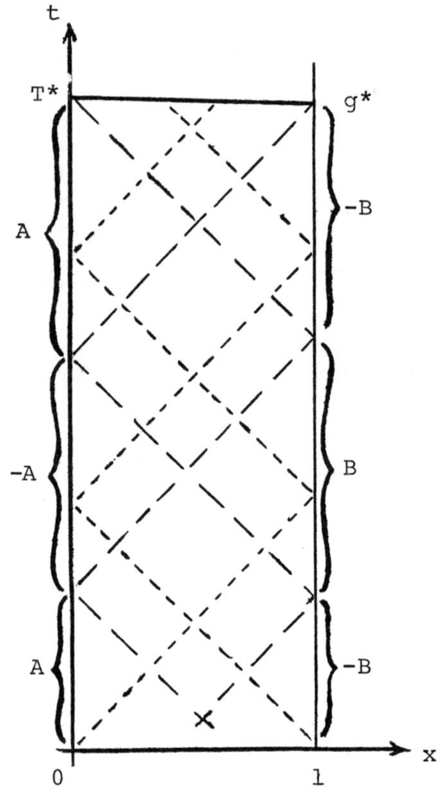

FIGURE 1

At this point there still remains the question of how to determine T*.

In the case of zero initial conditions

$$u(x,0) = u_t(x,0) = 0 \tag{11}$$

it is easy to find, with the aid of Figure 1, that

$$\int_0^1 u(x,T^*)dx = (A - B)T^* \tag{12}$$

Hence, in this case, $T^* = K/(A - B)$.

For nonzero initial conditions (2), we obtain by superposition

$$\int_0^1 u(x,T^*)dx = (A - B)T^* + I_1(T^*) \tag{13}$$

where $I_1(t)$ is the value of the integral for zero boundary conditions, which can be calculated in advance as it does not depend on the controls. In order to calculate $I_1(t)$, the solution $u_1(x,t)$ can be written as the superposition of two waves

$$u_1(x,t) = F(x-t) + G(x+t) \tag{14}$$

Here F and G are waves travelling in the positive, respectively, negative x-direction, and these waves get reflected at the boundaries with a change in sign. Hence

$$I_1(t) = \int_0^1 u_1(x,t)dx$$

will satisfy $I_1(t+2) = -I_1(t+1) = I_1(t)$ and be periodic.

The terminal time T* will therefore be determined by the condition

$$(A - B)T^* + I_1(T^*) = K \qquad (15)$$

and will be the smallest value of $T^* > 0$ satisfying this condition. Once we have determined T^*, we are able to determine the optimal controls $f^*(t)$ and $g^*(t)$.

§3. SINGULAR MANIFOLDS

According to the solution given above, *switching manifolds* are the instants $t = T^* - 1, T^* - 2, \cdots$, for both players "f" and "g."

Barriers appear in the following way. In equation (15), T^* is determined by a condition K = linear + periodic function, as illustrated in Figure 2. It is therefore possible that the smallest T^*, as a function of K, be discontinuous. It is also clear that there may be discontinuities of T^* considered as a function of the initial conditions (2); to make this statement precise we should decide which topology to use for the initial conditions, a matter which we do not want to pursue here further.

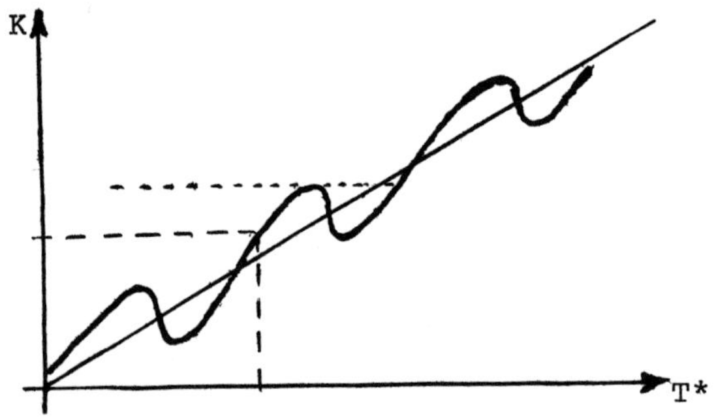

FIGURE 2

In order to have a *dispersal* or *efferent manifold* we change the terminal condition to

$$\int_0^1 |u(x,T)|dx = K \tag{16}$$

and keep the zero initial conditions (11). It is then clear that the time-minimizing player "f" may choose either to try to maximize u(x,T) or -u(x,T), and this choice may even be different and unrelated for different values of x. Hence we have a dispersal situation, from which more than one optimal play start.

On the other hand, player "g" should not commit himself to minimize u(x,T) or -u(x,T), until he is sure that his input cannot work to the advantage of "f" if this latter player switches his choice (from + to -, or conversely). For how long this situation for "g" may be the case, is a question which we do not intend to solve here.

A *universal* or *afferent manifold* appears clearly if the terminal time T* is fixed in advance and the cost is defined by

$$J = \int_0^{T^*} \int_0^1 |u(x,t)|dx\, dt \tag{17}$$

If, as before, "f" is minimizing, A > B and the initial conditions are zero, then it is easy to see that the optimal strategies are

$$g^*(t) = \pm B$$

(and there is no reason to change the sign)

$$\begin{aligned} f^*(t) &= 0 && \text{for } 0 < t < 1 \\ &= g^*(t-1) && \text{for } t > 1 \end{aligned}$$

(To see this one should consider the representation of u(x,t) as the superposition of the F(x-t) and G(x+t) waves.)

It is not difficult to see that this same "stationary" behavior is also reached in finite time, if the initial conditions are not zero. Hence all optimal paths end, after a sufficiently long time, in this *universal path*.

To find a *terminal manifold* with *usable* and *nonusable* parts, we refer to our first example. In Figure 2 it is shown how the terminal time T* is determined by the intersection of the given curve with the horizontal line at height K. If now the starting conditions are very near to an intersection where the curve is decreasing, this means that even with optimal play, "f" cannot avoid that the value of $\int u(x,t)dx$ drifts farther away from the terminal condition $\int u(x,t)dx = K$. Therefore, such points, where this condition is satisfied, are nonusable as terminal points.

REFERENCES

1. Isaacs, R., Differential Games, SIAM Series in Appl. Math., J. Wiley, New York, London, Sydney, 1965.

2. Roxin, E., Differential Games with Partial Differential Equations, Theory and Applic. of Differential Games, Proc. NATO Adv. Study Inst., Warwick, England (J. D. Grote, ed.), D. Reidel Publ. Co., Dordrecht, Boston, 1975, pp. 157-168.

3. Roxin, E., Dispersal Manifolds in Partial Differential Games, Proc. Internat. Symposium on Dynamical Systems, Gainesville, Florida, 1976, to appear.

SOME RESULTS ON THE STATIONARY OPTIMAL CONTROL OF A STOCHASTIC SYSTEM

Pan-Tai Liu

University of Rhode Island
Kingston, Rhode Island

§I. INTRODUCTION

In this paper, we study some stationary behavior of the optimal cost function V(x,t) of a stochastic control system over an infinite time interval. We shall show that as t approaches infinity, V(x,t) converges to $V_0(x)$, the optimal cost function of a corresponding deterministic (autonomous) control system in several ways. Our main results consist of three theorems in Section 3. Basically, the convergence is closely related to the stability of solutions of both systems and the stability properties will be discussed in Section 2.

Theorem 3.1 asserts the convergence of V(x,t) to $V_0(x)$ in the ordinary (deterministic) sense. The proof is carried out via the integration of Bellman's equation of dynamic programming of each system along the optimal solution of the other system. In a way, this theorem has similar significance to some results in [4] (see discussion at the end of Theorem 3.1). The martingale convergence theorem will then be used to prove stochastic convergence in Theorem 3.2.

Finally, we shall use the idea of a stable sequence of events in a probability space introduced by Renyi [6], and it will be shown in Theorem 3.3 that $V(x,t)$ converges to $V_0(x)$ as $t \to \infty$ under persistent random but "stable" perturbation on the state.

In the sequel, we shall assume that $f : R^n \times R^m \to R^m$, $g : R^n \times [0,\infty) \to R^n$, and $f_o : R^n \times R^m \to [0,\infty)$ satisfy the following assumptions:

(i) $\|f(x,u) - f(y,v)\| \le K_1(\|x-y\| + \|u-v\|)$

(ii) $\|f(x,u)\|^2 \le K_1{}^2(\|x\|^2 + \|u\|^2)$

(iii) $\|g(x,t)\|^2 \le \eta(t)(1 + \|x\|^2)$ where $\int_0^\infty \eta^2(t)dt < \infty$ and $\eta(t)$ is continuous

(iv) $g(x,t)$ is C' in x and measurable in t

(v) $f_o(x,u) \ge K_2(\|x\|^2 + \|u\|^2)$

(vi) $f(0,0) = 0$, $f_o(0,0) = 0$, and $g(0,t) = 0$ for all $t \ge 0$.

We now define the two systems.

System I

$$\frac{dx}{dt} = f(x,u), \; x(0) = x_0 \tag{1}$$

$$J_1 = \int_0^\infty f_o(x,u)dt \tag{2}$$

An admissible control u is a measurable function of t, i.e., $u = u(t)$ such that $J_1(u,x_0) < \infty$. We assume that an optimal control u^* which minimizes J_1 exists for all $x_0 \in R^n$ and can be synthesized, i.e., $u^*(t) = \hat{u}^*(x^*(t))$ where $x^*(t)$ is the solution to (1) corresponding to $u^*(t)$. For discussions of synthesis of an optimal control for nonlinear systems, see [2,5].

Let (Ω,F,P) be a probability triple and W_t an n-dimensional Wiener process measurable with respect to an increasing field of σ-algebra F_t, $t \geq 0$. We assume that $\sigma\left\{\bigcup_{t\geq 0} F_t\right\} = F$.

System II

$$dy_t = f(y,v)dt + g(y,t)dW_t \tag{3}$$

$$J_2 = E\left\{\int_0^\infty f_o(y,v)dt \,|\, y_o\right\} \tag{4}$$

An admissible control is a Markov policy $v = v(y,t)$, Lipschitz in y and measurable in t, satisfying $\|v(y,t)\| \leq C\|y\|$ and $J_2(v,y_o) < \infty$ for all $y_o \in R^n$. Because of assumptions (i)-(iii), (3) has a unique solution. We assume that an optimal control $v^*(y,t)$ which minimizes J_2 exists.

§2. SOME STABILITY PROPERTIES

Theorem 2.1 The solution $x(t,x_o,u)$ to (1) (henceforth written as $x(t)$) corresponding to an admissible control $u(t)$ and any initial condition x_o is asymptotically stable.

The proof of this theorem can be found in [2] and is omitted here. We proceed to prove the following theorem to System II.

Theorem 2.2 Let $y(t,x_o,v)$ (henceforth written as y_t) be the solution to (3) corresponding to an admissible control $v(y,t)$ and any (deterministic) initial condition y_o. Then y_t is asymptotically stable in the m.s.s., i.e.,

$$\lim_{t\to\infty} E\|y_t\|^2 = 0$$

Proof Suppose the contrary, i.e., $E\|y_t\|^2 \not\to 0$ as $t \to \infty$. Because of assumptions (v) and (vi) in Section 1 and the fact that $J_2 < \infty$, there exist two sequences $\{t_k\}$ and $\{\tau_k\}$, $k = 0, 1, 2, \cdots$, $\tau_k > t_k$ and $\ell > 0$, such that $E\|y_{t_k}\|^2 = 2\ell$,

$E\|y_{\tau_k}\|^2 = \ell$ and $E\|y_t\|^2 \in [\ell, 2\ell]$ for $t \in [t_\tau, \tau_k]$.

Next, we have from (3),

$$y_{\tau_k} - y_{t_k} = \int_{t_k}^{\tau_k} f(y_t, v(y_t, t))dt + \int_{t_k}^{\tau_k} g(y_t, t)dW_t \quad (5)$$

and hence, by using properties of Ito's integral

$$\begin{aligned} E\|y_{\tau_k} - y_{t_k}\|^2 &\le \int_{t_k}^{\tau_k} E\|f(y_t, v(y_t, t))\|^2 dt(\tau_k - t_k) \\ &\quad + \int_{t_k}^{\tau_k} E\|g(y_t, t)\|^2 dt \\ &\le \int_{t_k}^{\tau_k} K_1{}^2\, E\left\{\|y_t\|^2 + \|v(y_t, t)\|^2\right\} dt(\tau_k - t_k) \\ &\quad + \int_{t_k}^{\tau_k} \eta^2(t) E\|y_t\|^2 dt \\ &\le 2\ell(1 + C)K_1{}^2(\tau_k - t_k)^2 + 2\ell\eta_k^2(\tau_k - t_k) \quad (6) \end{aligned}$$

where $\eta_k^2 = \max_{t \in [t_k, \tau_k]} \eta^2(t)$. On the other hand, we have

$$E\|y_{t_k} - y_{\tau_k}\|^2 \ge E\|y_{t_k}\|^2 - E\|y_{\tau_k}\|^2 = \ell \quad (7)$$

Combining (6) and (7) and simplifying yield

$$2(1 + C)K_1{}^2(\tau_k - t_k)^2 + 2\eta_k^2 (\tau_k - t_k) \ge 1 \quad (8)$$

This implies that $\tau_k - t_k \ge m_1$ for all k, where m_1 is the (unique) positive root of the quadratic equation

$$2(1 + C)K_1{}^2 m^2 + 2\eta_k^2 m - 1 = 0 \quad (9)$$

But this contradicts the fact that $J_2 < \infty$, since

$$J_2 = E\left\{\int_0^\infty f_0(y_t, v(y_t, t))dt \,|\, y_o\right\}$$

$$\geq E\left\{\int_0^\infty K_2(1 + C^2)\|y_t\|^2 dt \,|\, y_o\right\}$$

$$\geq \sum_{k=1}^{p} \int_{t_k}^{\tau_k} E\|y_t\|^2 dt \geq pm_1 \, \ell \qquad (10)$$

where p can be arbitrarily large. The theorem is therefore proved.

§3. THE MAIN RESULTS

The optimal cost functions $V_o(x)$ and $V(x,t)$ are defined as

$$V_o(x) = \int_t^\infty f_0\left(x^*(s),\ \hat{u}^*\left(x^*(s)\right)\right)ds \qquad (11)$$

for System I, where $x^*(s) = x^*(s,x,u)$ is the solution to

$$\frac{dx}{ds} = f\left(x, u^*(s)\right),\ s \geq t,\ x(t) = x \qquad (12)$$

and

$$V(x,t) = E\left\{\int_t^\infty f_0\left(y_s^*, v^*\left(y_s^*, s\right)\right)ds \,|\, y_t^* = x\right\} \qquad (13)$$

for System II where y_s^*, $s \geq 0$, is the solution to (3) with $v = v^*(y,t)$. We remark that, obviously, $\hat{u}^*(0) = 0$, $v^*(0,t) = 0$, $V_o(0) = 0$ and $V(0,t) = 0$ for all $t \geq 0$.

We are now ready to prove our first theorem on the convergence of $V(x,t)$ to $V_o(x)$.

<u>Theorem 3.1</u> Assume that for all $(x,t) \in R^n \times [0,\infty)$,

(i) $V_o(x)$ and $V(x,t)$ are defined and C^2 in x and $V(x,t)$ is C^1 in t.

(ii) $V_o(x) \leq \alpha_1\|x\|^2$

(iii) $\left\|\frac{\partial^2 V_o}{\partial x^2}\right\| \le \alpha_2$ and $\left\|\frac{\partial^2 V}{\partial x^2}\right\| \le \alpha_3$

where α_i, i = 1, 2, 3, are positive constants. Then, we have, for all $x \in R^n$

$$\lim_{t\to\infty} V(x,t) = V_o(x) \tag{14}$$

<u>Proof</u> First, because of assumption (i) above, $V_o(x)$ and $V(x,t)$ must satisfy, respectively, Bellman's equations of dynamic programming (see [1]), i.e.,

$$\frac{\partial V_o}{\partial x} f(x,u) + f_o(x,u) \ge \frac{\partial V_o}{\partial x} f(x,\hat{u}^*(s)) + f_o(x,\hat{u}^*(x)) = 0 \tag{15}$$

for all $x \in R^n$ and $u \in R^m$ and

$$\frac{\partial V}{\partial t} + \frac{\partial V}{\partial x} f(x,v) + f_o(x,v) + \frac{1}{2}\sum_{i,j} g_i g_j \frac{\partial^2 V}{\partial x_i \partial x_j} \ge \frac{\partial V}{\partial t} + \frac{\partial V}{\partial x} f(x,v^*(x,t)) + f_o(x,v^*(x,t)) + \frac{1}{2}\sum_{i,j} g_i g_j \frac{\partial^2 V}{\partial x_i \partial x_j} = 0 \tag{16}$$

for all $x \in R^n$, $t \ge 0$, and $v \in R^m$.

Let us define

$$I_1(t) = \frac{1}{2}\int_t^\infty \sum_{i,j} g_i\big(x^*(s),s\big) g_j\big(x^*(s),s\big) \frac{\partial^2 V}{\partial x_i \partial x_j} \cdot \big(x^*(s),s\big)ds \tag{17}$$

It is easily seen that $I_1(t) < \infty$ since

$$I_1(t) \le \frac{\alpha_3}{2} \int_t^\infty \|g(x^*(s),s)\|^2 ds$$

$$\le \frac{\alpha_3}{2} \int_t^\infty \eta^2(s)\left(1 + \|x^*(s)\|^2\right) ds \qquad (18)$$

But $\lim_{s\to\infty} x^*(s) = 0$. So, for sufficiently large t, we have

$$1 + \|x^*(s)\|^2 \le 1 + \delta \text{ for all } s \ge t.$$

Thus,

$$I_1(t) \le \frac{\alpha_3}{2} (1 + \delta) \int_t^\infty \eta^2(s) ds < \infty \qquad (19)$$

We now let $v = u^*(x)$ in (16), integrate along $x^*(s)$, and obtain

$$0 \le \int_t^\infty (f_o(x^*(s),\hat{u}^*(x^*(s))) + \frac{\partial V}{\partial x} (x^*(s),s) f(x^*(s),\hat{u}^*(x^*(s)))$$

$$+ \frac{\partial V}{\partial t} [x^*(s),s] ds + I_1(t)$$

$$= V_o(x) - V(x,t) + I_1(t) \qquad (20)$$

Taking the limit in (20) as $t \to \infty$ yields

$$\limsup_{t\to\infty} V(x,t) \le V_o(x) \qquad (21)$$

Let us next define

$$I_2(x,t) = E\left\{\int_t^\infty \sum_{i,j} g_i(y_s^*,s) g_j(y_s^*,s) \cdot \frac{\partial V_o}{\partial y_i \partial y_j} (y_s^*) ds \,|\, y_t^* = x\right\} \qquad (22)$$

We now show that $I_2(x,t) < \infty$ in the following.

$$I_2(x,t) \le E\left\{\lim_{T\to\infty} \int_t^T \left| \sum_{i,j} g_i(y_s^*,s) g_j(y_s^*,s) \right.\right.$$

$$\left.\left. \frac{\partial^2 V_o}{\partial y_i \partial y_j}(y_s^*) \right| ds \,\middle|\, y_t^* = x\right\}$$

$$\le \alpha_2 \left[E\left\{ \lim_{T\to\infty} \int_t^T \eta^2(s) \|y_s^*\|^2 ds \,|\, y_t = x \right\} \right.$$

$$\left. + \int_t^\infty \eta^2(s) ds \right] \tag{23}$$

By using Faton's Lemma and continuing with (23), we have

$$I_2(x,t) \le \alpha_2 \left[\limsup_{T\to\infty} \int_t^T \eta^2(s) \; E\left\{ \|y_s^*\|^2 \,|\, y_t = x \right\} ds \right.$$

$$\left. + \int_t^\infty \eta^2(s) ds \right] \tag{24}$$

But, since $\lim_{s\to\infty} E\|y_s^*\|^2 = 0$, for sufficiently large t, we can have

$$E\left\{ \|y_s^*\|^2 \,|\, y_t^* = x \right\} \le \delta(x) \text{ for any } x \in R^n$$

Hence,

$$I_2(x,t) \le \alpha_2 \left\{ \limsup_{T\to\infty} \int_t^T \delta(x) \eta^2(s) ds + \int_t^\infty \eta^2(s) ds \right\} < \infty$$

Obviously, $\lim_{t\to\infty} I_2(x,t) = 0$ for all $x \in R^n$.

We now let $u = v^*(x,t)$ in (15) and integrate along y_s^* (adding and subtracting $I_2(x,t)$). We then obtain

$$0 \le E\left\{ \int_t^\infty f_o(y_s^*, v^*(y_s^*,s)) ds \,|\, y_s^* = x \right\}$$

$$+ E\left\{ \lim_{T\to\infty} \int_t^T \frac{\partial V_o}{\partial x}(y_s^*) f(y_s^*, v^*(x_s^*,s)) \right.$$

$$\left. + \frac{1}{2} \sum_{i,j} g_i(y_s^*,s) g_j(y_s^*,s) \frac{\partial^2 V_o}{\partial y_i \partial y_j}(y_s^*) ds \,|\, y_t^* = x \right\}$$

$$- I_2(x,t)$$

$$\leq V(x,t) + \limsup_{T\to\infty} E\left\{V_o(y_T^*) \mid y_t^* = x\right\} - V_o(x)$$

$$- I_2(x,t)$$

$$\leq \alpha_1 \lim_{T\to\infty} E\left\{\|y_T^*\|^2 \mid y_t^* = x\right\} + V(x,t) - V_o(x) - I_2(x,t)$$

$$= V(x,t) - V_o(x) - I_2(x,t) \qquad (25)$$

Taking the limit in (25) as $t \to \infty$ yields

$$\liminf_{t\to\infty} V(x,t) \geq V_o(x) \qquad (26)$$

Combining (21) and (26), we finally obtain (14) and the theorem is proved.

In [4], Fleming showed, among other things, that as the "noise intensity" goes to zero, the optimal cost function of a stochastic control system approaches that of a deterministic control system to which the original system is reduced. The nature of the problem there was somewhat different from ours. For one thing, Fleming considered a control problem in a cylinder, i.e., $(x,t) \in Q \times [0,T]$, where $Q \subset R^n$. However, Theorem 3.1, which asserts, heuristically speaking, that when the stochastic term "dies away" as time goes on (owing to the fact that $\int_0^\infty \eta^2(t)dt < \infty$), a similar convergence property holds. The analogy between the implications of both results is obvious.

The next theorem on the stochastic convergence of $V(x,t)$ is more general than Theorem 3.1.

Theorem 3.2 Let $z(\omega) \in R^n$ be any random variable with $E\|z(\omega)\|^2 < \infty$. Then for any $A \in F$, we have

$$\lim_{t\to\infty} \int_A E\{V(z(\omega),t)|F_t\}dP = \int_A V_o(z(\omega))dP \tag{28}$$

Proof Since $\lim_{t\to\infty} V(z(\omega),t) = V_o(z(\omega))$, w.p.1, there exist t_o and $\mu > 0$ such that

$$|V(z(\omega),t) - V_o(z(\omega))| \le \mu\|z(\omega)\|^2, \text{ w.p.1} \tag{29}$$

for all $t \ge t_o$. Hence, we have $E\{V_o(z(\omega))\} < \infty$ as well as $E\{V(z(\omega),t)\} < \infty$. From Lebesgue convergence theorem, it then follows that

$$\begin{aligned}&\lim_{t\to\infty} \int_A E\{|V(z(\omega),t) - V_o(z(\omega))| \,|F_t\}dP \\ &\le \lim_{t\to\infty} \int_\Omega |V(z(\omega),t) - V_o(z(\omega)|dP = 0\end{aligned} \tag{30}$$

But, we can then write

$$\begin{aligned}&\lim_{t\to\infty} \int_A E\{V(z(\omega),t)|F_t\}dP \\ &= \lim_{t\to\infty} \int_A E\{V(z(\omega),t) - V_o(z(\omega))|F_t\}dP \\ &\quad + \lim_{t\to\infty} \int_A E\{V_o(z(\omega))|F_t\}dP \\ &= \int_A \lim_{t\to\infty} E\{V_o(z(\omega))|F_t\}dP = \int_A V_o(z(\omega))dP\end{aligned} \tag{31}$$

The last step follows because of the martingale convergence theorem (see [7]) and the fact that $\sigma(F_\infty) = F$. The theorem is thus proved.

For any $A \in F$, it can be shown [8] that there exists $A_t \in F_t$ such that $\lim_{t\to\infty} P(A_t) = P(A)$ and $\lim_{t\to\infty} P(A_t \Delta A) = 0$.

To dwell upon the implication of Theorem 3.2, we write

$$\frac{1}{P(A_t)} \int_A E\{V(z(\omega)t)|F_t\}dP = \frac{1}{P(A_t)}\left[\int_{A_t} E\{V(z(\omega),t)|F_t\}dP\right.$$
$$+ \int_{A\sim A_t} E\{V(z(\omega),t)|F_t\}dP$$
$$\left.- \int_{A_t\sim A} E\{V(z(\omega),t)|F_t\}dP\right] \quad (32)$$

We now take the limit as $t \to \infty$ in (30). Because of the uniform integrability of $E\{V(z(\omega),t)|F_t\}$ (see [8]), the second and third integrals on the right hand side of (30) approach zero. We then have

$$\lim_{t\to\infty} \frac{1}{P(A_t)} \int_{A_t} V(z(\omega),t)dP = \frac{1}{P(A)} \int_A V_o(z(\omega))dP \quad (33)$$

or, equivalently,

$$\lim_{t\to\infty} E\{V(z(\omega),t)|A_t\} = E\{V_o(z(\omega))|A\} \quad (34)$$

Note that $V_o(z(\omega))$ represents the optimal cost function for System I with random initial condition $z(\omega)$ at t. As for $V(z(\omega),t)$, it can be shown, with some additional assumptions, that

$$E\{V(z(\omega),t)|A_t\} = \min_{v\in\vartheta} E\left\{\int_t^\infty f_0(y_s,v_s)ds|A_t\right\}$$

where $y_t = z(\omega)$ and ϑ is the class of nonanticipative controls (this will be discussed in a forthcoming paper). The random initial condition $z(\omega)$, which is not necessarily F_t measurable, could be attributed to outside disturbances. The implication of (32) in terms of the optimum cost functions is thus clear.

In the third theorem, we shall use the notion of a stable sequence of events, introduced by Renyi [6] to obtain further results on the convergence property of $V(\cdot,t)$. Let us first define it in the following.

Definition A sequence of events $\{A_n\}$, $A_n \in F$, $n = 1, 2, \cdots$, is said to be stable if $\lim_{n\to\infty} P(A_n \cap B) = Q(B)$ exists for every $B \in F$ and $0 < Q(\Omega) < 1$.

As is shown in [6], if $\{A_n\}$ is a stable sequence of events, then $Q(B)/Q(\Omega)$ is a probability measure on F, which is absolutely continuous with respect to P and Q(B) can be represented in the form

$$Q(B) = \int_B \lambda \, dP \tag{35}$$

where $\lambda = \lambda(\omega)$ is a random variable such that $0 \le \lambda(\omega) \le 1$, w.p.1.

Theorem 3.3 Suppose $V_o(z(\omega))$ is bounded, w.p.1. Let $\{A_n\}$ be a stable sequence of events and $\{t_n\}$ an increasing sequence such that $t_n \to \infty$ as $n \to \infty$. Then, we have

$$\lim_{n\to\infty} \int_{A_n \cap B} E\{V(z(\omega),t_n) | F_{t_n}\} dP = \int_B V_o(z(\omega)) dQ \tag{36}$$

Proof First, we have

$$\lim_{n\to\infty} \int_{A_n \cap B} E\{V(z(\omega),t_n) | F_{t_n}\} dP$$

$$= \lim_{n\to\infty} \int_{A_n \cap B} E\{V(z(\omega),t_n) - V_o(z(\omega)) | F_{t_n}\} dP$$

$$+ \lim_{n\to\infty} \int_{A_n \cap B} \left[E\{V_o(z(\omega)) | F_{t_n}\} - V_o(z(\omega))\right] dP$$

$$+ \lim_{n\to\infty} \int_{A_n \cap B} V_o(z(\omega)) dP \tag{37}$$

Let L_1, L_2, and L_3 represent the first, second, and third limit on the right hand side of (35), respectively. From Lebesgue's convergence theorem, it easily follows that $L_1 = L_2 = 0$ since $\lim_{n\to\infty} V(z(\omega),t_n) = V_o(z(\omega))$ and $\lim_{n\to\infty} E\{V_o(z(\omega))|F_{t_n}\} = V_o(z(\omega))$, w.p.l.

As for L_3, we note that since $V_o(z(\omega))$ is bounded, it can be approximated by a sequence of simple functions $V_o^k(\omega)$, i.e., $\lim_{n\to\infty} V_o^k(z(\omega)) = V_o(z(\omega))$, w.p.l, and

$$V_o^k = \sum_{i=1}^{r} \gamma_i^k I_{A_i^k}$$

where A_i^k's are disjoint sets in F and $I_{A_i^k}$ is the indicator function of A_i^k. Let k be so large that $|V_o^k - V_o| < \varepsilon$, w.p.l. We then have

$$\int_{A_n \cap B} V_o(z(\omega))dP - \int_B V_o(z(\omega))dQ$$

$$= \int_\Omega \left(V_o(z(\omega)) - V_o^k(\omega)\right) I_{A_n \cap B}\, dP$$

$$+ \int_\Omega V_o^k(\omega)(I_{A_n \cap B} - I_B\lambda)\, dP$$

$$+ \int_\Omega \left(V_o^k(\omega) - V_o(z(\omega))\right) I_B\lambda\, dP \tag{38}$$

Since V_o^k is simple, by choosing n sufficiently large, we have

$$\left|\int_\Omega V_o^k(\omega)\left(I_{A_n \cap B} - I_B\lambda\right)dP\right| < \varepsilon$$

and, also, since $|V_o^k - V_o| < \varepsilon$, w.p.l,

$$\left|\int_\Omega \left(V_o(z(\omega)) - V_o^k(\omega)\right) I_{A_n \cap B} dP\right| < \varepsilon$$

and

$$\left|\int_\Omega \left(V_o^k(\omega) - V_o(z(\omega))\right) I_B\lambda\, dP\right| < \varepsilon$$

Thus, for such n,

$$\left| \int_{A_n \cap B} V_o(z(\omega))dP - \int_B V_o(z(\omega))dQ \right| < 3\varepsilon \tag{39}$$

This means $L_3 = \int_B V_o(z(\omega))dQ$ and (34) follows. The theorem is thus proved.

A special case of (34) is when $B = \Omega$. We have

$$\lim_{n\to\infty} \int_A E\{V(z(\omega),t_n)|F_{t_n}\}dP = \frac{1}{Q(\Omega)} \int_\Omega V_o(z(\omega))dQ \tag{40}$$

and therefore

$$\lim_{n\to\infty} \frac{1}{P(A_n)} \int_{A_n} E\{V(z(\omega),t_n)|F_{t_n}\}dP$$

$$= \frac{1}{Q(\Omega)} \int_\Omega V_o(z(\omega))dQ \tag{41}$$

or, equivalently,

$$\lim_{n\to\infty} E\{E\{V(z(\omega),t_n)|F_{t_n}\}|A_n\} = \frac{1}{Q(\Omega)} \int_\Omega V_o(z(\omega))dQ \qquad (40)$$

If $A_n \in F_{t_n}$ for all n, then $P(A_n) \to P(A)$ for some $A \in F$. Then (40) can easily be shown to reduce to (28).

In general, we do not necessarily have $A_n \in F_{t_n}$. As in Theorem 3.2, $E\{V(z(\omega),t_n)|F_{t_n}\}$ could represent the optimum cost function with random initial condition $z(\omega)$ at t_n. Now, suppose the state of the system is subject to some persistent disturbances which occur at t_n, n = 1, 2, ⋯, and induce a stable sequence of events $\{A_n\}$, n = 1, 2, ⋯, in the probability space. $E\{E\{V(z(\omega),t_n)|F_{t_n}\}|A_n\}$ could then be interpreted as the expectation of the optimum cost function under such disturbances. Thus, in spite of the persistent but "stable" disturbances, the behavior of the stochastic system under the optimum control will settle down to a steady state condition as far as the cost function is concerned.

REFERENCES

1. Baudarel, R., J. Delmas, and P. Guichet, Dynamic Programming and Its Application to Optimal Control, Academic Press, New York, 1971.

2. Brunovsky, P., On the Optimal Stabilization of Nonlinear Systems, Czech. Math Journal, 18 (93) 1968, Praha, 278-293.

3. Doob, J. L., Stochastic Processes, Wiley, New York, 1953.

4. Fleming, W., Stochastic Control for Small Noise Intensities, SIAM J. on Control, 9, 1971, 473-517.

5. Markus, L. and B. Lee, Foundation of Optimal Control Theory, Wiley, New York, 1967.

6. Renyi, A., Foundation of Probability, Holden Day, San Francisco, Calif., 1970.

7. Rosenblatt, M., Random Processes, Springer-Verlag, New York, 1974.

8. Yeh, J., Stochastic Processes and the Wiener Integral, New York, 1973.

THE GENERALIZED PURSUIT PROBLEM WHERE THE PURSUER USES BANG-BANG CONTROLS

N. Levitt and H. Sussmann

Rutgers - State University of New Jersey
New Brunswick, New Jersey

ABSTRACT

We consider the pursuit problem where the pursuer and the evader both move on a differentiable manifold X^n and where the pursuer is constrained so that his trajectory must be a concatenation of segments of trajectories of finitely many vector fields $\pm V_i$ which are fixed in advance and which possess a certain generic property. We show that such a pursuer, having "memory" but not "foresight" can "catch" any evader whose trajectory y(t) has the property that $\dot{y}$ is always within the convex hull of $\pm \alpha V_i$ for some positive constant $\alpha < 1$.

We generalize this to the case when the pursuer moves on X^n and the evader on Y^m, and the pursuer's goal is to force the pair (x, y) into $W \subseteq X \times Y$.

§1. INTRODUCTION

We consider the "pursuit" problem and certain generalizations in a situation where both players move on a smooth manifold, and where the pursuer must utilize "bang-bang" controls; that is, there are fixed vector fields $V_1, \cdots, V_k$ on the smooth manifold X^n on which the pursuer moves, and the trajectory of the pursuer is a concatenation of partial trajectories of the various V_i, traversed in either the positive or negative direction. In other words, the control space for the pursuer is to be the discrete space $U = \{f_1, r_1, f_2, r_2, \cdots, f_k, r_k\}$ and the pursuer's system is given by $\dot{x} = P(x, u)$, $x \in X^n$, $u \in U$ with $P(x, f_i) = V_i(x)$, $P(x, r_i) = -V_i(x)$. The pursuer controls his trajectory by picking a piecewise constant function $\mathbb{R} \to U$, i.e., by switching from one vector field to another (taking sense of direction into account) at discrete intervals.

We shall study the case where the vector fields $V_1, \ldots, V_k$ possess a certain generic property for k-tuples of vector fields on X^n, viz:

If L is the Lie algebra generated by $V_1, \cdots, V_k$, then for any $x \in X^n$ and any tangent vector $v \in T_x X^n$ there is a vector field $W \in L$ such that $W(x) = v$. The genericity of this property is proved by Sussmann [2]. The property has a number of important consequences for the problem at hand.

We prove in this paper that the pursuer in these circumstances can always catch any evader who is, in some sense "slow" enough ("slowness" residing in the nature of the evader's velocity vector, without reference to acceleration and other higher order derivatives).

Let us describe more precisely the pursuit evasion game we have in mind. The pursuer moves as indicated on the (compact) manifold X^n. The evader moves on the smooth manifold Y^m. There is a subspace $W \subseteq X^n \times Y^m$, which we shall assume to be a submanifold, and roughly, the pursuer wins when $(x(t), y(t)) \in W$. For example, in the "pure" pursuit game where both pursuer and evader move on X^n (i.e., $X^n = Y^m$) and where the pursuer is trying to catch the evader in the obvious sense, then $W \subseteq X^n \times X^n$ is merely the diagonal submanifold $W = \{(x,y) | x = y\}$. We allow the evader to move by the rule $\dot{y} = E(y,z,t)$, $z \in Z$, assuming only that $z = z(t)$ so that the differential equation has a unique absolutely continuous solution $y(t)$. The game starts at time $t = 0$ with $x(0) = x_0 \in X^n$ and $y(0) = y_0 \in Y^m$. Suppose the evader uses the strategy $z = z(t)$ leading to a trajectory $y = y(t)$. A *winning strategy* for the pursuer is a piecewise constant function $u : [0,t) \to U$, $T > 0$, with the property that the limit $\lim_{t \to T^-} x(t) = x_T \in X^n$ exists, and that $(x_T, y(T)) \in W$. We say that the pursuit game is winnable for the pursuer, iff

1. given any x_0, y_0 and given any admissible evasion strategy $z(t)$, there is a piecewise constant function $U_z(t)$ which is a winning strategy
2. the selection of $u(t)$ on the interval $[0,s]$, $s \leq t$ is determined solely by the restriction of z to $[0,s]$

If we had only required (1) then we would only have a winnable "interception" game for the pursuer; that is, the pursuer would win, but his winning strategy would depend on knowing $z(t)$ for all t. Requirement (2) says, in effect, that the pursuer has "memory" but not "foresight," that is, the strategy for the pursuer up to time S is dependent only on the strategy of the evader up to time S.

Definition 1.1 If $\{P_1, \cdots, P_r\}\{Q_1, \cdots, Q_s\}$ are two sets of vector fields on M^n, we say that $\{Q_1, \cdots, Q_s\}$ is *slower* than $\{P_1, \cdots, P_r\}$ if there exists an α, $0 < \alpha < 1$ such that for all $x \in M$, all $1 \leq i \leq s$, $Q_i(x)$ is in the convex hull of the vectors $\alpha P_1(x)$, $-\alpha P_1(x)$, $\cdots$, $\alpha P_r(x)$, $-\alpha P_r(x)$.

<u>Definition 1.2</u> Given vector fields $\{P_1, \cdots, P_r\}$ on M and a system $\dot{x} = E(x,t,z)$, we say that E is slower than P_1, $\cdots$, P_r if there is an α, $0 < \alpha < 1$ so that the vector $E(x,t,z)$ is in the convex hull of $\alpha P_1(x)$, $-\alpha P_1(x)$, $\cdots$, $\alpha P_r(x)$, $-\alpha P_r(x)$ for all possible t, z, x.

We now state our theorems. First, we deal with the "pure" pursuit problem where the pursuer uses bang-bang controls for V_1, $\cdots$, V_k on X^n and the evader uses the rule $\dot{x} = E(x,z,t)$ [here X^n need not be compact].

<u>Theorem 1</u> The "pure" pursuit game is winnable if E is slower than V_1, $\cdots$, V_k.

This generalizes to the following problem where the pursuer moves on X^n with bang-bang controls for V_1, $\cdots$, V_k and the evader moves on Y^m using $\dot{y} = E(y,z,t)$, where the "winning" submanifold is $W \subset X^n \times Y^m$, and where X^n is compact.

<u>Theorem 2</u> The pursuit game is winnable if

(1) the projection map $X^n \times Y^m \to Y^m$ carries W onto Y

(2) there are vector fields P_1, $\cdots$, P_r on Y^m, Q_1, $\cdots$, Q_r on X^n so that Q_1, $\cdots$, Q_r is slower than V_1, $\cdots$, V_k, $E(y,z,t)$ is slower than P_1, $\cdots$, P_r

(3) P_1, $\cdots$, P_r has the generic property described in the introduction

(4) for $(x,y) \in W$, the vector $(Q_i(x), P_i(y)) \in T_xX \times T_yY = T_{(x,y)}X \times Y$ is an element of $T_{(x,y)}W$.

Theorem 1 for X^n compact is a special case of Theorem 2 because we can set $Y^m = X^n$, $W = \mathrm{diag}(X \times X)$ and $P_i = Q_i = \beta V_i$, (β slightly smaller than 1), and the hypothesis of 1 clearly implies the hypothesis of 2 in this special case. However, we shall prove Theorem 1 separately to begin with, inasmuch as it serves, more or less, as a lemma for the proof of Theorem 2.

§2. PROOF OF THEOREM 1

First we note that the property assumed of the set of vector fields $V_1, \cdots, V_k$ has some immediate and striking consequences. We first quote a well-known fact:

<u>Lemma 2.1</u> If $V_1, \cdots, V_k$ is a set of vector fields on X^n for which the generic property holds, then for any two points $x_1, x_2 \in X^n$, there is a path from x_1 to x_2 which is a concatenation of partial trajectories of the V_i, and their negatives.

In other words, given $x_1, x_2 \in X^n$, there is a piecewise constant function $u : [0,T] \to U$ such that the trajectory $x(t)$ determined by $\dot{x} = P(x,u,t)$ and $x(0) = x_1$ has $x(T) = x_2$.

Let $(\mathbb{R}^k)^\infty = \lim_m (\mathbb{R}^k)^m$ (with the weak topology; see Levitt [1]).

Note that the equation $\dot{x} = P(x,u,t)$ has the effect of determining a map $e_x : (\mathbb{R}^k)^\infty \to X^n$, given any initial point $x \in X^n$. That is, a point in $(\mathbb{R}^k)^\infty$ is a point in $(\mathbb{R}^k)^r = [(a_1^r, \cdots, a_k^r) \cdots (a_1^1, \cdots, a_k^1)]$. We identify this point with a piecewise constant map $[0, \sum_{i,j}|a_i^j|] \to U$ in an obvious way. That is, the map determined by $(a_1^r, \cdots, a_k^r, \cdots, a_1^1, \cdots, a_k^1)$ is equal to f_i on the interval from

$$\sum_{s<m} \sum_{q=1}^{k} |a_q^s| + \sum_{q=1}^{i-1} |a_q^m|$$

to

$$\sum_{s<m} \sum_{q=1}^{k} |a_q^s| + \sum_{q=1}^{i} |a_q|$$

provided $a_i^m > 0$ and is equal to r_i on the interval if $a_i^m < 0$.

Thus $e_x(p)$ is the endpoint

$$x\left(\sum_{s=1}^{r} \sum_{q=1}^{k} |a_q^s|\right)$$

of the trajectory determined by $p \in (\mathbb{R}^k)^\infty$ and the initial condition $x(0)$.

Let $N(t) \subseteq (\mathbb{R}^k)^\infty$ be the set of $(a_1^r, \cdots, a_k^r) \cdots (a_1^1, \cdots, a_k^1)$ such that $\sum_{j,i} |a_j^i| < t$.

<u>Lemma 2.2</u> Let $V_1, \cdots, V_k$ be a set of vector fields on X^n for which the generic property holds, and let $x \in X$. Then for any t, there is an open neighborhood A of x so that $A \subseteq e_x(N(t))$.

<u>Proof</u> If S is a set of points of X^n, let $e_S(N(t))$ denote the union of the $e_x(N(t))$, for x in S. It is clear that $e_S(N(t))$ is the set of all points that can be reached from some y in S by means of a trajectory which is a piecewise concatenation of integral curves of the V_i, in such a way that the total time is less than t. We have

a. If S is open, then $e_S(N(t))$ is open. Indeed, pick an arbitrary y in $e_S(N(t))$. Then there is an a in N(t) and an x in S, such that $e_x(a) = y$. The map F which takes z to $e_z(a)$ is clearly a diffeomorphism. Therefore, the image F(S) is open. This image is obviously a subset of $e_S(N(t))$, which contains y. So y is an interior point of $e_S(N(t))$. Since y was arbitrary, assertion (a) follows.

We now prove:

b. The interior of $e_x(N(t))$ is open for every x and every strictly positive t.

Indeed, let M be a submanifold of X^n. Consider the property:

(P) There is t' such that $0 < t' < t$ and that M is a subset of $e_x(N(t'))$.

Manifolds with property (P) clearly exist (e.g., pick t' arbitrarily, and then pick y in $e_x(N(t'))$, and let M be the zero-dimensional manifold $\{y\}$). Let Q be the set of manifolds that have property (P). Let M be an element of Q having the largest possible dimension. We claim that all the vector fields V_i are tangent to M. Indeed, assume that one of the V_i, say V_1, is not tangent to M. Then there is a point y in M such that $V_1(y)$ is not tangent to M at y. Consider the map G defined by: G(z,s) = the value at s of the integral curve of V_1 which passes through z at time 0. Then G maps $M \times \mathbb{R}$ to X^n, and the differential of G at (y,0) has rank $r = 1 + \dim M$. So G maps a neighborhood U of (y,0) onto an r-dimensional submanifold N of X^n. We can assume U to be a product $V \times [-\varepsilon,\varepsilon]$, where V is a neighborhood of y in M, and ε can be replaced by any smaller number, if so desired. In particular, let $\varepsilon < t-t'$, where t' is such that M is a subset of $e_x(N(t))$. Then it is clear that N is a subset of $e_x(N(t''))$, where $t'' = t' + \varepsilon$. Since $t'' < t$, it follows that N has property (P) as well. This contradicts the hypothesis that M was an element of Q of maximal dimension.

So, we have shown that the V_i are tangent to M. But the set of vector fields tangent to a given submanifold is a Lie algebra. Therefore, every V in the Lie algebra generated by the V_i is tangent to M. This implies that M is n-dimensional, hence open in X^n. So (b) is proved.

The desired conclusion now follows easily. Let $U \neq \phi$ be an open subset of X^n which is contained in $e_x(N(t/2))$. The existence of U follows from (b). By (a), $e_U(N(t/2))$ is open. It is clear that $e_U(N(t/2))$ is contained in $e_x(N(t))$. Moreover, x belongs to $e_U(N(t/2))$, because we can pick an element y of U, and a trajectory which steers x to y in time less than t/2. If we then reverse the trajectory, we see that y can be steered into x in time less than t/2.

So x is an interior point of $e_x(N(t))$. This completes the proof.

Now let α be a fixed strictly positive constant. Consider an absolutely continuous trajectory $h : [0,T] \to X$ such that $\dot{h} \in$ convex hull $(\pm\alpha V_1, \pm\alpha V_k)$ at all points.

<u>Lemma 2.3</u> Let $h : [0,T] \to X^n$ be any such trajectory. Then for any neighborhood A of h(T) there is a piecewise constant function $u : [0,\tau] \to U$, $\tau = \alpha T$ such that, in the trajectory x(t) determined by ϕ, the rule $\dot{x} = P(x,u,t)$, and the initial condition $x(0) = h(0)$, we have $x(\tau) \in A$.

<u>Proof</u> Consider the initial value problem

$$(*) \qquad \dot{x}(t) = v_1(t)V_1(x(t)) + \cdots + v_k(t)V_k(x(t)),\ x(0) = h(0)$$

where the v_i are measurable real-valued functions on [0,T] that satisfy $|v_1(t)| + \cdots + |v_k(t)| \leq \alpha$ for all t. It follows from Filippov's Lemma that h is a solution of (*) for some choice of the measurable functions v_i. Let S be the subset of $\mathbb{R}^k$ whose elements are those k-tuples $(v_1, \cdots, v_k)$ that satisfy $|v_1| + \cdots + |v_k| \leq \alpha$. It is well known that, if the set F of S-valued measurable functions on [0,T] is given the topology of weak convergence, then the map which to each function in F assigns the corresponding solution of (*) is continuous. Moreover, the set B of piecewise constant elements of F whose values are extreme points of S is dense in F. Hence there is a choice of functions v_i such that $(v_1, \cdots, v_k)$ is in B, and that the corresponding trajectory y(t) satisfies $y(T) \in A$. Now reparametrize by letting $x(t) = y(t/\alpha)$, for $0 \leq t \leq \tau$. It is clear that this trajectory satisfies the desired properties.

Now we complete the proof of Theorem 1. Let α be as given, and choose α_1 such that $\alpha < \alpha_1 < 1$. Let the evader's trajectory in X^n be given by y(t). Let x(0) be the pursuer's starting point. By Lemma 1, the pursuer may choose a trajectory which brings him to y(0) at time T_1. Consider a neighborhood A of $y(0) = x(T_1)$ such that any point of A may be reached by the pursuer in time $\varepsilon < (\alpha_1 - \alpha)T_1$. (By Lemma 2.2, such a neighborhood exists.) Consider the evader's trajectory y(t) from y(0) to $y(T_1)$ and consider its "reverse," i.e., $\hat{y}(t) = y(T_1 - t)$. By Lemma 2.3 we may choose a trajectory w(t) which is a concatenation of partial trajectories of the $\pm V_i$'s, with w(t) defined on $[0,\tau]$, $\tau < \alpha_1 T_1$, so that $w(\tau) \in A$. Choose a trajectory for pursuer from y(0) to $w(\tau)$ which is completed in time $\sigma < (\alpha_2 - \alpha_1)T_1$. Follow this by a trajectory which is

$w(t)$ "run in reverse" from $w(\tau)$ to $w(0) = y(T_1)$ for an interval of length τ. This concatenation determines a trajectory $x(t)$ for the pursuer from $t = T_1$ to $t = T_2$ where $(T_2 - T_1) < \alpha_1 T_1$ and were $x(T_2) = y(T_1)$.

We now reiterate this procedure, that is, we find, as above, a further trajectory for the pursuer from $t = T_2$ to $t = T_3$ so that $x(T_3) = y(T_2)$ and $(T_3 - T_2) < \alpha_1(T_2 - T_1)$.

Now consider the sequence $\{T_i\}$; this must have a finite limit since $\Sigma(T_{i+1} - T_i)$ is majorized by the geometric series $\sum_{n=0}^{\infty} \alpha_2^n T_1$. By the continuity of $y(t)$, moreover, the distance $\rho(y(T_{i+1}), y(T_i))$ must go to zero as $i \to \infty$. Clearly $\lim_{t\to T}(x(t)) = \lim_{t\to T}(y(t)) = y(T)$ so the evader is "caught" at time T.

Moreover, it is clear from the proof that the pursuer's choice of strategy on the interval $[0,S]$, $S < T$, can be determined by the evader's trajectory on $[0,S]$ (in fact on $[0,S-\varepsilon]$ for some small enough ε).

§3. PROOF OF THEOREM 2

Consider now the situation described in the hypothesis of Theorem 2. The pursuer's winning strategy is described as follows.

First, consider the manifold Y^m; the evader's trajectory is given by $y(t)$. Pick a point x_1 in X^n so that $(x_1, y_0) \in W$. Find a trajectory $x(t)$ for the pursuer $0 \leq t \leq T_1$ so that $x(T_1) = x_1$ (by virtue of Lemma 2.1). Now a path which is a concatenation of segments of trajectories of the $\pm P_i$'s will briefly, be called a $(P_1, \cdots, P_r)$-trajectory. By the proof of Theorem 1, there is a $(P_1, \cdots, P_r)$-trajectory from $y(0)$ to $y(t_1)$ which is traversed from T_1 to T_2 where $(T_1 - T_2) < \beta T_1$

for some fixed constant $\beta < 1$. We follow this by a $(P_1, \cdots, P_r)$ trajectory from time T_2 to T_3 starting with $y(T_2)$, ending at $y(T_3)$ with $(T_3 - T_2) < \beta^2 T_1$, and so forth. Note that each choice of the segment from $y(T_i)$ to $y(T_{i+1})$ is made at time T_{i+1} when the endpoint of the trajectory previously chosen is $y(T_i)$. As in the proof of Theorem 1, we have a $(P_1, \cdots, P_r)$-trajectory $w(t)$, defined on some finite interval $[T_1,T]$, so that

$$\lim_{t\to T} w(t) = \lim_{t\to T} y(t) = y(T)$$

Now note that each $(P_1, \cdots, P_r)$-trajectory in Y^m corresponds to a $(Q_1, \cdots, Q_r)$ trajectory in X^n, with initial point x_1. That is, for each segment of time (t, s) when the $(P_1, \cdots, P_r)$-trajectory runs along an orbit of $\pm P_i$, we have the $(Q_1, \cdots, Q_r)$-trajectory run along $\pm Q_i$. Since the point $(s_1, y(0)) \in W$, and since $(\pm P_i(x), \pm Q_i(y)) \in T_{x,y}W$ for $x,y \in W$, it follows that if $u(t)$ is the trajectory in X^n corresponding to $w(t)$, then $(u(t),w(t))$ is a trajectory in W.

Lemma 3.1 There is a trajectory $x(t)$, $T_1 \le t < T$, of $V_1, \cdots, V_k$ so that $x(T_i) = u(T_i)$.

Proof Assume $x(t)$ has been determined for $t \in [T_1,T_{i-1}]$ satisfying the condition. By the proof of Theorem 1, there is a $(V_1, \cdots, V_k)$-trajectory $\bar{x}(t)$ from $u(T_{i-1})$ to $u(T_i)$ defined on $[T_{i-1},\tau]$, $\tau < T_i$. But we extend $\bar{x}(t)$ to x by going out along the V_1 trajectory through $u(T_i)$ for the interval between τ and $(T_i + \tau)/2$ and then back along $-V_1$ for the interval between $(T_i + \tau)/2$ and T_i.

Thus we have a $(V_1, \cdots, V_k)$-trajectory $x(t)$ from time T_1 to T which, concatenated with the earlier path from $x(0)$ to x_1 during $[0,T_1]$ gives the pursuer's strategy. Obviously, $\lim_{t\to T} x(t) = \lim_{t\to T} u(t) = x_T$, but note that

$$(x_T, y(T)) = \lim_{t\to T}(x(t), y(T))$$

$$= (\lim_{t\to T} u(t), \lim_{t\to T} w(t)) \in W$$

so $x(t)$ is a winning strategy for the pursuer.

Note also that the choice of $x(t)$ involves no foreknowledge on the part of the pursuer because at time T_i he knows the evader's path from $y(T_{i-1})$ to $y(T_i)$, thus he 'computes' $w(t)$ from $t = T_i$ to $t = T_{i+1}$, thus obtaining $u(t)$, and consequently allowing the computation of $x(t)$ at the same time.

Remark The reader will easily see that at each point T_i the pursuer might have a small subsequent interval to do his "computation" during which he "pursues;" but, of course, the interval available for such computation shrinks to zero as $i \to \alpha$.

Remark The assumption of Theorem 2 that X^n be compact may be replaced by the assumption that $V_i(x)$ is uniformly bounded for some complete Riemannian metric.

REFERENCES

1. Levitt, N., Homotopy and Continuous Reachability, Topology, 15 (1976), pp. 55-67.

2. Sussmann, H., Some Properties of Vector Fields that are Invariant Under Small Perturbations, J. Diff. Eq., 20 (1976), pp. 292-315.

A MINIMUM PRINCIPLE FOR SYSTEMS GOVERNED BY ITO DIFFERENTIAL EQUATIONS WITH MARKOV JUMP PARAMETERS

N. U. Ahmed and H. W. Wong

University of Ottawa
Ottawa, Canada

ABSTRACT

We consider the following Ito differential equation,

$$d\xi(t) = f(\xi(t),t,y(t),u(\xi(t),t,y(t)))dt + g(\xi(t),t,y(t))dw$$

in R^n, where y is a homogeneous additive Markov chain with a finite state space and w is a vector Wiener process. The problem is to choose a control law u(x,t,y) so as to minimize the cost functional

$$J(u) = \int_0^{T\wedge\tau} f_o(x(t),t,y(t),u(t,x(t),y(t)))dt$$

where τ is the first exit time of the process ξ from a given open bounded and connected set $\Omega \subset R^n$. A computational algorithm is developed.

§1. INTRODUCTION

The problem stated in the abstract is equivalent to the optimal control problem of a system of parabolic differential equations with first boundary conditions (Lemma 1.1). In section 2 we consider the existence of (weak) solutions of a system of general linear parabolic equations (Theorem 2.2) covering the system equivalent to the stochastic problem. In section 3 necessary conditions of optimality (Theorem 3.5, Corollary 3.6) for the general problem are developed. In the same section Theorem 3.8 gives the optimality conditions for the stochastic problem as a special case. The optimality condition for a linear quadratic regulator problem as given by Wonham [13] and Sworder and Kazangey [6,12] follows from a direct application of Theorem 3.8 (see section 4, Corollary 4.1). In section 5 we present an algorithm for computing the optimal controls on the basis of the optimality conditions developed in this paper.

Similar problems were considered by Krasvoskii and Lidskii [7,9] on an infinite time interval. Sworder [11] gave a maximum principle for the fixed time linear problem without terminal cost. For the linear regulator problem, Wonham [13, p. 192] computed the optimal control law explicitly. Recently, Rishel [10] considered the Mayer's problem (terminal cost) for a system governed by ordinary differential equations with Markov jump parameters contained in the coefficient f. In this paper, we consider a Lagrange problem (integral cost) for a stochastic differential system with the drift coefficient containing Markov jump parameters (as in Rishel's model) plus a diffusion term which also depends on the jump process.

In Rishel's problem the coefficient f is assumed to be C^1 with respect to both the state and the control. In this paper we consider both f and g (diffusion coefficient) measurable in t and x but Gâteaux differentiable in the control variable (in the space of controls). Rishel studies in detail the continuity and differentiability properties of the conditional expectation of the cost function. Here we use the concept of weak solutions as given in Ladyženskaja, Solonnikov, and Ural'ceva [8] with the conditional expectation of the cost function belonging to $V_2(Q_T)$, or equivalently $L_\infty((0,T); L_2(\Omega)) \cap L_2((0,T); W_2^1(\Omega))$. Rishel uses dynamic programming approach to develop some interesting necessary and sufficient conditions of optimality consisting of integral and ordinary differential equations (Theorems 9, 11, 12 [10]). The approach taken in this paper is the variational approach involving adjoint systems. The optimality (only necessary) conditions consist of two parabolic systems (one being the adjoint) with first boundary conditions and certain inequalities.

The optimality conditions (Theorem 3.5 and Corollary 3.6) are improvements of those obtained by Ahmed and Teo [1] in which necessary conditions are given for one parabolic equation with Cauchy condition instead of a system of parabolic equations with first boundary conditions. Also, in this paper the coefficients and their corresponding Gâteaux derivatives are not required to be bounded in $R^N \otimes R^N$ or in R^N.

When the Markov chain y in the Ito equation (1.1) is absent, the optimality conditions (3.8) and (3.9) coincide with those given by Ahmed and Teo [2] who consider fixed time problems.

It would be interesting to consider control laws based on partial information as in Fleming [4]. Indeed it has been observed that when u(t) is adapted to the sigma-algebra $\sigma\{y(t)\} \times \sigma\{\hat{\xi}(t)\}$ ($\hat{\xi}$ the observable part of ξ), the necessary condition is easily obtained from Theorem 3.5. However, when

$u(t)$ is adapted to $\sigma\{\hat{y}(t)\} \times \sigma\{\hat{\xi}(t)\}$ ($\hat{y}$ the observable part of y), it is not so straightforward.

Consider $N = \{z_1, z_2, \cdots, z_N\}$ a finite set of points in R^m and $\{y(t), t \in [0,T]\}$ a homogeneous additive Markov chain with state space N, and transition matrix P with elements

$$p_{ik}(t) = \text{prob}\{y(t) = z_k \mid y(o) = z_i\}$$

$(i, k = 1, 2, \cdots, N)$ and with initial distribution P_o. Define

$$q_{ik} \triangleq \lim_{t \downarrow o} \frac{P_{ik}(t)}{t}, \qquad i \neq k$$

$$q_{ii} \triangleq \lim_{t \downarrow o} \frac{P_{ii}(t) - 1}{t} = -\sum_{k \neq i} q_{ik}$$

It is known that P satisfies the differential equation

$$\frac{dP}{dt} = QP, \; P(o) = P^o \quad \text{and} \quad Q = \{q_{ik}\}$$

Denote the time interval $[0,T]$ by I. Let $w(t)$, $t \in I$ be the standard Wiener process with values in R^d. Suppose U is any closed convex set in R^r. Consider the stochastic differential system in R^n given by

$$d\xi = f(\xi,t,y,u)dt + g(\xi,t,y)dw$$

$$\xi(o) = \xi_o, \text{ with initial distribution } \pi \tag{1.1}$$

where u is any measurable function on I with values in U. Let Ω be an open bounded and connected set in R^n with boundary $\partial\Omega$. Denote the cylinder $(0,T) \times \Omega$ by Q_T and the lateral boundary $(0,T) \times \partial\Omega$ by S_T and let τ be the first time the process ξ hits the set $\partial\Omega$. Define U to be the class of admissible control laws consisting of functions u defined on $\bar{\Omega} \times I \times N$

with values in U so that for each $z \in N$, $u(\cdot,\cdot,z)$ is measurable on $\Omega \times I$. For any function h defined on $\Omega \times I \times N$ we will use $h(t,x,k)$ for $h(t,x,z_k)$. Let $f_o : R^n \times I \times N \times U \to R^1$ so that for each arbitrary but fixed point $z \in N$ and $u \in U$, f_o is measurable in $(x,t) \in R^n \times I$, and for each $(x,t,z) \in R^n \times I \times N$, f_o is continuous in u on U.

The problem is to find a control law $u_o \in \mathcal{U}$ so that the cost functional

$$J(u) = E\int_o^{\tau\wedge T} f_o(\xi(t),t,y(t),u(t))dt \tag{1.2}$$

attains its minimum at $u(t) = u_o(\xi(t),t,y(t))$ subject to the dynamic constraints (1.1). Call this problem P_o.

The optimization problem P_o can be transformed into an equivalent problem involving a system of partial differential equations of parabolic type with first boundary conditions. Define

$$\varphi(x,s,k) = E\left\{\int_s^{T\wedge\tau} f_o(\xi(\theta),\theta,y(\theta),u(\theta,y(\theta))d\theta \,\middle|\, \xi(s) = x, y(s) = z_k\right\} \tag{1.3}$$

and denote by ϕ and $F(u)$, the vectors $(\varphi(\cdot,\cdot,1), \cdots, \varphi(\cdot,\cdot,N))'$ and $(f_o(\cdot,\cdot,1,u), \cdots, f_o(\cdot,\cdot,N,u))'$. Redefine $\mathcal{U}$ to be the class of admissible controls consisting of functions $\{u(x,t)\}$ where $u(x,t) = (v_1(x,t,1), v_2(x,t,1), \cdots, v_r(x,t,1); \cdots; v_1(x,t,N), v_2(x,t,N), \cdots, v_r(x,t,N))'$ defined and measurable on Q_T with $v(x,t,k) = \{v_i(x,t,k), i = 1, \cdots, r\}$, $k = 1, \cdots, N$, taking values in U. In other words u is a measurable function on Q_T with values in U^N (N copies of U) contained in R^{rN}.

Using the convention of repeated indices for summation, we define the operator $A(u)$, for $u \in \mathcal{U}$, by

$$A(u)\phi = A_{ij}\phi_{x_i x_j} + B_i(u)\phi_{x_i} + C\phi \tag{1.4}$$

or in the divergence form,

$$A(u)\phi = (A_{ij}\phi_{x_i})_{x_j} - \bar{B}_i(u)\phi_{x_i} + C\phi \tag{1.4'}$$

where $\bar{B}_i(u) \triangleq \partial A_{ij}/\partial x_j - B_i(u)$ and

$$\begin{aligned} A_{ij} &= \operatorname{diag}(a_{ij}^1, a_{ij}^2, \cdots, a_{ij}^N),\ a_{ij}^k(x,t) \\ &= \frac{1}{2}(gg')_{ij}(x,t,k) \end{aligned}$$

$$\begin{aligned} B_i(u) &= \operatorname{diag}(f_i^1, f_i^2, \cdots, f_i^N),\ f_i^k(x,t) \\ &= f_i(x,t,k,v(x,t,k)) \end{aligned}$$

$$C = Q$$

Lemma 1.1 Suppose for each control $u \in \mathcal{U}$ the system (1.1) has a solution ξ. Then the optimization problem P_o is equivalent to the following problem P_1:

$$P_1 \begin{cases} \phi_t + A(u)\phi + F(u) = 0, & (x,t) \in \Omega \times [0,T) \\ \phi(x,t) = 0, & (x,t) \in \partial\Omega \times (0,T) \\ \phi(x,T) = 0, & x \in \Omega \end{cases} \tag{1.5}$$

$$J(u) = \int_\Omega (\phi(x,0)P^o)\pi(dx) = \min_{u \in \mathcal{U}}$$

where ϕP^o denotes the scalar product in R^N.

Note that the generator $A(u)$ can be easily derived using the principle of dynamic programming.

By virtue of Lemma 1.1 we have been able to convert the original stochastic optimization problem P_o into an optimal control problem of a deterministic parabolic system with first boundary conditions. Thus from this point on, we will consider the equivalent problem P_1.

§2. EXISTENCE OF SOLUTIONS OF A CLASS OF GENERAL PARABOLIC SYSTEMS AND ASSOCIATED ADJOINTS

In this section we present sufficient conditions for the existence of weak solution of the following more general problem

$$\phi_t + [A_{ij}(u)(x,t)\phi_{x_i} + A_j(u)(x,t)\phi]_{x_j} + B_i(u)(x,t)\phi_{x_i}$$

$$+ C(u)(x,t)\phi + F(u)(x,t) = 0, \quad (x,t) \in \Omega \times [0,T)$$

$$\phi(x,t)|_{S_T} = 0$$

$$\phi(x,T) = \phi_T(x), \quad x \in \Omega \tag{2.1}$$

and its adjoint problem. The result will cover the system (1.5) and its associated adjoint problem. First, we introduce some notations.

The Euclidean inner product in R^N will be denoted by vw, the Euclidean norm by $|v|$, the integral $\int_\Omega vw\, dx\, dt$ by $\langle v,w\rangle_\Omega$ and the integral $\int_{Q_T} vw\, dx\, dt$ by $\langle v,w\rangle$. Matrix norms are defined as $|\cdot|$,

$$|A|^2 = \sum_{k,\ell=1}^{N} |A_{k\ell}|^2 \quad \text{and} \quad |v_x|^2 = \sum_{k=1}^{N} \sum_{i=1}^{n} \left(\frac{\partial v_k}{\partial x_i}\right)^2$$

The symbols $R^N \otimes R^N$ will denote the space of $N \times N$ matrices and "'" will denote matrix transpose. The complements of a set Σ will be denoted by $C\Sigma$.

Define the following function spaces,

$W_2^{1,0}(Q_T)$ is the Hilbert space of R^N-valued functions with scalar product

$$(\varphi,\psi)_{W_2^{1,0}(Q_T)} = \int_{Q_T}[\varphi\psi + \varphi_{x_i}\psi_{x_i}]dx\, dt$$

$V_2(Q_T)$ will denote the Banach space consisting of all elements in $W_2^{1,0}(Q_T)$ with the norm defined by

$$|\varphi|_{Q_T} = \operatorname*{ess\,sup}_{0\le t\le T}\|\varphi(\cdot,t)\|_{2,\Omega} + \|\varphi_x\|_{2,Q_T}$$

where

$$\|\varphi_x\|_{2,Q_T} = \left(\int_{Q_T}|\varphi_x|^2 dx\, dt\right)^{1/2}$$

$L_{q,r}(Q_T)$ is the Banach space consisting of all R^N-valued functions on Q_T with a finite norm

$$\|\varphi\|_{q,r,Q_T} = \left[\int_I\left(\int_\Omega|\varphi(x,t)|^q dx\right)^{r/q} dt\right]^{1/r}$$

where $q \ge 1$ and $r \ge 1$.

$W_2^1(\Omega)$ is the Banach space consisting of all elements of $L_q(\Omega,R^N)$ having first order generalized derivatives that are square integrable with the norm

$$\|\psi\|_{W_2^1(\Omega)} = \sum_{j=o}^{1} \langle\langle\ \psi\ \rangle\rangle_{2,\Omega}^{(j)}$$

where

$$\langle\langle\ \psi\ \rangle\rangle_{2,\Omega}^{(j)} = \sum_{(j)} \|D_x^j\psi\|_{2,\Omega}$$

The symbol D_x^j denotes any derivative of $\psi(x)$ with respect to x of order j, where $\sum_{(j)}$ denotes the summation over all possible derivatives of ψ of order j.

$\overset{o}{W}{}^1_2(\Omega)$ is the subspace of $W^1_2(\Omega)$ in which the set of all functions that are infinitely differentiable and finite in Ω is dense.

The following assumptions will be made on the coefficients of (2.1):

(A1) $A_{ij}(u) : \Omega \times I \to R^N \otimes R^N$ satisfies for every $u \in \mathcal{U}$ the parabolicity condition

$$\nu \sum_{k=1}^{N} \sum_{i=1}^{n} \left(\xi_i^k\right)^2 \le A_{ij}(u)\xi_i\xi_j \le \mu \sum_{k=1}^{N} \sum_{i=1}^{n} \left(\xi_i^k\right)^2$$

for all $\xi_i = \left(\xi_i^1, \xi_i^2, \cdots, \xi_i^N\right)'$, $i = 1, 2, \cdots, n$, belonging to R^N uniformly with respect to $(x,t) \in Q_T$ for some positive constants ν and μ.

(A2) $A_i(u)$, $B_i(u)$, $C(u) : \Omega \times I \to R^N \otimes R^N$ are measurable functions with norms

$$\| |A_i(u)|^2;\ |B_i(u)|^2;\ |C(u)| \|_{q,r,Q_T} \le \mu_1$$

for every $u \in \mathcal{U}$ and q, r satisfying

$$\begin{aligned} &1/r + n/2q = 1 \\ &q \in (n/2, \infty],\ r \in [1, \infty] \text{ for } n \ge 2 \\ &q \in [1, \infty],\ r \in [1, 2] \text{ for } n = 1 \end{aligned} \tag{2.2}$$

(A3) The free term $F(u) : \Omega \times I \to R^N$ has a finite norm

$$\|F(u)\|_{q_1,r_1,Q_T} \le \mu_2$$

for every $u \in \mathcal{U}$ and

$$\begin{aligned} &1/r_1 + n/2q_1 = 1 + n/4 \\ &q_1 \in [2n/n{+}2, 2],\ r_1 \in [1, 2] \text{ for } n \ge 3 \\ &q_1 \in (1, 2],\quad r_1 \in [1, 2) \text{ for } n = 2 \\ &q_1 \in [1, 2],\quad r_1 \in [1, 4/3] \text{ for } n = 1 \end{aligned} \tag{2.3}$$

A generalized (or weak) solution ϕ of (2.1) is defined to be an element in $V_2(Q_T)$ that satisfies $\phi(x,t)|_{S_T} = 0$ and that for any element $\eta(x,t) \in W_2^{1,1}(Q_T)$, $\eta(x,0) = 0$, the following integral identity holds,

$$\int_{Q_T} \Big\{-\phi\eta_t - A_{ij}(u)\phi_{x_i}\eta_{x_j} - A_j(u)\phi\eta_{x_j} + B_j(u)\phi_{x_j}\eta + C(u)\phi\eta + F(u)\eta\Big\} dx\, dt = -\int_\Omega \phi\eta\, dx\,\Big|_{t=T} \tag{2.4}$$

For proving the existence of generalized solution for the parabolic system (2.1), we need an energy estimate.

Lemma 2.1 (Energy inequality) Suppose $\phi \in V_2(Q_T)$, with $\phi(x,t)|_{S_T} = 0$ satisfies for almost all t_1, $t_2 \in [0,T]$ the following inequality

$$\frac{1}{2}\int_\Omega |\phi(x,t)|^2 dx\Big|_{t=t_1}^{t=t_2} - \int_{t_1}^{t_2}\int_\Omega (A_{ij}(u)\phi_{x_i} + A_j(u)\phi)\phi_{x_j} + (B_i(u)\phi_{x_i} + C(u)\phi + F(u))\phi\, dx\, dt \geq 0 \tag{2.5}$$

and $A_{ij}(u)$, $A_j(u)$, $B_i(u)$, $C(u)$, $F(u)$ satisfy (A1)-(A3). Then there exists a constant c depending on n, ν, μ, μ_1, and q so that

$$|\phi|_{Q_T} \leq c[\|\phi(x,T)\|_{2,\Omega} + \|F(u)\|_{2,Q_T}]$$

Proof Let $\tau \in [t_1,t_2]$. From (2.5)

$$\frac{1}{2}\int_\Omega |\phi(x,t)|^2 dx\Big|_{t=\tau}^{t=t_2} - \int_\tau^{t_2}\int_\Omega A_{ij}(u)\phi_{x_i}\phi_{x_j}\, dx\, dt + \int_\tau^{t_2}\int_\Omega \Big\{- A_j(u)\phi\phi_{x_j} + B_j(u)\phi_{x_j}\phi + C(u)\phi\phi + F(u)\phi\Big\}\, dx\, dt \geq 0$$

Using the assumption (A1) and Cauchy's inequality in the above expression, we arrive at

$$\int_\Omega |\phi(x,\tau)|^2 dx + \nu\int_\tau^{t_2}\int_\Omega |\phi_x|^2 \, dx\, dt \le \int_\Omega |\phi(x,t_2)|^2 \, dx$$
$$+ 2\int_\tau^{t_2}\int_\Omega \left\{ \left(\frac{1}{\nu}|A_j(u)|^2 + \frac{1}{\nu}|B_j(u)|^2 + |C(u)| \right) |\phi|^2 \right.$$
$$\left. + |F(u)| \right\} dx\, dt$$

By taking ess sup with respect to τ over $[t_1,t_2]$, we obtain

$$\mathrm{Min}(1,\nu)\,|\phi|^2_{Q_{t_1,t_2}} \le \|\phi(\cdot,t_2)\|^2_{2,\Omega} + 2\int_{Q_{t_1,t_2}} \mathcal{D}|\phi|^2 dx\, dt$$
$$+ 2\int_{Q_{t_1,t_2}} |F(u)\phi|\, dx\, dt \qquad (2.5')$$

where

$$\mathcal{D} = \frac{1}{\nu}\sum_{j=1}^{n} |A_j(u)|^2 + |B_j(u)|^2 + |C(u)|$$

and

$$|\phi|_{Q_{t_1,t_2}} = \operatorname*{ess\,sup}_{t_1\le\tau\le t_2} \|\phi(\cdot,\tau)\|_{2,\Omega} + \left[\int_{t_1}^{t_2}\int_\Omega |\phi_x|^2 dx\, dt\right]^{1/2}$$

The rest of the proof can be carried out along the same line as in Ladyženskaja et al. [8, Lemma 2.1, p. 139].

<u>Theorem 2.2</u> If the conditions (A1)-(A3) are satisfied and $\phi_T(x) \in L_2(\Omega, R^N)$, then the system (2.1) has a generalized solution in $V_2(Q_T)$.

<u>Proof</u> Following Galerkin's method, we choose a fundamental system of functions $\psi_k(x)$, $k = 1, 2, \cdots$, in $W_2^1(\Omega)$ so that ψ_k vanishes on S_T for each k and that they are orthonormal in $L_2(\Omega, R^N)$. Let us denote

$$\phi^m(x,t) = \sum_{k=1}^{m} C_k^m(t)\psi_k(x)$$

where $C_k^m(t) = \int_\Omega \phi^m(x,t)\psi_k(x)dx$ are determined by a system of ordinary differential equations

$$\frac{d}{dt} C_k^m(t) = \sum_{\ell=1}^{m} \Big\{ \langle A_{ij}(u)(\psi_\ell)_{x_i}, (\psi_k)_{x_j} \rangle_\Omega + \langle A_j(u)\psi_\ell, \cdot (\psi_k)_{x_j} \rangle_\Omega - \langle B_i(u)(\psi_\ell)_{x_i}, \psi_k \rangle_\Omega - \langle C(u)\psi_\ell, \psi_k \rangle_\Omega \Big\} C_\ell^m(t) - \langle F(u), \psi_k \rangle_\Omega ; \; C_k^m(T) = \int_\Omega \phi_T^m(x)\psi_k(x)\, dx$$

From here on the proof is carried out in the same way as in Ladyženskaja et al. [8, p. 153].

Consider the adjoint problem to (2.1) defined by

$$-\psi_t + (A'_{ij}(u)\psi_{x_j} - B'_i(u)\psi)_{x_i} - A'_j(u)\psi_{x_j} + C'(u)\psi = 0, \quad (x,t) \in \Omega \times [0,T)$$

$$\psi(x,t)\big|_{S_T} = 0$$

$$\psi(x,0) = \psi_o(x) \tag{2.6}$$

Similar to the definition of generalized solution of (2.1), we define a generalized or weak solution ψ of (2.6) to be an element in $V_2(Q_T)$ that satisfies $\psi(x,t)\big|_{S_T} = 0$ and that for any element $\gamma(x,t) \in W_2^{1,1}(Q_T)$, $\gamma(x,T) = 0$, the following integral identity holds,

$$\int_{Q_T} \{\gamma_t\psi - \gamma_{x_i} A'_{ij}(u)\psi_{x_j} + \gamma_{x_i} B'_i(u)\psi - \gamma A'_j(u)\psi_{x_j} + \gamma C'(u)\psi\}dx\, dt = -\int_\Omega \gamma\psi dx\big|_{t=0}$$

The following theorem follows from Theorem 2.2 by simply reversing the flow of time.

Theorem 2.3 If the coefficients $A_{ij}(u)$, $A_j(u)$, $B_i(u)$, $C(u)$ of the adjoint problem (2.6) satisfy the assumptions (A1)-(A3) and that $\psi_o(x) \in L_2(\Omega, R^N)$, then (2.6) has a generalized solution in $V_2(Q_T)$.

Denote the adjoint operator corresponding to $A(u)$ in (1.4') by $A^*(u)$, defined as

$$A^*(u)\psi = (A'_{ij}\psi_{x_j} + \bar{B}'_i(u)\psi)_{x_i} + C'\psi$$

where $\bar{B}_i(u) = \partial A_{kj}/\partial x_j - B_i(u)$. Consider the adjoint problem to (1.5),

$$-\psi_t + A^*(u)\psi = 0, \quad (x,t) \in \Omega \times (0,T]$$
$$\psi(x,t)|_{S_T} = 0$$
$$\psi(x,0) = \psi_o(x) \underline{\Delta} P^o q(x), \quad x \in \Omega \tag{2.7}$$

where $\int_A q(x)dx = \pi(A)$. The following corollary is an immediate consequence of Theorems 2.2 and 2.3.

<u>Corollary 2.4</u> Suppose $\partial A_{ij}/\partial x_j$ is measurable, $i = 1, 2, \cdots, n$ and belong to $L_{q,r}(Q_T)$ with q, r satisfying (2.2) and A_{ij}, $B_i(u)$, C, $F(u)$ satisfy the assumptions (A1)-(A3) for each fixed but arbitrary $u \in \mathcal{U}$. Then the system (1.5) and its adjoint (2.7) have weak solutions in $V_2(Q_T)$.

§3. NECESSARY CONDITIONS FOR OPTIMALITY

Optimality conditions for the problem P_1 (see Lemma 1.1) will be presented under a more general setting. Assuming that the coefficients and free terms satisfy the hypothesis of Theorem 2.2, we consider the following problem P_2.

$$\phi_t + (A_{ij}(u)\phi_{x_i} + A_j(u)\phi)_{x_j} + B_i(u)\phi_{x_i}$$
$$+ C(u)\phi + F(u) = 0, \quad (x,t) \in \Omega \times [0,T)$$

$$\phi(x,t)|_{S_T} = 0$$

$$\phi(x,T) = \phi_T(x), \quad x \in \Omega \tag{3.1}$$

$$J(u) = \int_\Omega \phi(u)(x,o) P^o\pi(dx) = \min_{u \in \mathcal{U}}$$

where $\phi(u)$ denotes the generalized solution of the first boundary value problem (3.1) corresponding to $u \in \mathcal{U}$. The following lemma will be needed for the derivation of the optimality conditions.

Lemma 3.1 Suppose $A_{ij}^{(k)}$, $A_j^{(k)}$, $B_i^{(k)}$, $C^{(k)}$, $F^{(k)}$, satisfying the assumptions (A1)-(A3) independently of k, converge almost everywhere to $A_{ij}(u)$, $A_j(u)$, $B_i(u)$, $C(u)$, $F(u)$, respectively, on Q_T and there exist bounded measurable functions G $\in L_{q,r}(Q_T)$ and H $\in L_{q_1,r_1}(Q_T)$ such that $|A_j^{(k)}|$, $|B_i^{(k)}|$, $|C^{(k)}|$ $\leq |G|$ and $|F^{(k)}| \leq H$ a.e. Further, suppose that $\phi_T(x) \in L_2(\Omega, R^N)$. Let the sequence $\phi^k \in V_2(Q_T)$ be the generalized solution of the system (3.1) with coefficients $A_{ij}^{(k)}$, $A_j^{(k)}$, $B_i^{(k)}$, $C^{(k)}$, $F^{(k)}$. Then there exists a subsequence of the sequence $\{\phi^k\}$, again denoted by $\{\phi^k\}$, and an element $\phi \in V_2(Q_T)$ so that ϕ^k and $\phi_{x_i}^k$ converge weakly to ϕ and ϕ_{x_i}, respectively, in $L_2(Q_T)$ where $\phi = \phi^u$ is the weak solution of (3.1).

Proof It follows from the energy inequality in Lemma 2.1 that $|\phi^k|_{Q_T} \leq \beta$ uniformly for $k = 1, 2, \cdots$, where β is independent of k. Therefore, we can select a common subsequence from $\{\phi^k\}$, again denoted by $\{\phi^k\}$, such that ϕ^k and $\phi_{x_i}^k$ converge weakly to ϕ and ϕ_{x_i}, respectively, in $L_2(Q_T)$. To show that ϕ is a weak solution of (3.1), it suffices to consider $\eta \in \overset{o}{W}{}_2^{1,1}(Q_T)$ with $\eta(x,o) = 0$ and show that as $k \to \infty$,

$$<\phi_t - \phi_t^k + (A_{ij}(u)\phi_{x_i} + A_j(u)\phi)_{x_j} - (A_{ij}^k\phi_{x_i}^k + A_j^k\phi^k)_{x_j}$$
$$+ B_i(u)\phi_{x_i} - B_i^k\phi_{x_i}^k + C(u)\phi - C^k\phi^k + F(u) - F^k, \eta > 0 \quad (3.2)$$

This follows from dominated convergence theorem and the fact that if $f_k \to f_o$ weakly and $g_k \to g_o$ strongly, then $<f_k, g_k> \to <f_o, g_o>$. This completes the proof.

The proof of the necessary conditions for optimality of problem P_2 is based on approximating the system coefficients by their integral averages and the convergence properties of the resulting coefficients. We need the preparations below.

Following the method in Aronson [3] and Ahmed and Teo [1, 2], we let $F = 0$ in CQ_T (complement of Q_T in $R^n \times I$) and $\phi_T = 0$ in $C\Omega$ (complement of Ω in R^n). Let $A_{ij}^{(m)}(u)$, $A_j^{(m)}(u)$, $B_i^{(m)}(u)$, $C^{(m)}(u)$, $F^{(m)}(u)$ denote the integral averages of the corresponding coefficients formed with an averaging kernel whose support lies in $|x|^2 + t^2 < m^{-2}$ for integers $m \geq 1$ and define the operators

$$L^m(u)\phi = \phi_t + (A_{ij}^{(m)}(u)\phi_{x_i} + A_j^{(m)}(u)\phi)_{x_j}$$
$$+ B_i^m(u)\phi_{x_i} + C^{(m)}(u)\phi$$

Let ϕ_T^m denote the integral average of ϕ_T formed with a kernel whose support lies in $|x| < m^{-1}$. Let $\{\Omega^m\}$ be a sequence of open domains with smooth boundaries such that $\overline{\Omega^m} \subset \Omega^{m+1} \subset \overline{\Omega^{m+1}} \subset \Omega$ for all $m \geq 1$ and $\lim_{m\to\infty} \Omega^m = \Omega$. Let $\{g_m\}$ be a sequence of functions in $C_o^\infty(\Omega^m)$ (the space of infinitely differentiable functions defined and having compact support in Ω^m) so that $g_m = 1$ on $\overline{\Omega^{m-1}}$ and $0 \leq g_m(x) \leq 1$ for $x \in \Omega^m \backslash \Omega^{m-1}$. Then the sequence of boundary value problems

$$L^m(u)\phi + F^m = 0, \quad (x,t) \in Q_T^m = \Omega^m \times [0,T)$$

$$\phi(x,T) = \phi_T^m(x) g_m(x), \quad x \in \Omega^m$$

$$\phi(x,t) = 0, \quad (x,t) \in \partial\Omega^m \times [0,T) \tag{3.3}$$

with $F^m \equiv F^m(u)$ and $\partial\Omega^m$ sufficiently smooth, has a classical solution $\phi^m \in C^1(\overline{Q^m})$. For details on smooth solutions see, for example, Friedman [5, p. 65] and Ladyženskaja, Solonikov, Uralcera [8, p. 320).

<u>Remark 3.2</u> It follows from Lemma 3.1 that ϕ^m and $\phi^m_{x_i}$ converge weakly to ϕ and ϕ_{x_i}, respectively in $L_2(Q_T)$ where ϕ^m and ϕ are weak solutions of the systems (3.3) and (3.1).

Define the adjoint problem of (3.1) by

$$L^*(u_o)\psi = 0, \quad (x,t) \in \Omega \times (0,T]$$

$$\psi(x,t)\big|_{S_T} = 0$$

$$\psi(x,o) = \psi_o(x) \equiv P^o q(x), \; x \in \Omega, \; \pi(A) = \int_A q(x)dx \tag{3.4}$$

where $L^*(u_o)$ is the formal adjoint of the operator

$$L(u_o)\phi = \phi_t + (A_{ij}(u_o)\phi_{x_i} + A_j(u_o)\phi)_{x_j}$$

$$+ B_i(u_o)\phi_{x_i} + C(u_o)\phi$$

where u_o is an optimal control for the problem P_2 (existence assumed). The existence of generalized solution of (3.4) follows from Theorem 2.3. Similar to the integral averaged system (3.3), we define the following adjoint problems to (3.3),

$$L^{*m}(u_o)\psi^m = 0, \qquad (x,t) \in Q_T^m$$

$$\psi^m(x,t) = 0, \qquad (x,t) \in \partial\Omega^m \times (0,T]$$

$$\psi^m(x,o) = \psi_o^m(x)g_m(x), \quad x \in \Omega^m \tag{3.5}$$

It follows that (3.5) has a unique classical solution $\psi^m \in C^1(\overline{Q^m})$.

Remark 3.3 By the same arguments as in remark 3.2, ψ^m and $\psi^m_{x_i}$ converge weakly to ψ and ψ_{x_i}, respectively, in $L_2(Q_T)$ where ψ is the weak solution of (3.4).

Let $\phi(u)$ denote the generalized solution of (3.1) corresponding to an admissible control u and ψ denote the generalized solution of (3.4) corresponding to the optimal control u_o. We now state another lemma needed in deriving the optimality conditions.

Lemma 3.4 Suppose the coefficients and the free term in (3.1) satisfy the assumptions (A1)-(A3) and that there exist functions $G \in L_{q,r}(Q_T)$ and $H \in L_{q_1,r_1}(Q_T)$ so that $|A_i(u)|$, $|B_i(u)|$, $|C(u)| \le G$ a.e. and $|F(u)| \le H$ a.e. uniformly in $\mathcal{U}$. Let $\phi_T \in L_2(\Omega, R^N)$. Then

$$\lim_{m\to\infty} \langle L(u_o)(\phi(u_o) - \phi(u)), \psi^m\rangle =$$

$$- \int_\Omega (\phi(u_o) - \phi(u))\psi dx\Big|_{t=o} \tag{3.6}$$

for any $u \in \mathcal{U}$.

Proof We will follow a similar procedure as given in [1, p. 986]. Let $\phi^r(u)$, $\phi^r(u_o)$ be the solutions of problem (3.3) with m = r and $\phi(u)$, $\phi(u_o)$ the weak solutions of the problem

(3.1) both corresponding to the controls u and u_o, respectively. The following equalities can be easily verified.

(i) $$<L(u_o)(\phi(u_o) - \phi(u)), \psi^m> = <L(u_o)(\phi(u_o) - \phi(u) - \phi^r(u_o) + \phi^r(u)), \psi^m> + <L(u_o)(\phi^r(u_o) - \phi^r(u)), \psi^m>$$

(ii) $$<L(u_o)(\phi^r(u_o) - \phi^r(u)), \psi^m> = <\phi^r(u_o) - \phi^r(u), L^*(u_o)\psi^m> + \int_\Omega(\phi^r(u_o) - \phi^r(u))\psi^m dx\Big|_{t=0}^{t=T}$$

(iii) $$<\phi^r(u_o) - \phi^r(u), L^*(u_o)\psi^m> = <\phi^r(u_o) - \phi^r(u), (L^*(u_o) - L^{*m}(u_o))\psi^m> + <\phi^r(u_o) - \phi^r(u), L^{*m}(u_o)\psi^m>$$

(iv) $$<\phi^r(u_o) - \phi^r(u), L^{*m}(u_o)\psi> = -\int_\Omega(\phi^r(u_o) - \phi^r(u))\psi^m dx\Big|_{t=T}$$

The equality (ii) follows from integration by parts and (iv) follows from Green's formula. From (i)-(iv) we obtain

$$<L(u_o)(\phi(u_o) - \phi(u)), \psi^m> = <L(u_o)(\phi(u_o) - \phi(u) - \phi^r(u_o) + \phi^r(u)), \psi^m> + <\phi^r(u_o) - \phi^r(u), (L^*(u_o) - L^{*m}(u_o))\psi^m> - \int_\Omega(\phi^r(u_o) - \phi^r(u))\psi^m dx\Big|_{t=o} \quad (3.7)$$

By remark 3.2, $\phi^r(u_o)$, $\phi^r_{x_i}(u_o)$ converge weakly to $\phi(u_o)$, $\phi_{x_i}(u_o)$, respectively, in $L_2(Q_T)$. Using this fact and the boundedness properties of the coefficients and the free term (see the hypothesis of this lemma), one can show that the first term on the right hand side vanishes. By virtue of the convergence of $\phi^r(u_o)$, $\phi^r_{x_i}(u_o)$ mentioned above, together with the a.e. convergence of the integral averaged coefficients on Q_T and the dominated convergence theorem, the second term on the right hand side of (3.7) vanishes as r, m $\to \infty$. In a similar way, the first two terms on the right side of (3.7) vanishes as m $\to \infty$ and then r $\to \infty$, i.e., the order of taking limits is immaterial. It is also easily shown that as r, m $\to \infty$,

$$\int_\Omega (\phi^r(u) - \phi^r(u_o))\psi^m \, dx \to \int_\Omega (\phi(u) - \phi(u_o))\psi dt$$

and the proof is complete.

The following definition will be used in the optimality conditions. Let Γ be a map from a Banach space E_1 into a Banach space E_2. Γ_u is the strong Gâteaux derivative of Γ on E_1 and evaluated at $u \in E_1$ so that

$$\lim_{\varepsilon \downarrow 0} \left\| \frac{\Gamma(u + \varepsilon v) - \Gamma(u)}{\varepsilon} - \Gamma_u(v) \right\|_{E_2} = 0$$

for all $v \in E_1$. $\Gamma_u(v)$ is the Gâteaux differential of Γ at u in the direction v. The following assumptions will be made on the strong Gâteaux derivatives of the coefficients and free terms of (3.1).

(A4) The strong Gâteaux differentials $A_{iju}(v)$, $A_{ju}(v)$, $B_{ju}(v)$, $C_u(v)$, $F_u(v)$ exist at every point $u \in \mathcal{U}$. These differentials belong to the same function space as A_{ij}, A_j, B_j, C, F, respectively (see A1, A2, A3) for every $v \in \mathcal{U}$.

With these preparations, we now present the optimality conditions of the control problem P_2.

Theorem 3.5 Suppose that the coefficients $A_{ij}(u)$, $A_j(u)$, $B_j(u)$, $C(u)$, $F(u)$ in the systems (3.1) and (3.4) satisfy the hypothesis of Lemma 3.4 and that their strong Gâteaux derivatives satisfy the assumption (A4). Let $\phi_T \in L_2(\Omega, R^N)$. Then in order that $u_o \in \mathcal{U}$ be an optimal control, it is necessary that there exists a $\psi \in V_2(Q_T)$ so that

$$\int_{Q_T}\left\{\left[A_{iju_o}(v - u_o)\phi_{x_i}(u_o) + A_{ju_o}(v - u_o)\phi(u_o)\right]\psi_{x_j} - B_{iu_o}(v - u_o)\phi_{x_i}(u_o)\psi - C_{u_o}(v - u_o)\phi(u_o)\psi - F_{u_o}(v - u_o)\psi\right\}dx\,dt \le 0 \tag{3.8}$$

for all $v \in \mathcal{U}$.

Proof Multiplying the first equation in (3.1) by $\eta \in \overset{o}{W}{}_2^{1,1}(Q_T)$, where $\eta|_{S_T} = 0$, and integrating by parts, we obtain

$$\int_\Omega \phi_T(x)\eta(x,T)dx + \int_{Q_T}(-\phi(u)\eta_t - A_{ij}(u)\phi_{x_i}(u)\eta_{x_j} - A_j(u)\phi(u)\eta_{x_j} + B_j(u)\phi_{x_j}(u)\eta + C(u)\phi(u)\eta + F(u)\eta)\,dx\,dt = \int_\Omega \phi(u)(x,o)\eta(x,o)dx$$

Using the above formula, one can show that

$$\langle L(u_o)\phi(u_o) - L(u_o + \varepsilon v)\phi(u_o + \varepsilon v), \psi^m\rangle = \langle F(u_o + \varepsilon v) - F(u_o), \psi^m\rangle$$

Therefore,

$$\langle L(u_o)(\phi(u_o) - \phi(u_o + \varepsilon v)), \psi^m\rangle + \langle (L(u_o) - L(u_o + \varepsilon v))\phi(u_o + \varepsilon v), \psi^m\rangle = \langle F(u_o + \varepsilon v) - F(u_o), \psi^m\rangle \tag{3.9}$$

Letting $m \to \infty$, the first term on the left hand side converges to $\int_\Omega (\phi(u_o + \varepsilon v) - \phi(u_o))\psi dx|_{t=o}$ (≥ 0) by Lemma 3.4. Therefore, taking limits as $m \to \infty$ in the last equation, we arrive at

$$<(L(u_o) - L(u_o + \varepsilon v))\phi(u_o + \varepsilon v) + F(u_o) - F(u_o + \varepsilon v),\psi> \leq 0$$

Dividing both sides by $\varepsilon \in (0, 1]$, we obtain

$$\int_{Q_T} \left\{ \left[\frac{A_{ij}(u_o + \varepsilon v) - A_{ij}(u_o)}{\varepsilon}\right] \phi_{x_i}(u_o + \varepsilon v)\psi_{x_j} + \left[\frac{A_j(u_o + \varepsilon v) - A_j(u_o)}{\varepsilon}\right] \phi(u_o + \varepsilon v)\psi_{x_j} - \left[\frac{B_i(u_o + \varepsilon v) - B_i(u_o)}{\varepsilon}\right] \phi_{x_i}(u_o + \varepsilon v)\psi - \left[\frac{C(u_o + \varepsilon v) - C(u_o)}{\varepsilon}\right] \phi(u_o + \varepsilon v)\psi - \left[\frac{F(u_o + \varepsilon v) - F(u_o)}{\varepsilon}\right] \psi \right\} dx\, dt \leq 0$$

Note that the integrands here (except the last one) involve a matrix operation followed by a scalar multiplication in R^N. Next we take the limit in the last inequality as $\varepsilon \downarrow 0$. Since the strong Gâteaux differentials exist and are linear in $\mathcal{U}$ (assumption A4), it follows from the fact "$<f_k, g_k> \to <f_o, g_o>$ whenever $f_k \to f_o$ strongly and $g_k \to g_o$ weakly (in $L_2(Q_T)$)" that (3.8) holds for every $u \in \mathcal{U}$. Indeed, using the inequality 1.12 in Ladyženskaja et al. [6, p. 137], one can verify that

$$\left[\frac{B_i'(u_o + \varepsilon v) - B_i'(u_o)}{\varepsilon}\right]\psi \to B_{ju_o}'(v)\psi$$

strongly in $L_2(Q_T, R^n)$. Further, we know that $\phi_{x_i}(u_o + \varepsilon v) \to \phi_{x_i}(u_o)$ weakly in $L_2(Q_T, R^N)$. Therefore the third term in the above inequality converges to the corresponding term in the inequality (3.8) as $\varepsilon \downarrow 0$. Similar conclusions hold for the other terms. This completes the proof.

Note that for a fixed u_o the integrand in (3.8) is linear in $(v - u_o)$. Thus, there exists a function $G : Q_T \times U^N \times R^N \times (R^N \otimes R^N) \times R^N \times (R^N \otimes R^N) \to R^{rN}$ such that the left hand side of (3.8) is equal to

$$\int_{Q_T} G(x,t,u_o,\phi(u_o), \phi_x(u_o), \psi(u_o),\psi_x(u_o)) \cdot (v - u_o)dx\, dt$$

Using Theorem 3.5 one can obtain the following pointwise necessary condition.

Corollary 3.6 Suppose the hypotheses of Theorem 3.5 are satisfied, then for every regular point $(x,t) \in Q_T$ corresponding to the optimal control u_o

$$G(x,t,u_o(x,t), \phi(u_o)(x,t), \phi_x(u_o)(x,t), \psi(u_o)(x,t), \psi_x(u_o)(x,t)) \cdot (v - u_o(x,t)) \le 0 \tag{3.10}$$

for all $v \in U^N$.

Proof The proof is given as in Ahmed and Teo [1, p. 992].

The necessary condition (3.8) leads to a gradient method for computing the optimal control. The algorithm is presented in section 5. The pointwise necessary condition (3.10) gives an expression for the optimal control u_o in terms of ϕ, ϕ_x, ψ, ψ_x. In the linear quadratic problem considered in the following section, (3.10) gives rise to ordinary differential equations.

Remark 3.7 Problem P_o can be modified to the case $\Omega = R^n$ and $J(u) = \int_o^T f_o(\xi,t,y,u)dt$ as in [11]. The optimality conditions remain the same as in (3.8) and (3.10) where now $\phi(u_o)$ and ψ are weak solutions in $V_2^{1,1/2}(R^n \times I)$ (as defined in Ladyženskaja et al. [8, p. 171]) of the corresponding Cauchy problems.

The optimality condition for the original optimization problem P_1 as stated in Lemma 1.1 follows from Theorem 3.5. This is presented below.

Theorem 3.8 Suppose the following assumptions for the system (1.5) are satisfied.

(i) A_{ij} satisfies the assumption (A1).

(ii) For each $u \in \mathcal{U}$, $\bar{B}_i(u) \underline{\Delta} \partial A_{ij}/\partial x_j - B_i(u)$ and $F(u)$ are measurable functions on Q_T with values in $R^N \otimes R^N$ and R^N, respectively. Further, there exist functions $G \in L_{q,r}(Q_T)$ and $H \in L_{q_1,r_1}(Q_T)$ so that $|\bar{B}_i(u)| \leq G$ and $|F(u)| \leq H$ a.e. uniformly in $\mathcal{U}$.

(iii) The strong Gâteaux differentials $\bar{B}_{iu}(v)$ and $F_u(v)$ exist for every $u \in \mathcal{U}$ and are linear in $v \in \mathcal{U}$. These differentials belong to the same function space as $\bar{B}_i(u)$ and $F(u)$.

(iv) $\phi_T \in L_2(\Omega)$.

Let ψ be the weak solution of the adjoint system (2.7) corresponding to the control u_o. Then in order that u_o be the optimal control (existence assumed), it is necessary that

$$\int_{Q_T} [\bar{B}_{iu_o}(v - u_o)\phi_{x_i}(u_o) - F_{u_o}(v - u_o)]\psi \, dx \, dt \le 0 \quad (3.11)$$

for all $v \in \mathcal{U}$. Further, for each regular point $(x,t) \in Q_T$ corresponding to the measurable function represented by the integrand in (3.11),

$$[\bar{B}_{iu_o}(x,t)(v - u_o(x,t))\phi_{x_i}(u_o)(x,t)$$
$$- F_{u_o}(x,t)(v - u_o(x,t))]\psi(x,t) \le 0 \quad (3.12)$$

for all $v \in \mathcal{U}^N$.

§4. EXAMPLES

In this section we apply the necessary condition (3.12) (Theorem 3.8) to a linear quadratic regulator problem and discuss some examples. Consider the completely controllable linear dynamical system

$$dx(t) = [A(t)x(t) + B(t)u(t) + C(t)]dt$$
$$+ D(t)dw(t), \ t \in [0, T] \quad (4.1)$$

where $\{A(t), B(t), C(t), D(t)\}$ takes the value $\{A^m(t), B^m(t), C^m(t), D^m(t)\}$ when a homogeneous additive Markov chain $y(t)$ takes the value m, $m = 1, 2, \cdots, N$. $y(t)$ has the differential transition matrix $Q = \{q_{ij}\}$. The problem P_3 will be to choose a control $u(t,x,m)$ so as to minimize the quadratic cost criteria,

$$J(u) = E\int_0^T [x'Rx + u'Su + L(y)] \, dt \quad (4.2)$$

where $R \geq 0$, $S > 0$ depend on y and t, L(y) is a scalar function.

Corollary 4.1 The optimal control in the problem P_3 is given by

$$u^m = - S^{m^{-1}} B^{m'} \left(\frac{1}{2} H^m + P^m x\right)$$

and the conditional expected cost is given by

$$\phi^m = E\left\{\int_t^T [x'Rx + u'Su + L(y)]\, dx\, d\tau \,\middle|\, x(t) = x,\ y(t) = m\right\}$$

$$= p^m + H^{m'} x + x'P^m x$$

where p^m, H^m, P^m satisfy the following scalar, vector and matrix Riccati differential equations,

$$\dot{p}^m = - \operatorname{trace}(D^{m'} P^m D^m) - C^{m'} H^m + \frac{1}{4} H^{m'} B^m S^{m^{-1}} B^{m'} H^m$$

$$- q_{mi} p^i - L(m)$$

$$\dot{H}^m = (P^m B^m S^{m^{-1}} B^{m'} - A^{m'}) H^m - 2P^{m'} C^m - q_{mi} H^i$$

$$\dot{P}^m = - P^m A^m - A^{m'} P^m + P^m B^m S^{m^{-1}} B^{m'} P^m - R^m - q_{mi} P^i \quad (4.3)$$

Proof By remarks 3.7, we can apply the optimality condition (3.12). Noting that $\psi(x,t) > 0$ for every $(x,t) \in R^N \times I$, we obtain

$$u^m = - \frac{1}{2} S^{m^{-1}} B^{m'} \phi_x^m, \quad m = 1, 2, \cdots, N \quad (4.4)$$

Assuming a quadratic form

$$\phi^m(x,t) = p^m(t) + H^m(t) x + x'P^m(t) x \quad (4.5)$$

Substituting it into (1.5) and grouping terms with the same

power in x, we arrive at (4.3). The expression for the optimal control u^m is easily obtained from (4.4) and (4.5).

The result of a linear quadratic regulator problem considered by Wonham [13, p. 192] follows as a special case of Corollary 4.1. A housing policy problem was considered by Kzangey and Sworder [6, pp. 1120-1128]. Another problem of control of plant dynamics, repair and inventory was also considered by the same authors [12, p. 342], who used the dynamic programming approach. The differential equations they obtain are covered in (4.3) of Corollary 4.1.

§5. COMPUTATION ALGORITHM

The pointwise necessary condition (3.10) can be used to obtain an expression of an extremal control in terms of ϕ, ϕ_x, ψ, and ψ_x. The resulting computation problem is converted into solving the two point boundary value problem (3.1) and (3.4) with u replaced by the expression of extremal control as obtained above. In general, this procedure is difficult to handle. In this section we use the necessary condition to obtain a gradient method for computation purposes.

Let G be as defined in section 3. From the necessary condition (3.8), it is clear that

$$\int_{Q_T} G(x,t,v,\phi(v),\ \phi_x(v),\ \psi(v),\ \psi_x(v))v\ dx\ dt$$

has to be maximized with respect to $v \in \mathcal{U}$. The function G will provide the gradient of the cost functional J of the problem P_2. The following theorem is needed to justify the subsequent algorithm.

Theorem 5.1 Suppose the hypotheses of Theorem 3.5 are satisfied. Then the gradient $J_u(v)$ of the cost functional J evaluated at u in the direction v is given by

$$J_u(v) = \int_{Q_T} G(x,t,u,\phi(u),\phi_x(u),\psi(u),\psi_x(u))v\, dx\, dt \tag{5.1}$$

Let $G^i = G(x,t,u^i,\phi(u^i),\phi_x(u^i),\psi(u^i),\psi_x(u^i))$. Then the sequence of controls $\{u^i\}$ generated by the algorithm

$$u^{i+1} \triangleq \begin{cases} u^i - \rho_i G^i/|G^i| & \text{for } \mathcal{U} \subset L_\infty \\ u^i - \rho_i G^i & \text{for } \mathcal{U} \subset L_2 \end{cases} \tag{5.2}$$

converges for any sequence of sufficiently small positive numbers ρ_i to an extremal control.

Proof Replacing u_o by $u \in \mathcal{U}$ in (3.9) and taking the limit as $m \to \infty$, we obtain from Lemma 3.4

$$\int_\Omega [\phi(u + \varepsilon v) - \phi(u)]\psi(u)dx\Big|_{t=o} = <(L(u + \varepsilon v)$$

$$- L(u))\phi(u + \varepsilon v),\psi(u)> + <F(u + \varepsilon v) - F(u),\psi(u)>$$

The expression (5.1) now follows from dividing both sides of the above equality by ε and taking limit as $\varepsilon \downarrow 0$ (same as in Theorem 3.5). Using Taylor's expansion

$$J(u^{i+1}) = J(u^i) + J_{u^i}(u^{i+1} - u^i) + o(\|u^{i+1} - u^i\|) \tag{5.3}$$

Assuming $\|u^{i+1} - u^i\|$ to be small, it follows from the Hölder's inequality that $J(u^{i+1}) - J(u^i)$ is minimized when

$$u^{i+1} = \begin{cases} u^i - \rho_i G^i/|G^i|, & \mathcal{U} \subset L_\infty \\ u^i - \rho_i G^i, & \mathcal{U} \subset L_2 \end{cases} \tag{5.4}$$

for a sufficiently small positive number ρ_i.

Note that for a fixed but arbitrary $u \in \mathcal{U}$ ($\subset L_\infty$ or L_2), $G^i v$ considered as a function on Q_T to R is in L_1. It follows that the integral $J_{u^i}(v)$ is well defined. Also note that if $\mathcal{U}$ is considered to be a w* compact subset of L_∞ (or L_2), then J_{u^i} is a w* continuous functional on $\mathcal{U}$. Consequently, J_{u^i} has a maximum and a minimum on $\mathcal{U}$.

Using the iteration scheme (5.4), it follows from (5.3) that

$$J(u^{i+1}) = J(u^i) - \rho_i \|J_{u^i}\| + o(\rho_i) < J(u^i) \tag{5.5}$$

where

$$\|J_{u^i}\| = \begin{cases} \int_{Q_T} |G^i| dx\, dt, & \mathcal{U} \subset L_\infty \\ \int_{Q_T} |G^i|^2 dx\, dt, & \mathcal{U} \subset L_2 \end{cases}$$

This completes the proof.

Note that Gv is required to be L_1. Hence the function spaces for $A_{ju}(v)$, $B_{ju}(v)$, $\cdots$, $F_u(v)$ (and hence the corresponding coefficients, see (A4)) and the control space $\mathcal{U}$ is chosen so that $Gv \in L_1$.

Based on Theorem 5.1, we now present the algorithm for computing the optimal control.

Algorithm 5.2

Step 1: Provide an initial guess $u^o \in \mathcal{U}$ and a small positive number ρ_o.

Step 2: Solve equations (3.1) and (3.4) to obtain $\phi(u^i)$ and $\psi(u^i)$, $i = 0, 1, 2, \cdots$.

Step 3: Compute $J(u^i) = \int_\Omega \phi(u^i)(x,o)\psi_o(x)dx$.

Step 4: If $J(u^i) < J(u^{i-1})$, let $\rho_i = \rho_{i-1}$. If $J(u^i) > J(u^{i+1})$, let $\rho_i = \rho_{i-1}/a$, $0 < a < 1$. If $|J(u^i) - J(u^{i-1})| < \varepsilon$ for some small preassigned number ε, iteration stops.

Step 5: Obtain the next admissible control by (5.2) and repeat step 2.

ACKNOWLEDGMENT

The work presented in this paper was supported by the National Research Council of Canada under Grant No. A-7109.

REFERENCES

1. Ahmed, N. U. and K. L. Teo, Necessary Conditions for Optimality of Cauchy Problems for Parabolic Differential Systems, SIAM J. Control, vol. 13, no. 5 (1975), pp. 981-993.

2. Ahmed, N. U. and K. L. Teo, Optimal Control of Stochastic Ito Differential Systems with Fixed Terminal Time, Advances in Appl. Prob., vol. 7, no. 1 (1975), pp. 154-178.

3. Aronson, D. G., Non-negative Solutions of Linear Parabolic Equations, Ann. Scuola Norm. Sup. Pisa, 22 (1968), pp. 607-694.

4. Fleming, W. H., Optimal Control of Partially Observable Diffusions, SIAM J. Control, 6 (1968, pp. 194-214.

5. Friedman, A., Partial Differential Equations of Parabolic Type, Prentice-Hall, Englewood Cliffs, N. J., 1974.

6. Kazangey, T. and D. D. Sworder, Effective Federal Policies for Regulating Residential Housing, Proc. 1971 Summer Computer Simulation Conference, pp. 1120-1128.

7. Krasvoskii, N. N. and E. A. Lidskii, Analytic Design of Controllers in Stochastic Systems with Velocity Limited Controlling Action, Appl. Math. Mech., 25 (1961), pp. 627-643.

8. Ladyženskaja, O. A., V. A. Solonnikov, and N. N. Ural'ceva, Linear and Quasi-linear Equations of Parabolic Type, Translation of Mathematical Monographs, Amer. Math. Society, Providence, 1968.

9. Lidskii, E. A., Optimal Control of Systems with Random Properties, Appl. Math. Mech., 27 (1963), pp. 33-45.

10. Rishel, R., Dynamic Programming and Minimum Principles for Systems with Jump Markov Disturbances, SIAM J. Control, 13 (1975), pp. 338-371.

11. Sworder, D. D., Feedback Control of a Class of Linear Systems with Jump Parameters, IEEE Trans. Automatic Control, AC-14 (1969), pp. 9-14.

12. Sworder, D. D. and T. Kazangey, Optimal Control, Repair and Inventory Strategies for a Linear Stochastic System, IEEE Trans. Systems, Man, and Cybernetics, SMC-2, No. 3 (1972), pp. 342-347.

13. Wonham, W. M., Random Differential Equations in Control Theory, Probability Methods in Applied Mathematics, A. T. Bharucha-Reid, ed., Academic Press, New York, 1970, pp. 191-199.

CONTROLLABILITY OF NONLINEAR SYSTEMS WITH RESTRAINED CONTROLS TO CLOSED CONVEX SETS

Ethelbert N. Chukwu and Jan M. Gronski

Cleveland State University
Cleveland, Ohio

ABSTRACT

The study deals with the nonlinear control system

$$\dot{x} = f(t,x,u) \tag{1}$$

on $[t_0,\infty) \times E^n \times E^m$, where f is continuous and the control function $u : [t_0,\infty) \to E^m$ are measurable and $u(t) \in \Omega(t)$ for $t \geq t_0$. Here Ω is a continuous multifunction on $[t_0,\infty)$ taking its values in nonempty, compact subsets of E^m. First, we characterize the G-controllability of (1) by means of a growth condition. As a consequence it is shown that all points of E^n can be attained from G, a closed and convex subset of E^n, if and only if (1) is G-controllable. Finally, (1) is specialized to

$$\dot{x} = A(t)x + k(t,u) \tag{2}$$

and its perturbation

$$\dot{x} = A(t)x + k(t,u) + g(t,x,u) \tag{3}$$

If G is symmetric about 0 and is positively invariant with respect to solutions of $\dot{x} = A(t)x$ then (2) is G-controllable if and only if (3) is approximate G-controllable provided g is integrably bounded.

§1. INTRODUCTION

This paper considers the nonlinear control system

$$\frac{dx}{dt} = f(t,x,u) \tag{1}$$

on $[t_0,\infty) \times E^n \times E^m$, where f is continuous and the control functions $u : [t_0,\infty) \to E^m$ are measurable and $u(t) \in \Omega(t)$ for $t \geq t_0$. Here Ω is a continuous multifunction on $[t_0,\infty)$ taking values in nonempty, compact subsets of E^m which contain the origin. We shall also deal with special cases of (1), namely

$$\frac{dx}{dt} = A(t)x + k(t,u) \tag{2}$$

and its perturbation

$$\frac{dx}{dt} = A(t)x + k(t,u) + g(t,x,u) \tag{3}$$

where $k(t,0) \equiv 0$, $g(t,x,0) \equiv 0$ and f and g are continuous for $t \geq t_0$, $x \in E^n$ and $u \in E^m$.

Let G be a subset of E^n. The system (1) is G-controllable if for any $x_0 \in E^n$ there exists a measurable function $u : [t_0,\infty) \to E^m$ and a time $t_1 \in [t_0,\infty)$ such that $u(t) \in \Omega(t)$ for $t \geq t_0$ and the solution of

$$\dot{x} = f(t,x,u(t)) \text{ a.e. on } [t_0,t_1]$$
$$x(t_0) = x_0 \tag{4}$$

satisfies $x(t_1) \in G$. The system is approximately G-controllable if given any $x_0 \in E^n$ and any $\varepsilon > 0$ there exists a measurable $u : [t_0,\infty) \to E^n$ and a time $t_1 \in [t_0,\infty)$ such that $u(t) \in \Omega(t)$ for $t \geq t_0$ and the solution $x(t)$ of (4) satisfies

$$d(x(t_1),G) < \varepsilon$$

where $d(x(t_1),G) = \inf\{|x(t_1) - p| : p \in G\}$. We say that the system (1) can attain every point of E^n from G if given any

$x_1 \in E^n$, there exists a measurable $u : [t_0,\infty) \to E^n$, a time $t_1 \in [t_0,\infty)$ and an $x_0 \in G$ such that $u(t) \in \Omega(t)$ for $t \geq t_0$ and the solution of (4) satisfies $x(t_1) = x_1$.

First, we shall characterize the G-controllability of (1) by means of a growth condition. This then extends such a characterization of (2) in Chukwu and Silliman [1]. As an immediate consequence it is shown that the system (1) can attain every point of E^n from G if and only if (1) is G-controllable.

Let X(t) denote the fundamental matrix solution of $\dot{x} = A(t)x$ such that $X(t_0)$ is the identity matrix. When dealing with the systems (2) and (3) we shall assume as basic that

$$X^{-1}(t)G \subseteq X^{-1}(t')G \text{ for all } t_0 \leq t \leq t' \tag{5}$$

where G is a closed and convex target set. It is clear that condition (5) which first appeared in [1] implies that

$$X(t',t)G \subseteq G$$

where $X(t')X^{-1}(t) = X(t',t)$ is a fundamental matrix of $\dot{x} = A(t)x$ with $X(t,t) = I$, the identity matrix. That is, G is positively invariant with respect to solutions $x(t')$ of $\dot{x} = A(t)x$, $t \leq t'$. This assumption is obviously weaker than the stability of the set G with respect to $x = A(t)x$, [2].

Recall that the function $h(t,x,u)$ is integrably bounded on $[t_0,\infty) \times \Omega$ if there exists a function $\lambda(t)$ which is integrable on $[t_0,\infty)$ and such that $|h(t,x,u)| \leq \lambda(t)$ for $t \geq t_0$, $x \in E^n$, $u \in \Omega(t)$. For the class of perturbations g where $X^{-1}(t)g(t,x,u)$ is integrably bounded on $[t_0,\infty) \times \Omega$ it was shown in [1] that if (5) holds and if the set

$$\{h(t,u) + g(t,x,u) : u \in \Omega(t)\}$$

is convex, then (3) is G-controllable if and only if (4) is G-controllable. We now remove this convexity assumption, and impose a Lipschitz condition in x on g to deduce that (3) is G-controllable if and only if (4) is approximately G-controllable.

It is interesting to compare the above result with the complete controllability treatment of Dauer [3]. In [3] condition (5) is missing and instead, $X^{-1}(t)$ is assumed to be uniformly bounded on $[t_0,\infty)$, a hypothesis which was fundamental in the treatment (see also [5, p. 134]). This assumption can be omitted when dealing with null-controllability. Since G is convex and contains zero the question naturally arises whether there is a class of systems which satisfy (5) for which controllability into G is not an immediate consequence of null-controllability. An example is given to show that there is indeed such a class. Since the invariance of G with respect to $\dot{x} = A(t)x$ is far weaker than the asymptotic stability of the zero solution of $\dot{x} = A(t)x$, it is clear that condition (5) is not very restrictive. Indeed, it has appeared in another guise in [6] as a necessary and sufficient condition for the continuity of the minimal time function for the system

$$\dot{x} = Ax + u$$

with target G.

<u>Proposition 1</u> Let G be a fixed convex subset of E^n which contains zero. Assume that in (1), $f(t,x,0) = 0$ and that the set

$$f(t,x,\Omega) = \{f(t,x,u(t)) : u(t) \in \Omega(t)\}$$

is convex for every $x \in E^n$ and $t \geq t_0$. Then (1) is

G-controllable if and only if the following growth condition holds: For every $\varepsilon > 0$ and each nonzero vector $\eta \in E^n$ there exists a measurable function u, satisfying $u(t) \in \Omega(t)$ for all $t \geq t_0$ and there exist $p \in G$ and $t_1 \geq t_0$ such that

$$\eta^T\left\{\int_{t_0}^{t_1} f(t,x(t),u(t))dt - p\right\} \geq \varepsilon \tag{6}$$

Proof Let $\varepsilon > 0$ and let a nonzero $\eta \in E^n$ be given. Choose $x_0 \in E^n$ so that

$$\eta^T(-x_0) \geq \varepsilon$$

Now assume that (1) is G-controllable and let $t_1 \geq t_0$ and $u(t) \in \Omega(t)$ be such that the solution $x(t) = x(t;t_0,x_0,u)$ of (1) satisfies $x(t_1) \in G$. Then for some $p \in G$

$$p = x_0 + \int_{t_0}^{t_1} f(t,x(t),u(t))dt$$

so that

$$-x_0 = \int_{t_0}^{t_1} f(t,x(t),u(t)dt - p$$

It follows from this that

$$\eta^T(-x_0) = \eta^T\left\{\int_{t_0}^{t_1} f(t,x(t),u(t))dt - p\right\} \geq \varepsilon$$

Conversely, consider the set

$$s(t) = \int_{t_0}^{t} f(\tau,x(\tau),\Omega(\tau))d\tau - G \tag{7}$$

The integral evidently contains zero and is convex. Since G also contains zero and is convex, so does s(t). Because G is fixed, s(t) is increasing in t; that is, $s(t) \subseteq s(t')$ for

$t_0 \le t \le t'$. Therefore $\bigcup_{t \ge t_0} s(t)$ is convex and contains zero. We now assume that for every $\varepsilon > 0$ and each nonzero vector $\eta \in E^n$, there exists a bounded measurable u, satisfying $u(t) \in \Omega(t)$, for all $t \ge t_0$, $p \in G$ and $t_1 \ge t_0$ such that (6) holds. It follows from [7, Lemma 4, p. 7] that

$$\bigcup_{t \ge t_0} s(t) = E^n$$

Take any $x_0 \in E^n$, there exists $t_1 \ge t_0$ such that

$$-x_0 \in s(t_1) = \int_{t_0}^{t_1} f(\tau, x(\tau), \Omega(\tau))d\tau - G$$

Therefore, for some u, $u(\tau) \in \Omega(\tau)$, and some $p \in G$, we have

$$p = x_0 + \int_{t_0}^{t_1} f(\tau, x(\tau), u(\tau)d\tau = x(t_1; t_0, x_0, u)$$

But then this implies that $x(t_1) \in G$, concluding the proof.

Remark If G is a symmetric about 0, we can assume s in (7) to be of the form

$$s(t) = \int_{t_0}^{t} f(\tau, x(\tau), \Omega(\tau))d\tau + G \tag{8}$$

Our proof of Proposition 1 has then inadvertently shown a little more than the proposition had announced: If (6) is satisfied, the system (1) can attain every point of E^n from G. A natural question to ask is whether the converse holds. Let us denote by $A(G, t_1)$ the set of all points x in E^n such that there exists a point in G which can be steered to x in time t_1. Let

$$A(G) = \bigcup_{t \ge t_0} A(G, t)$$

We observe that the system (1) can attain every point of E^n

from G if and only if $A(G) = E^n$. The next proposition will answer the question posed above.

<u>Proposition 2</u> Let G be a fixed convex subset of E^n which is symmetric about zero. Assume that in (1), $f(t,x,0) = 0$ and that the set

$$f(t,x,\Omega) = \{f(t,x,u(t)) : u(t) \in \Omega(t)\}$$

is convex for every $x \in E^n$ and $t \geq t_0$. Then system (1) can attain every point of E^n from G if and only if condition (6) is satisfied.

<u>Proof</u> We need to show that condition (6) is true if and only if

$$A(G) = E^n$$

But then

$$A(G) = \bigcup_{t \geq t_0} s(t)$$

where s(t) is as in (8), and we have shown that condition (6) implies $A(G) = E^n$. For the converse, for any nonzero vector η and for any $\varepsilon > 0$, choose $x_0 \in E^n$ such that $\eta^T(-x_0) \geq \varepsilon$. Assume $A(G) = E^n$. Then $x_0 \in A(G)$ implies there exists a $t_1 \geq t_0$, $p \in G$ and $u : I \to E^m$ with $u(t) \in \Omega(t)$ such that

$$x_0 = p + \int_{t_0}^{t_1} f(s,x(s),u(s))ds$$

$$= -(-p) + \int_{t_0}^{t_1} f(s,x(s),u(s))ds$$

Since G is symmetric $p' = -p \in G$. Hence

$$\eta^T x_0 = \eta^T \left\{ \int_{t_0}^{t_1} f(t,x(t),u(t)) - p' \right\} \geq \varepsilon$$

which concludes the proof.

Corollary 1 Suppose G and f are as in Proposition 2, then system (1) can attain every point of E^n from G if and only if it is G-controllable.

The following results which are proved in [1] are needed in the sequel. Proposition 3 is generalized above in Proposition 1.

Proposition 3 Let G be a fixed closed and convex subset of E^n which is symmetric about 0, and assume that $X^{-1}(t)G \subseteq X^{-1}(t')G$ for all t',t with $t_0 \le t \le t'$. Further assume that the set

$$\{k(t,u) : u(t) \in \Omega(t),\ t \ge t_0\}$$

is convex and that $k(t,0) = 0$. A necessary and sufficient condition for (2) to be G-controllable is the following growth condition: For every $\varepsilon > 0$ and each nonzero vector $\eta \in E^n$ there exists $u : I \to E^m$, $u(t) \in \Omega(t)$, $t \ge t_0$, $p \in G$ and $t \ge t_0$ such that

$$\eta T\left\{\int_{t_0}^{t} X^{-1}(s)k(s,u(s))ds - X^{-1}(t)p\right\} \ge \varepsilon$$

The above proposition can be improved: The convexity of $\{k(t,u) : u(t) \in \Omega(t),\ t \ge t_0\}$ can be removed. It was only used to show that

$$\int_{t_0}^{t} X^{-1}(s)k(s,\Omega(s))ds$$

is convex. But the above integral is convex by Theorem 1 of [8].

Proposition 4 Assume that $X^{-1}(t)g(t,x,u)$ is integrably bounded on $[t_0,\infty) \times \Omega$ and that (5) holds. Suppose (2) is G-controllable. Let $x_0 \in E^n$. Then there exists a time $t_1 \geq t_0$ with the following property: Given any continuous function $w : [t_0,\infty) \to E^n$ there exists $u : I \to E^n$, $u(t) \in \Omega(t)$, $t \geq t_0$ and a solution ψ of

$$\dot{x} = A(t)x + k(t,u(t)) + g(t,w(t),u(t)), \quad x(t_0) = x_0$$

which satisfies $\psi(t_1) \in G$.

Theorem 1 Let G be closed, convex and symmetric about 0. Assume that $g(t,x,0) = 0$, g satisfies a Lipschitz condition in x and $X^{-1}(t)g(t,x,u)$ is integrably bounded on $[t_0,\infty) \times \Omega$. Assume that X and G satisfy the condition, $X^{-1}(t)G \subseteq X^{-1}(t')G$ for all $t_0 \leq t \leq t'$. Then (2) is G-controllable if and only if (3) is approximately G-controllable.

Proof Suppose (2) is G-controllable and let $x_0 \in E^n$. It follows from Proposition 4 that there exist a $t_1 \geq t_0$, a sequence of measurable functions $u_1, u_2, \cdots$, with $u_i(t) \in \Omega(t)$ for all $t \geq t_0$, and a sequence of continuous functions $w_i : [t_0,\infty) \to E^n$, $i = 1, 2, \cdots$ such that

$$w_{i+1}(t) = X(t)\left[x_0 + \int_{t_0}^{t} X^{-1}(s)[k(s,u_i(s)) + g(s,w_i(s),u_i(s)]ds\right]$$

$$w_{i+1}(t_0) = x_0 \qquad w_{i+1}(t_1) \in G$$

for each $i = 1, 2, \cdots$. Also the sequence $\{w_i\}$ can be chosen so that it is uniformly bounded on $I = [t_0,t_1]$. Let ρ be such that $w_i(t) \in S^n_\rho(x_0)$ for $i = 1, 2, \cdots$, where $S^n_\rho(x_0)$ is a ball of radius ρ and center x_0. It now follows that there exists a $p(t) \in L^1(I)$ such that

$$\lim_{i\to\infty} \int_{t_0}^{t} X^{-1}(s)[k(s,u_i(s)) + g(s,w_i(s),u_i(s)]ds$$
$$= \int_{t_0}^{t} p(s)ds$$

for each $t \in I$. Now set $\bar{w}(t) = X(t)\left[x_0 + \int_{t_0}^{t} p(s)ds\right]$ then

$$\lim_{i\to\infty} w_i(t) = \bar{w}(t) \text{ for all } t \in I$$

Note that $\bar{w}(t_0) = x_0$ and because G is closed $\bar{w}(t_1) \in G$. Let

$$R(t,x) = \{X^{-1}(t)[k(t,u(t)) + g(t,x,u(t)) : u(t) \in \Omega(t)\}$$

R is a multifunction which is nonempty, closed and continuous in t and x. It is uniformly bounded on the compact set $I \times S_\rho^n(x_0)$. Let $H(t,x)$ denote the closed convex hull of the set $R(t,x)$. It follows from a theorem of Wazewski's [9] that the limit $\bar{w}$ of the convergent sequence of solutions w_i of

$$\dot{x}(t) \in A(t)x + R(t,x(t))$$
$$\dot{x}(t_0) = x_0 \qquad x(t_1) \in G$$

is a solution of

$$\dot{x}(t) \in A(t)x + H(t,x(t))$$
$$x(t_0) = x_0 \qquad x(t_1) \in G$$

for all $(t,x) \in I \times S_\rho^n(x_0)$.

Since $R(t,x)$ is continuous and bounded for all $t \in I$, $x \in S_\rho^n(x_0)$, we have from [10, Theorem 3] that given $\varepsilon > 0$ there exists an absolutely continuous function ϕ satisfying equation

$$\dot{x}(t) \in A(t)x(t) + R(t,x(t)) \text{ a.e. on } I$$

$$x(t_0) = x_0$$

with

$$\max_{t \in I} |\phi(t) - \bar{w}(t)| < \varepsilon$$

From Filippov's Lemma [11, p. 78] there exists a measurable function u with $u(t) \in \Omega(t)$, $t \geq t_0$ such that

$$\dot{\phi}(t) = A(t)\phi(t) + k(t,u(t)) + g(t,\phi(t),u(t)) \text{ a.e. on } I$$

with $\phi(t_0) = x_0$ and

$$\max_{t \in I} |\phi(t) - \bar{w}(t)| < \varepsilon$$

But since $\bar{w}(t_1) \in G$ this last inequality implies that

$$d(\phi(t_1),G) < \varepsilon$$

proving approximate G-controllability.

For the converse, assume that (2) is not G-controllable. Then by Proposition 3, there exists $\varepsilon > 0$, some $\eta \neq 0$, $\eta \in E^n$ such that for all $u : [t_0,\infty) \to E^n$, all $p \in G$, all $t \geq t_0$

$$\eta^T\left\{\int_{t_0}^{t} X^{-1}(s)k(s,u(s))ds - X^{-1}(t)p\right\} < \varepsilon$$

It follows from the hypotheses in $X^{-1}(t)g(t,x,u)$ that there exists a number N such that

$$\left|\int_{t_0}^{t} X^{-1}(s)g(s,w(s),u(s))ds\right| \leq N$$

for each continuous $w : [t_0,\infty) \to E^n$, each $u(s) \in \Omega(s)$ and each $t \geq t_0$. Let $\lambda > 0$ and choose $x_0 \in E^n$ such that

$$-\eta^T x_0 > \varepsilon + N + \lambda$$

Assume (3) is approximately G-controllable. Then for the x_0 and for λ above, there exists a measurable $u : [t_0, \infty) \to E^n$, such that $u(t) \in \Omega(t)$ for all $t \geq t_0$, a time $t_1 \in [t_0, \infty)$, and an $q \in E^n$ such that the solution $x(t)$ of (3) satisfies

$$x(t_1) - p = q$$

for some $p \in G$ where $|q| < \lambda / |X^{-1}(t_1)|$. From the variation of parameter formula,

$$X(t_1)\left[x_0 + \int_{t_0}^{t_1} X^{-1}(s)[k(s,u(s)) + g(s,x(s),u(s)]ds\right]$$
$$- p = q$$

from which

$$-x_0 = \int_{t_0}^{t_1} X^{-1}(s)[k(s,u(s)) + g(s,x(s),u(s))\, ds]$$
$$- X^{-1}(t_1)[p + q]$$

From this follows the estimate

$$-\eta^T x_0 < \varepsilon + N + |X^{-1}(t_1)||q| < \varepsilon + N + \lambda$$

This contradicts the choice of x_0, proving the converse.

Example 1

$$\dot{\underline{x}} = \begin{bmatrix} -2 & 0 \\ 0 & -1 \end{bmatrix} \underline{x} + \begin{bmatrix} 1 & 1 \\ 1 & 2 \end{bmatrix} \underline{u} + \begin{bmatrix} e^{-3t}\sin u_1 x_1 \\ e^{-2t}\cos(x_1 x_2 u_2 + \pi/2) \end{bmatrix}$$

$$u : |u_i| \leq 1, \; t = 0$$

$$G = \{(x_1, x_2) : |x_1| \leq 1 \; |x_2| \leq 1\}$$

Note that by Corollary 2.2 of [1] the base system is G-controllable. The function $g(t,x,\underline{u}) = [(e^{-3t}\sin u_1x_1,\ e^{-2t}\cos(x_1x_2u_2 + \pi/2)]^T$ satisfies $g(t,x,0) = 0$ and is Lipschitz. $X^{-1}(t)g(t,\underline{x},\underline{u}) = [(e^{-t}\sin x_1u_1,\ e^{-t}\cos(x_1x_2u_2 + \pi/2)]^T$ is integrably bounded on $[0,\infty) \times \Omega$. Set G is closed, convex and symmetric about 0 and $e^{-At}G = \{(x_1,x_2) : -e^{2t} \le x_1 \le e^{2t},\ -e^t \le x_2 \le e^t\}$ satisfies $e^{-At}G \subseteq e^{-At'}G$ for $0 \le t \le t'$. By Theorem 1, the system is approximately G-controllable.

Example 2 Consider the system

$$\dot{x}_1 = -2x_1 + u_1 + u_2 + e^{-3t-u^2}\sin u_1x_1$$

$$\dot{x}_2 = -x_2 + e^{-2t}\cos(x_1x_2u_2 + \pi/2)$$

where $|u_i| \le 1$, $t = 0$; $G = \{(x_1,x_2) : |x_1| \le 1\ |x_2| \le 1\}$

Here

$$A = \begin{bmatrix} -2 & 0 \\ 0 & -1 \end{bmatrix} \quad k(t,u) = \begin{bmatrix} 1 & 1 \\ 0 & 0 \end{bmatrix} u$$

$$g = (e^{-3t-u^2}\sin u_1x,\ e^{-2t}\cos x_1x_2u_2)^T$$

We note that the controllability space $\{A|B\} = \text{span}\ \{[1,0]^T\} = \text{span}\ [B,\ AB]$. Thus our system is not null controllable. But

$$\{A|B\} + \bigcup_{t>0} e^{-At}G = E^2,$$

since $e^{-At}G = \{(x_1,x_2) : -e^{2t} \le x_1 \le e^{2t},\ -e^t \le x_2 \le e^t\}$

It follows from Corollary 2.2 of [1] that the base system is G-controllable. Since g and $X^{-1}(t)g(t,\underline{x},u)$ satisfy all the required conditions and since

$$e^{-At}G \subseteq e^{-At'}G \text{ for } 0 \le t \le t'$$

we conclude that the above system is approximately G-controllable.

REFERENCES

1. Chukwu, E. N. and S. D. Silliman, Complete Controllability to a Closed Target Set, J. Optimization Theory Appl., to appear. Technical Report CSUMD-42, Department of Mathematics, Cleveland State University, Cleveland, Ohio 44115, August 1975.

2. Yoshizawa, T., Stability of Sets and Perturbed System, Funkcialaj Ekuacioj, 5 (1962), pp. 33-69.

3. Dauer, J. P., Approximate Controllability of Nonlinear Systems with Restrained Controls, J. Math. Anal. Appl, 46 (1974) pp. 126-131.

4. Dauer, J. P., Erratum: Approximate Controllability of Nonlinear Systems with Restrained Controls, to appear.

5. Dauer, J. P., On Controllability of Systems of the Form $\dot{x} = A(t)x + g(t,u)$, J. Optimization Theory and Appl, 11 (1973), pp. 132-138.

6. Hsu, B. H., On the Continuity of the Minimal Time Function, Technical Report, Case Western Reserve University, Cleveland, Ohio (1973).

7. LaSalle, J. P., The Time-Optimal Control Problem, Contributions to Nonlinear Oscillation, vol. 5, Princeton University Press, Princeton, New Jersey, 1960.

8. Neustadt, L. W., The Existence of Optimal Controls in the Absence of Convexity Conditions, J. Math. Anal. Appl., 7 (1963), pp. 110-117.

9. Wazewski, "Sur une généralisation de la notion des solutions d'une équation au contingent, Bull. Acad. Polon Sci. Ses. Sci. Math Astronom. Phys., 10 (1962), pp. 11-15.

10. Filippov, A. F., Classical Solutions of Differential Equations with Multivalued right-hand side," SIAM J. Control, 5 (1967), pp. 607-627.

11. Filippov, "On Certain Questions in the Theory of Optimal Control, SIAM J. Control, 1 (1962), pp. 76-84.

A THREAT-RECIPROCITY CONCEPT FOR PURSUIT/EVASION

H. J. Kelley

Analytical Mechanics Associates, Inc.
Jericho, New York

§1. INTRODUCTION

Combat between fighter aircraft of roughly equal performance is difficult to analyze even when the roles of pursuer and evader are clearly determined, but the hardest decisions arise when there is role ambiguity. This situation is examined in the following, assuming, for the sake of argument, that acceptable approximations to optimal strategies have already been worked out for the one-on-one case with the combatants designated pursuer and evader, each in turn. The boundaries separating successful pursuit from successful evasion in the role-designated strategy results, the so-called barrier surfaces, may be thought of as generalized capture envelopes, although one must bear in mind that these are surfaces in the <u>joint</u> pursuer/evader state space.

§2. THE DRAW REGION

Attention is directed to what might be termed the "draw" region of the joint state space, in which both participants have the option of disengaging, i.e., there is sufficient margin of one kind or another to permit <u>either</u> vehicle electing the role of evader to get away safely. This margin could take the form of separation distance for a vehicle with superior maximum speed, energy difference for a craft with higher ceiling, or favorable angular geometry for a more maneuverable vehicle. It is not clear, even for a simulator environment where the heroic is commonplace, that seizing the initiative and taking on the pursuer's role is best in this draw region for either player. Maneuvering for position without actually launching an attack is attractive, and a rational approach to this could be based on each player attempting to drive the point representing the current state in the joint state space toward that player's generalized capture envelope.

§3. STRATEGIES IN THE DRAW SPACE

Describe the generalized capture envelope by $Q_1 = 1$, where Q_1 is a function of the joint state and $Q_1 < 1$ implies that Player 1 can capture his opponent by assuming the role of pursuer. Values of $Q_1 > 1$ are of interest, with the interpretation of Q_1 as a sort of generalized closest-approach distance. If Q_1 exceeds unity, however slightly, Player 2 can insure that this continues by playing pure evasion. If there is more than a thin margin, however, a compromise strategy in the draw space that attempts to reduce Q_2, his own generalized miss as a potential attacker, while maintaining a defensive margin $Q_1 > 1$, is a logical consideration.

If one assumes (rashly) differentiability of the Q-functions, then each of the players may wish to extremize a linear combination of the miss rates $\dot{Q}_1$ and $\dot{Q}_2$ with respect to his controls under some assumption about the other player's actions. Generally, the weighting factors in the linear combinations chosen for decision making by the two players will be different, depending on degrees of conservatism in defensive margins. If each player were to insist on nonnegativity of the other's miss rate, the resulting control policies would both come out pure evasion. A less defensive policy could be obtained rationally, either by extrapolation of an opponent's nonoptimal control histories or by relaxing the constraint on the opponent's miss rate, thus leading to a less dominant weighting factor on the term in the linear combination. An analogous scheme applied to the increments ΔQ_i over some finite time interval in the "plan-ahead" spirit has greater appeal but is more difficult to treat. Smoothness considerations alone may require this formulation, however—recall that there are jump discontinuities in the value function for the "homicidal chauffeur" game.

One expects that there may be some equilibrium situations in which $\dot{Q}_1 = \dot{Q}_2 = 0$ and the linear combinations chosen for extremization by the two players are identical. These should be relatively easy to determine and perhaps illuminating as concerns the structure of stand-offs.

§4. TWO-ON-ONE

If Vehicles 1 and 2 are in coalition against Vehicle 3, there are then three terms in each of the two linear combinations extremized by the two players; however, one player now has two sets of controls available. This should, of course, increase his effectiveness, except in presumably temporary situations when his two vehicles are in nearly identical states. If one of the two team vehicles is at once the more threatening and the more threatened, the situation will momentarily have the character of one-on-one, with the second team vehicle available for redeployment. When the two team-mates pose equal threats, the coalition is really working; the composite threat as seen by the opponent will be characterized by discontinuous partial derivatives of max $[Q_{31}, Q_{32}]$ which make decisions awkward. The arrangement in which the team vehicles divide responsibility, one as the more threatening and the other as the more threatened (bait?) has interesting aspects also.

§5. CONCLUDING REMARKS

The threat-reciprocity concept proposed for consideration may seem somewhat academic in that barrier surfaces, or extended capture envelopes, are not within easy computational reach for air-to-air engagements with realistic modelling. Nonetheless, the role-determination and threat-reciprocity problems appear to be central in air-combat decision making and deserve more attention than so far accorded. A notable exception is the investigation of Ref. 1, which uses 2-D constant-speed vehicle models in a differential-game formulation.

While the gaming approach does not seem attractive for the immediate future with realistic point-mass or "energy" vehicle models in a 3-D setting, the determination of barriers with suboptimal pursuit and evasion logic is within reach via simulation. In spite of the approximations entailed in obtaining and representing such surfaces for threat-reciprocity study, some useful insights seem possible from such an investigation.

REFERENCE

1. G. J. Olsder and J. V. Breakwell, Role Determination in an Aerial Dogfight, U. S. Navy Workshop in Differential Games, U. S. Naval Academy, Annapolis, Maryland, July 31-August 3, 1973; also in International Journal of Game Theory, Vol. 3, No. 1, 1974, pp. 47-66.

A NOTE ON THE EXISTENCE OF SYNTHESIS OF SADDLE POINTS IN DIFFERENTIAL GAMES

Richard C. Scalzo

University of Illinois at Chicago Circle
Chicago, Illinois

and

Colby College
Waterville, Maine

§1. INTRODUCTION

Ever since the concepts of value and saddle points for differential games were made precise, attempts have been made to find an elementary proof of the existence of value and synthesis of saddle points using Isaac's condition. In this paper I offer an elementary proof of the existence of value and of the synthesis of saddle points, using the Isaac's condition, for differential games with state variable $x \in R^1$. The basis of the proof is the fact that, given the Isaac's condition, one player controls the location of the terminal state of the game in time T_0. Furthermore, the proof extends in an obvious manner to differential games as formulated by Elliott and Kalton, as well as to stochastic games. All that is required is the Isaac's condition hold and that the game is one for which we have complete information. Throughout this paper the notation used conforms to that of [1].

§2. ASSUMPTIONS

In this section we set forth the assumptions to be used throughout this paper. We consider a differential game associated with

$$\frac{dx}{dt} = f(t,x,y,z), \quad x \in R^1, \quad t \in [t_0, T_0] \tag{2.1}$$

$$x(t_0) = x_0 \tag{2.2}$$

The control functions $y(t)$, $z(t)$ are Lebesgue measurable on $[t_0, T_0]$ and

$$y(t) \in Y \subset R^{q_1}, \quad z(t) \in Z \subset R^{q_2} \tag{2.3}$$

where Y and Z are compact. The payoff is given by

$$P(y,z) = g(x(T_0)) \tag{2.4}$$

We assume

(A1) $f(t,x,y,z)$ is continuous in (t,x,y,z) on $[t_0,T_0) \times R^1 \times Y \times Z$.

(A2) There is a nonnegative Legesgue measurable function $k(t)$ such that $\int_{t_0}^{T_0} k(t) < \infty$ and $|f(t,x,y,z)| \leq k(t)(1 + |x|)$ for all $(t,x,y,z) \in [t_0,T_0] \times R^1 \times Y \times Z$.

(A3) For every $R > 0$, there is a nonnegative Lebesgue measurable function $k_R(t)$ such that $\int_{t_0}^{T_0} k_R(t)dt < \infty$, and $|f(t,x,y,z) - f(t,\bar{x},y,z)| \leq k_R(t)|x - \bar{x}|$ for all (t,y,z) in $[t_0,T_0] \times Y \times Z$ and all $|x| \leq R$, $|\bar{x}| \leq R$.

(I) For every unit vector $p \in R^1$,

$$\min_z \max_y p \cdot f(t,x,y,z) = \max_y \min_z p \cdot f(t,x,y,z)$$

Remarks Since $p \in R^1$, (I) is equivalent to

$$\max_y \min_z f(t,x,y,z) = \min_z \max_y f(t,x,y,z)$$

and

$$\min_y \max_z f(t,x,y,z) = \max_z \min_y f(t,x,y,z).$$

Finally, we assume

(B_2') $g(x)$ is continuous for $x \in R^1$.

§3. A FUNDAMENTAL LEMMA

In this section we consider the differential game associated with

$$\frac{dx}{dt} = f(y,z), \quad x \in R^1, \quad t \in [t_0,T_0] \tag{3.1}$$

and (2.2)-(2.4), (A1)-(A3), (B_2'), and (I) hold.

Denote by $\bar{y}_1$ a point in Y such that

$$\min_z f(\bar{y}_1,z) = \min_z \max_y f(y,z)$$

by $\bar{z}_1$ a point in Z such that

$$\max_y f(y,\bar{z}_1) = \min_z \max_y f(y,z)$$

by $\bar{y}_2$ a point in Y such that

$$\max_z f(\bar{y}_2,z) = \min_y \max_z f(y,z)$$

and by $\bar{z}_2$ a point in Z such that

$$\min_y f(y,\bar{z}_2) = \min_y \max_z f(y,z)$$

From the definitions of $\bar{y}_1$, $\bar{y}_2$, $\bar{z}_1$, and $\bar{z}_2$ it follows that if $\tilde{\beta}_1 \le \tilde{\beta}_2$, where $\tilde{\beta}_1$, $\tilde{\beta}_2$ are as below, then

$$\min_z \max_y f(y,z) \le f(\bar{y}_1,z) \le \max_y \max_z f(y,z), \text{ for all } z,$$

$$\min_y \min_z f(y,z) \le f(\bar{y}_2,z) \le \min_y \max_z f(y,z), \text{ for all } z,$$

$$\min_y \min_z f(y,z) \le f(y,\bar{z}_1) \le \min_z \max_y f(y,z), \text{ for all } y$$

and

$$\min_y \max_z f(y,z) \le f(y,\bar{z}_2) \le \max_y \max_z f(y,z), \text{ for all } y.$$

Let $\delta = T_0 - t_0/n$, $n = 1, 2, \cdots$. Partition $[t_0,T_0]$ with $\{t_0 + j\delta;\ 0 \le j \le n\}$ and set $I_k = [t_{k-1},t_k)$, $1 \le k \le n-1$, $I_n = [t_{n-1},T_0]$.

Define δ-strategies as in [1] and set

$$\tilde{\beta}_0 = \min_y \min_z f(y,z), \quad \tilde{\beta}_1 = \min_y \max_z f(y,z)$$

$$\tilde{\beta}_2 = \min_z \max_y f(y,z), \quad \tilde{\beta}_3 = \max_y \max_z f(y,z).$$

Let K be a positive constant greater than

$$\max\{|\tilde{\beta}_3 - \tilde{\beta}_2|,\ |\tilde{\beta}_3 - \tilde{\beta}_1|,\ |\tilde{\beta}_0 - \tilde{\beta}_1|,\ |\tilde{\beta}_0 - \tilde{\beta}_2|\}.$$

We shall prove

<u>Lemma 3.1</u> Consider the differential equation (3.1), (2.2), (2.3), and suppose that the system satisfies (A1)-(A3) and (I). Then if $\tilde{\beta}_1 \le \tilde{\beta}_2$, and $x_0 + \tilde{\beta}_1(T_0-t_0) \le x \le x_0 + \tilde{\beta}_2(T_0-t_0)$, there is a lower δ-strategy $\tilde{\Gamma}_\delta$ for y such that if Δ_δ or Δ^δ is any δ-strategy for z, and x (t) corresponds to $(\Delta_\delta,\tilde{\Gamma}_\delta)$ or $(\Delta^\delta,\tilde{\Gamma}_\delta)$, then $|x_\delta(T_0) - x| \le K\delta$. If $\tilde{\beta}_2 \le \tilde{\beta}_1$, and $x_0 + \tilde{\beta}_2(T_0-t_0) \le x \le x_0 + \tilde{\beta}_1(T_0-t_0)$, there is a lower δ-strategy for z, $\tilde{\Delta}_\delta$, such that for any δ-strategy Γ_δ or Γ^δ for y, if $x\delta(t)$ corresponds to $(\tilde{\Delta}_\delta,\Gamma_\delta)$ or $(\tilde{\Delta}_\delta,\Gamma^\delta)$, then $|x_\delta(T_0) - x| \le K\delta$.

Proof For convenience assume that $\tilde{\beta}_1 \leq \tilde{\beta}_2$ and that $x_0 + \tilde{\beta}_1(T_0 - t_0) \leq x \leq x_0 + \tilde{\beta}_2(T_0 - t_0)$. From this it follows that there is a β such that $\tilde{\beta}_1 \leq \beta \leq \tilde{\beta}_2$ and that $x = x_0 + \beta(T_0 - t_0)$. Construct $\tilde{\Gamma}_\delta$ as follows:

Set $\tilde{\Gamma}_{\delta,1} = \bar{y}_1$. Next suppose that $\Delta^{\delta,1}(\bar{y}_1) = z_1$, then $f(\bar{y}_1,z_1) \geq \tilde{\beta}_2 \geq \beta$. If $f(\bar{y}_1,z_1) > \beta$, then set $\tilde{\Gamma}_{\delta,2}(z_1,\bar{y}_1) = \bar{y}_2$, and if $f(\bar{y}_1,z_1) = \beta$, then $\tilde{\Gamma}_{\delta,2}(z_1,\bar{y}_1) = \bar{y}_1$. Proceeding inductively, if

$$\frac{1}{k-1}\sum_{j=1}^{k-1} f(\bar{y}_j,z_j) > \beta,\ \bar{y}_j \in \{\bar{y}_1,\bar{y}_2\}$$

then set

$$\tilde{\Gamma}_{\delta,k}(z_1,\ \bar{y}_1,\ \cdots,\ z_{k-1}) = \bar{y}_2$$

and if

$$\frac{1}{k-1}\sum_{j=1}^{k-1} f(\bar{y}_j,z_j) \leq \beta,\ \bar{y}_j \in \{\bar{y}_1,\bar{y}_2\},$$

then set

$$\tilde{\Gamma}_{\delta,k}(z_1,\ \bar{y}_1,\ \cdots,\ z_{k-1},\ \bar{y}_{k-1}) = \bar{y}_1$$

Let $x_\delta(t)$ be the trajectory corresponding to $(\Delta^\delta,\tilde{\Gamma}_\delta)$, and proceed by induction to show that $|x_\delta(T_0) - x| \leq K\delta$.

$k = 1$: Clearly $0 \leq f(\bar{y}_1,z_1) - \beta < K$, hence

$$0 \leq x_\delta(t_0 + \delta) - x_0 - \beta\delta < K\delta$$

$k = j - 1$: There are two cases. First assume that

$$\text{(a)}\quad \frac{1}{j-1}\sum_{r=1}^{j-1} f(\bar{y}_r,z_r) > \beta,\ \bar{y}_r \in \{\bar{y}_1,\bar{y}_2\}$$

and that

$$0 \le x_\delta(t_0 + (j - 1)\delta) - x_0 - \beta(j - 1)\delta < K\delta$$

$$= \sum_{r=1}^{j-1} f(\bar{y}_r, z_r)\delta - \beta(j - 1)\delta < K\delta$$

Then by definition of $\tilde{\Gamma}_\delta$, $\bar{y}_j = \bar{y}_2$ and

$$\sum_{r=1}^{j} f(\bar{y}_r, z_r)\delta - \beta j\delta = \sum_{r=1}^{j-1} f(\bar{y}_r, z_r)\delta$$

$$- \beta(j - 1)\delta + f(\bar{y}_j, z_j)\delta - \beta\delta$$

But clearly

$$[f(\bar{y}_j, z_j) - \beta]\delta \le \left[\sum_{r=1}^{j-1} f(\bar{y}_r, z_r) - \beta(j - 1)\right]\delta$$

$$+ \left[f(\bar{y}_j, z_j) - \beta\right]\delta$$

$$\le \left[\sum_{r=1}^{j-1} f(\bar{y}_r, z_r) - \beta(j - 1)\right]\delta$$

and from this it follows that

$$|x_\delta(t_0 + j\delta| - x_0 - \beta j\delta| \le K\delta \tag{3.2}$$

Next, assume that

(b) $\frac{1}{j-1} \sum_{r=1}^{j-1} f(\bar{y}_r, z_j) \le \beta,\ \bar{y}_r \in \{\bar{y}_r \in \{\bar{y}_1, \bar{y}_2\}\}$

and

$$-K\delta \le x_\delta(t_0 + (j - 1)\beta\delta) - x_0 - \beta(j - 1)\delta \le 0$$

Then by definition of $\tilde{\Gamma}_\delta$, $\bar{y}_j = \bar{y}_1$, and in a fashion similar to that of case (a) we get (3.2). This concludes the proof of Lemma 3.1.

Remark The above lemma establishes the fact that one player controls the terminal state of the game, in the set $[x_0 + \tilde{\beta}(T_0 - t_0), x_0 + \bar{\beta}(T_0 - t_0)]$, where $\tilde{\beta} = \min\{\tilde{\beta}_1, \tilde{\beta}_2\}$,

$\bar{\beta} = \max\{\tilde{\beta}_1, \tilde{\beta}_2\}$. Next we establish

Lemma 3.2 Consider the differential equation with control given by (3.1), (2.2), (2.3) and suppose that the equation satisfies (A1)-(A3) and (I). Then if $\tilde{\beta}_1 \le \tilde{\beta}_2$ there is a lower δ-strategy for z, Δ_δ^*, such that if Γ^δ is any upper δ-strategy for y and $x_\delta(t)$ is the trajectory corresponding to $(\Delta_\delta^*, \Gamma^\delta)$, then

$$x_0 + \tilde{\beta}_1(t - t_0) - K\delta \le x_\delta(t) \le x_0 + \tilde{\beta}_2(t - t_0) + K\delta \tag{3.3}$$

for all $t \in [t_0, T_0]$.

Proof Set $\Delta_{\delta,1}^* \equiv z_1^* = \bar{z}_1$. Then define $\Delta_{\delta,k}^*$ as follows:

$$\Delta_{\delta,k}^*(y_1, z_1^*, \cdots, y_{k-1}, z_{k-1}^*) \equiv z_k^* = \bar{z}_1$$

if

$$\sum_{j=1}^{k-1} f(y_j, z_j^*) \ge (k - 1)\tilde{\beta}_1,$$

and

$$\Delta_{\delta,k}^*(y_1, z_1^*, \cdots, y_{k-1}, z_{k-1}^*) \equiv z_k^* = \bar{z}_2$$

if

$$\sum_{j=1}^{k-1} f(y_j, z_j^*) < (k - 1)\tilde{\beta}_1$$

For $t \in I_1$, it is clear that (3.3) holds. Now assume that (3.3) holds for all $t \in I_{k-1}$ and consider some $t \in I_k$, then

Case 1 $\sum_{j=1}^{k-1} f(y_j, z_j^*) \ge (k - 1)\tilde{\beta}_1$

Then $z_k^* = \bar{z}_1$, and so $f(y_k, z_k^*) \le \tilde{\beta}_2$, and since (3.3) holds for $t \in I_{k-1}$, we have

$$x_0 + \tilde{\beta}_1(k-1)\delta \le x_0 + \sum_{j=1}^{k-1} f(y_j, z_j^*)\delta$$

$$\le x_0 + \tilde{\beta}_2(k-1)\delta + K\delta$$

Hence, we have

<u>Subcase 1a</u> $\tilde{\beta}_1 \le f(y_k, z_k^*) \le \tilde{\beta}_2$. Then

$$x_0 + \tilde{\beta}_1(t - t_0) \le x_0 + \sum_{j=1}^{k-1} f(y_j, z_j^*)\delta + f(y_k, z_k^*)(t - t_{k-1})$$

$$\le x_0 + \tilde{\beta}_2(t - t_0) + K\delta$$

<u>Subcase 1b</u> $f(y_k, z_k^*) < \tilde{\beta}_1$. Then

$$x_0 + \tilde{\beta}_1(t - t_0) - K\delta \le x_0 + \sum_{j=1}^{k-1} f(y_j, z_j^*)\delta$$

$$+ f(y_k, z_k^*)(t - t_{k-1}) \le x_0 + \tilde{\beta}_2(t - t_0) + K\delta$$

Thus for Case 1, we have (3.3) for $t \in I_k$.

<u>Case 2</u> $(k-1)\tilde{\beta}_1 \le \sum_{j=1}^{k-1} f(y_j, z_j^*) \le (k-1)\tilde{\beta}_2$, then $z_k^* = \bar{z}_1$, and so $f(y_k, z_k^*) \le \tilde{\beta}_2$. Next we have

$$x_0 + \tilde{\beta}_1(k-1)\delta \le x_0 + \sum_{j=1}^{k-1} f(y_j, z_j^*)\delta \le x_0 + \tilde{\beta}_2(k-1)\delta$$

and·

<u>Subcase 2a</u> $\tilde{\beta}_1 \le f(y_k, z_k^*) \le \tilde{\beta}_2$, then

$$x_0 + \tilde{\beta}_1(t - t_0) \le x_0 + \sum_{j=1}^{k-1} f(y_j, z_j^*)\delta + f(y_k, z_k^*)(t - t_{k-1})$$

$$\le x_0 + \tilde{\beta}_2(t - t_0)$$

Subcase 2b $f(y_k, z_k^*) < \tilde{\beta}_1$, then

$$x_0 + \tilde{\beta}_1(t - t_0) - K\delta \le x_0 + \sum_{j=1}^{k-1} f(y_j, z_j^*)\delta$$

$$+ f(y_k, z_k^*)(t - t_{k-1}) \le x_0 + \tilde{\beta}_2(t - t_0).$$

Thus for Case 2, (3.3) holds for $t \in I_k$.

Case 3 $\sum_{j=1}^{k-1} f(y_j, z_j^*) < (k - 1)\tilde{\beta}_1$.

The situation in this case is similar to that in Case 1. This completes the proof of Lemma 3.2.

Remark There is a corresponding lemma in case $\tilde{\beta}_2 \le \tilde{\beta}_1$, and a δ-strategy Γ_δ^* for y.

For the sake of completeness, we state

Lemma 3.3 Consider the differential equation with controls associated with (3.1), (2.2)-(2.3) and assume that (A1)-(A3) and (I) hold. If Γ_1^δ is the constant strategy associated with $\bar{y}_1$, then there is a lower δ-strategy Δ_δ^1 for z such that if $x_\delta(t)$ is the trajectory corresponding to $(\Delta_\delta^1, \Gamma_1^\delta)$ then $|x_\delta(T_0) - x| \le K\delta$, where x is any preassigned point in the interval $[x_0 + \tilde{\beta}_2(T_0 - t_0), x_0 + \tilde{\beta}_2(T_0 - t_0)]$.

We are now in a position to prove

Theorem 3.1 Let G be a differential game associated with (3.1), (2.2)-(2.4), (A1)-(A3), (B_2') and (I). Then G has value.

Proof Let $M_1 = \max\{g(x);\ x_0 + \tilde{\beta}_1(T_0 - t_0) \le x \le x_0 + \tilde{\beta}_2(T_0 - t_0)\}$ when $\tilde{\beta}_1 \le \tilde{\beta}_2$, and $m_1 = \min\{g(x);\ x_0 + \tilde{\beta}_2(T_0 - t_0) \le x \le x_0 + \tilde{\beta}_1(T_0 - t_0)\}$ when $\tilde{\beta}_2 \le \tilde{\beta}_1$.

We have the following claims.

<u>Claim 1</u> If $\tilde{\beta}_1 \le \tilde{\beta}_2$, then $V(t_0, x_0) = M_1$.

<u>Claim 2</u> If $\tilde{\beta}_2 \le \tilde{\beta}_1$, then $V(t_0, x_0) = m_1$.

We prove only Claim 1, the proof of Claim 2 being similar.
Proof of Claim 1: Since $\tilde{\beta}_1 \le \tilde{\beta}_2$, by Lemma 3.1, there is a $\tilde{\Gamma}_\delta$ such that

$$P[\Delta^\delta, \tilde{\Gamma}_\delta\} \ge M_1 - \gamma(K\delta) \text{ for all } \Delta^\delta$$

where $\gamma(s)$ is a modulus of continuity of $g(x)$. Thus

$$\inf_{\Delta^\delta} P[\Delta^\delta, \tilde{\Gamma}_\delta] \ge M_1 - \gamma(K\delta)$$

and

$$V_\delta = \sup_{\Gamma_\delta} \inf_{\Delta\delta} P[\Delta^\delta, \tilde{\Gamma}_\delta] \ge M_1 - \gamma(K\delta) \tag{3.4}$$

Suppose now that $V_\delta > M_1 + \gamma(K\delta)$. This means that there is a $\bar{\Gamma}_\delta$ such that

$$\inf_{\Delta^\delta} P[\Delta^\delta, \bar{\Gamma}_\delta] > M_1 + \gamma(K\delta)$$

But if z plays Δ^*_δ, then with $\Delta^*_\delta = \Delta^\delta_*$ as in Lemma 3.2,

$$P[\Delta^*_\delta, \bar{\Gamma}_\delta] > M_1 + \gamma(K\delta)$$

which contradicts the fact that the trajectory corresponding to $(\Delta^*_\delta, \bar{\Gamma}_\delta)$ satisfies

$$x_0 + \tilde{\beta}_1(T_0 - t_0) - K\delta \le x_\delta(T_0) \le x_0 + \tilde{\beta}_2(T_0 - t_0) + K\delta$$

Hence,

$$V_\delta \le M_1 + \gamma(K\delta) \tag{3.5}$$

Equations (3.4) and (3.5) taken together give

$$M_1 - \gamma(K\delta) \le V_\delta \le M_1 + \gamma(K\delta) \tag{3.6}$$

Since every lower δ-strategy is also an upper δ-strategy, it follows in exactly the same manner that

$$M_1 - \gamma(K\delta) \le V^\delta \le M_1 + \gamma(K\delta) \tag{3.7}$$

Now from (3.6) and (3.7) it follows that G has value.

The next theorem requires a different definition of value than that found in [1]. We adopt the notation $\bar{V}_\delta$, $\bar{V}^\delta$ and $\bar{V}$ for this different definition. We now state

Theorem 3.2 Let $\bar{G}$ be the differential game associated with (2.1)-(2.4), (A1)-(A3), (B_2'), and (I). Then $\bar{G}$ has value $\bar{V}$.

Proof In a manner similar to Fleming we replace the dynamics (2.1), (2.2) by an approximation.

$$x_k(t) = x_{k-1} + \int_{t_{k-1}}^{t} f(t_{k-1}, x_{k-1}, y(x), z(s))ds \tag{3.8}$$

$$x_1(t_0) = x_0,\ 1 \le k \le n \tag{3.9}$$

It then follows from Theorem 3.1 that the game associated with

$$x_n(t) = x_{n-1} + \int_{t_{n-1}}^{t} f(t_{n-1}, x_{n-1}, y(s), z(s))ds \tag{3.10}$$

and (2.4) played on $[t_{n-1}, T_0]$ has value V. Thus set

$$\bar{V}^{\delta,n}(t_{n-1},x_{n-1}) = V(t_{n-1},x_{n-1}) = \bar{V}_{\delta,n}(t_{n-1},x_{n-1})$$

and proceed as in [1], to define $\bar{V}^{\delta,n-j}$, $\bar{V}_{\delta,n-j}$. Since at each stage, we have $\bar{V}^{\delta,n-j} = \bar{V}_{\delta,n-j}$, the game $\bar{G}$ has value $\bar{V}$.

Now a straightforward argument shows that $\bar{V}$ as defined above agrees with Fleming's value.

Remark It should be noted that Δ_δ^*, $\tilde{\Gamma}_\delta$ as defined in Lemma 3.1 and Lemma 3.2, respectively, form a saddle point for G as in Theorem 3.1.

§4. EXISTENCE OF SYNTHESIS

For the sake of clarity in this section we consider only games associated with (3.1), (2.2)-(2.4).

Keeping Theorem 3.1 in mind, suppose that $\tilde{\beta}_1 \le \tilde{\beta}_2$ and that $\bar{x} \in [x_0 + \tilde{\beta}_1(T_0 - t_0), x_0 + \tilde{\beta}_2(T_0 - t_0)]$ and $g(\bar{x}) = M_1$, and that $|\bar{x}| \le |\tilde{x}|$ whenever $g(\tilde{x}) = M_1$ and $\tilde{x} \in [x_0 + \tilde{\beta}_1(T_0 - t_0), x_0 + \tilde{\beta}_2(T_0 - t_0)]$. Then if $\bar{\beta}$ is such that $x_0 + \bar{\beta}(T_0 - t_0) = \bar{x}$, define

$$\tilde{y}(t_0,x_0,t,x) = \begin{cases} \bar{y}_1 & \text{if } x + \bar{\beta}(T_0 - t) \ge \bar{x} \\ \bar{y}_2 & \text{if } x + \bar{\beta}(T_0 - t) < \bar{x} \end{cases}$$

and

$$z^*(t_0,x_0,t,x) = \begin{cases} \bar{z}_1 & \text{if } x + \tilde{\beta}_2(T_0 - t) \ge x_0 + \tilde{\beta}_2(T_0 - t_0) \\ \bar{z}_2 & \text{if } x + \tilde{\beta}_1(T_0 - t) < x_0 + \tilde{\beta}_1(T_0 - t_0) \end{cases}$$

It is easy to see that $y^*(t_0,x_0,t,x)$ and $z^*(t_0,x_0,t,x)$ are measurable in (t,x), and that

Theorem 4.1 Let G be the differential game associated with (3.1), (2.2)-(2.4), (A1)-(A3), (B_2'), and (I). Suppose that $\tilde{\beta}_1 \le \tilde{\beta}_2$. Then for each (t_0, x_0), a synthesis of the saddle point $(\Delta_\delta^*, \tilde{\Gamma}_\delta)$ is given by $\tilde{y}(t_0,x_0,t,x)$, $z^*(t_0,x_0,t,x)$.

The treatment of the case in which we replace (3.1) by (2.1) is a straightforward extension of Theorem 4.1.

Remark It is an easy matter now to get switching theorems from Theorem 4.1.

REFERENCES

1. Friedman, A., Differential Games, Wiley Interscience, 1971.

2. Fleming, W. H., A Note on Differential Games of Prescribed Duration, Contributions to the Theory of Games, vol. III, Ann. Math. Stud. No. 39, Princeton University Press, Princeton, New Jersey, 1957, pp. 407-416.

3. Fleming, W. H., The Convergence Problem for Differential Games, J. Math. Anal. Appl., 3, 102-116, 1961.

4. Fleming, W. H., The Convergence Problem for Differential Games II, Advances in Game Theory, Ann. Math. Stud. No. 52, Princeton University Press, Princeton, New Jersey, 1964, pp. 195-216.

A STOCHASTIC OPTIMAL CONTROL MODEL FOR A PROBLEM IN RESOURCE ECONOMICS

Mou-Hsiung Chang and Jon G. Sutinen

University of Alabama at Huntsville
Huntsville, Alabama

and

University of Rhode Island
Kingston, Rhode Island

§1. INTRODUCTION

The purpose of this paper is to formally develop a minimum principle for the stochastic optimal control of leasing the lands of the Outer Continental Shelf (OCS).

Recently, the Department of Interior began leasing larger areas of the OCS for oil and gas exploitation in an attempt to increase the domestic production of petroleum. A major problem in the leasing process is the determination and selection of the optimal leasing arrangements for the exploration, development and production activities. Formal specification of this problem is as follows.

Let us assume the existence of a central authority whose goal is to maximize the discounted stream of expected benefits from the leasing of lands on the outer continental shelf for the purpose of exploiting suspected deposits of oil and gas. The lease is for a fixed duration, beginning in time period $t = 0$ and expiring at the end of $t = T$. The authority is responsible for negotiating and enforcing the terms of the lease, and we assume the associated negotiation and enforcement costs (i.e., the transaction costs) are negligible. To achieve its goal, the authority is faced with choosing an optimal set of leasing arrangements subject to the constraints imposed by the marketplace and by the exhaustible nature of the resource. That is, the authority selects from a set of feasible arrangements (assumed made available to it through a competitive bidding process), a specific royalty rate r(t), a fixed payment R(t), and levels of inputs K(t), for each time period covered by the lease.

If x(t) and p(t) denote the stock of the petroleum and market price of output of petroleum, respectively, then x(t) and p(t) satisfy the following scalar stochastic differential equations:

$$\dot{x}(t) = -F(x(t),K(t)),\ x(o) = x_o$$
$$dp(t) = f(x(t),p(t),K(t))dt + \sigma(p(t))dW_t,$$
$$p(o) = p_o \qquad (1)$$

where x_o and p_o are random variables with given probability distribution functions. W_t is the standard Wiener process. F(x,K) is the production function, where $F(x,K) \geq 0$, $F(0,K) = 0$, $\partial F/\partial x$, $\partial F/\partial K > 0$ and $\partial^2 F/\partial K^2 < 0$, whereas $\partial f/\partial x$, $\partial f/\partial K < 0$ and $\partial\sigma/\partial p < 0$. One physical constraint is that total extraction cannot exceed the available petroleum stock, i.e.,

$$\int_o^T F(x(t),K(t))dt \leq x_o$$

with probability 1. However, this is always satisfied since we have assumed that $F(0,K) = 0$

The feasible set of leasing arrangements from which the authority chooses is assumed to be offered (at least partially) by a large number of identical forms. Under these assumptions, the authority's objective is to choose a deterministic admissible (control) (r(t),R(t),K(t)) so that

$$E\left\{\int_0^T V_1(\pi_1(t))\,dt\right\} \qquad (2)$$

is maximized, where V_1 is the authority's concave objective function, i.e.,

$$\frac{\partial V_1}{\partial \pi_1} > 0 \quad \text{and} \quad \frac{\partial^2 V_1}{\partial \pi_1^2} < 0$$

and $\pi_1(t)$ is the revenue received by the authority at time t, i.e., $\pi_1(t) = r(t)p(t)F(x(t),K(t)) + R(t)$.

On the other hand, the representative firm has a concave benefit function, $V_2(\pi_2(t))$, where $\partial V_2/\partial \pi_2 > 0$ and $\partial^2 V_2/\partial \pi_2^2 < 0$. The firm's net revenue is given by

$$\pi_2(t) = (1 - r(t))p(t)F(x(t),K(t)) - cK(t) - R(t)$$

where c is rental price of the variable input K. Due to the fact that bidding is conducted in an open auction-like setting, competition will induce the applicants to increase their respective bids to the point where any element of an above competitive return will be eliminated. However, the firm will be willing to accept terms of a lease only if it will be made no worse off than if it employed its assets elsewhere in the economy (where, we assume, it can earn an income I(t) at time t). Therefore, the following constraint should also be satisfied:

$$E\left\{\int_0^T V_2(\pi_2(t))dt\right\} = E\left\{\int_0^T V_2(I(t))dt\right\} = \text{constant} \qquad (3)$$

In this paper, we shall find necessary conditions, in the form of a minimum principle, for a set of deterministic optimal arrangements (controls) $(r^*(t),R^*(t),K^*(t))$. Based on the minimum principle obtained, we shall also find a set of first order necessary conditions from which we can prove the optimal royalty rate $r^*(t)$ is neither zero nor one.

§2. ASSUMPTIONS

In addition to those assumptions made in Section 1, which are due to the economic nature of the problem, we shall also assume that the following technical mathematical conditions are satisfied:

(A1) $F_K(x,K)$, $f_x(x,p,K)$, $f_p(x,p,K)$, and $\sigma_p(p)$ exist and are jointly continuous in their arguments

(A2) $|f(x,p,K)| \le C_1(1 + |p| + |x| + |K|)$

$|F(x,K)| \le C_2(1 + |x| + |K|)$

$|\sigma(p)| \le C_3(1 + |p|)$

(A3) $|f(\tilde{x},\tilde{p},K) - f(x,p,K)| \le K_1(|\tilde{x} - x| + |\tilde{p} - p|)$

$|F(\tilde{x},K) - F(x,K)| \le K_2|\tilde{x} - x|$

$|\sigma(\tilde{p}) - \sigma(p)| \le K_3|\tilde{p} - p|$

where C_i and K_i, $i = 1, 2, 3$, are some constants. These conditions are essential to the existence and uniqueness of a solution to (1) (see [1]) and analysis in the sequel.

A three-dimensional vector function, $u(t) = (r(t),R(t),K(t))$, is said to be an admissible control if

(i) $r(t)$, $R(t)$, and $K(t)$ are measurable functions and $(r(t),R(t),K(t)) \in [0,1] \times [0,\bar{R}] \times [0,\bar{K}] \subset \mathbb{R}^3$;

(ii) system (1) has a unique solution $(x(t),p(t))$ with $E\{x^2(t)\} < \infty$ and $E\{p^2(t)\} < \infty$ for all $t \in [0,T]$; and

(iii) constraint (3) is satisfied.

3. MINIMUM PRINCIPLE

Suppose that $u^* = (r^*(t),R^*(t),K^*(t))$ is an optimal control to the control problem (1)-(3) with corresponding trajectory $(x^*(t),p^*(t))$. Let us consider the following linear stochastic adjoint system:

$$\begin{aligned}
\dot{\psi}_1(t) &= \psi_1(t)F_x(x^*(t),K^*(t)) - \psi_2(t)f_x(x^*(t),p^*(t),K^*(t)) \\
&\quad + V_1'(\pi_1^*(t))r^*(t)p^*(t)F_x(x^*(t),K^*(t)) \\
&\quad + \lambda V_2'(\pi_2^*(t))(1 - r^*(t))p^*(t)F_x(x^*(t),K^*(t)), \\
&\quad \psi_1(T) = 0
\end{aligned}$$

$$\begin{aligned}
d\psi_2(t) &= -\psi_2(t)f_p(x^*(t),p^*(t),K^*(t))dt - \psi_2(t)\sigma_p(p^*(t))dW_t \\
&\quad + V_1'(\pi_1^*(t))r^*(t)F(x^*(t),K^*(t))dt \\
&\quad + \lambda V_2'(\pi_2^*(t))(1 - r^*(t))F(x^*(t),K^*(t))dt, \\
&\quad \psi_2(T) = 0
\end{aligned} \tag{4}$$

where λ is a constant, $\pi_1^*(t) = r^*(t)p^*(t)F(x^*(t),K^*(t)) + R^*(t)$ and $\pi_2^*(t) = (1 - r^*(t))p^*(t)F(x^*(t),K^*(t)) - cK^*(t) - R^*(t)$. The solution $(\psi_1(t),\psi_2(t))$ to (4) has been proved to exist by Fleming [2].

We have the following lemma which is essential to the proof of Theorem 2.

<u>Lemma 1</u> Assume that assumptions (A1)-(A3) are satisfied. Let $u^* = (r^*(t),R^*(t),K^*(t))$ be an optimal control with

corresponding trajectory $(x^*(t),p^*(t))$ and $u = (r(t),R(t),K(t))$ be any admissible control with corresponding trajectory $(x(t), p(t))$. Then there exists a constant λ and a stochastic process $(\psi_1(t),\psi_2(t))$ satisfying (4) such that

$$E\left\{\int_0^T V_1(\pi_1^*(t))dt\right\} - E\left\{\int_0^T V_1(\pi_1(t))dt\right\}$$

$$= E\left\{\int_0^T [-\psi_1(t)F(x^*(t),K(t)) + \psi_2(t)f(x^*(t),p^*(t),K(t))\right.$$

$$- V_1(r(t)p^*(t)F(x^*(t),K(t)) + R(t)$$

$$\left. - \lambda V_2((1-r(t))p^*(t)F(x^*(t),K(t)) - cK(t) - R(t))]dt\right\}$$

$$- E\left\{\int_0^T [-\psi_1(t)F(x^*(t),K^*(t)) + \psi_2(t)f(x^*(t),p^*(t),K^*(t))\right.$$

$$\left. - V_1(\pi_1^*(t)) - \lambda V_2(\pi_2^*(t))]dt\right\}$$

$$+ \int_0^T \theta(|r^*(t) - r(t)| + |R^*(t) - R(t)|$$

$$+ |K^*(t) - K(t)|)dt$$

where $\theta(\cdot)$ is a quantity such that $\lim_{h\to 0+} \frac{\theta(h)}{h} = 0$.

Proof See appendix for proof.

Now, based on Lemma 1, we have the following necessary condition for an optimal control $(r^*(t),R^*(t),K^*(t))$ in the form of a minimum principle.

Theorem 2 (Minimum Principle) Assume that (A1)-(A3) are satisfied. Let $(r^*(t),R^*(t),K^*(t))$ be an optimal control with optimal trajectory $(x^*(t),p^*(t))$ to control problem (1)-(3). Then there exists a nonzero $\lambda(t)$ and a nonzero stochastic process $(\psi_1(t),\psi_2(t))$ satisfying (4) such that

$$E\{-\psi_1(t)F(x^*(t),K^*(t)) + \psi_2(t)f(x^*(t),p^*(t),K^*(t)) - V_1(\pi_1^*(t)) - \lambda V_2(\pi_2^*(t))\}$$
$$\leq E\{-\psi_1(t)F(x^*(t),K(t)) + \psi_2(t)f(x^*(t),p^*(t),K(t)) - V_1(r(t)p^*(t)F(x^*(t),K(t)) + R(t)) - \lambda V_2((1-r(t))p^*(t)F(x^*(t),K(t)) - cK(t) - R(t))\} \quad (5)$$

for all admissible $(r(t),R(t),K(t))$ and almost all $t \in [0,T]$.

<u>Proof</u> Suppose that the conclusion of the theorem were false, then there exists an admissible control $v(t) = (r(t),R(t),K(t))$ and a measurable subset A of $[0,T]$ with $mA \neq 0$ such that

$$E\{-\psi_1(t)F(x^*(t),K^*(t)) + \psi_2(t)f(x^*(t),p^*(t),K^*(t)) - V_1(\pi_1^*(t)) - V_2(\pi_2^*(t))\}$$
$$> E\{-\psi_1(t)F(x^*(t),K(t)) + \psi_2(t)f(x^*(t),p^*(t),K(t)) - V_1(r(t)p^*(t)F(x^*(t),K(t)) + R(t)) - \lambda V_2((1 - r(t))p^*(t)F(x^*(t),K(t)) - cK(t) - R(t))\} \quad (6)$$

for all $t \in A$.

If (6) holds, then for an arbitrary, but sufficiently small positive number α, there is a nonzero-measure subset $A(\alpha)$ of A, such that

$$E\{-\psi_1(t)F(x^*(t),K^*(t)) + \psi_2(t)f(x^*(t),p^*(t),K^*(t)) - V_1(\pi_1^*(t)) - V_2(\pi_2^*(t))\}$$

$$- E\{-\psi_1(t)F(x^*(t),K(t)) + \psi_2(t)f(x^*(t),p^*(t),K(t))$$
$$- V_1(r(t)p^*(t)F(x^*(t),K(t)) + R(t))$$
$$- \lambda V_2((1 - r(t))p^*(t)F(x^*(t),K(t)) - cK(t) - R(t))\}$$
$$> \alpha \quad \text{for all } t \in A(\alpha)$$

Now let us define a new admissible control $\tilde{u}(t) = (\tilde{r}(t),\tilde{R}(t),\tilde{K}(t))$ by

$$\tilde{u}(t) = \begin{cases} (r(t),R(t),K(t)) & \text{if } t \in A(\alpha) \\ (r^*(t),R^*(t),K^*(t)) & \text{if } t \notin A(\alpha) \end{cases}$$

Note that $u(t)$ is admissible, because both $(r(t),R(t),K(t))$ and $(r^*(t),R^*(t),K^*(t))$ are admissible and $A(\alpha)$ is a measurable set. Since $\tilde{u}(t)$ and $(r^*(t),R^*(t),K^*(t))$ are both bounded and measurable, it is no restriction to suppose that

$$\sup_{t\in A(\alpha)} \|\tilde{u}(t) - (r^*(t),R^*(t),K^*(t))\| = b < \infty \text{ for some } b,$$

where $\|(r,R,K)\| = |r| + |R| + |K|$. Then, from Lemma 1, we have

$$E\left\{\int_0^T V_1(\pi_1^*(t))dt - \int_0^T V_1(\tilde{\pi}_1(t))dt\right\}$$
$$= E\left\{\int_0^T [-\psi_1(t)F(x^*(t),\tilde{K}(t)) + \psi_2(t)f(x^*(t),p^*(t),\tilde{K}(t))\right.$$
$$- V_1(\tilde{r}(t)p^*(t)F(x^*(t),\tilde{K}(t)) + \tilde{R}(t))$$
$$- \lambda V_2((1 - \tilde{r}(t))p^*(t)F(x^*(t),\tilde{K}(t))$$
$$\left. - c\tilde{K}(t) - \tilde{R}(t))]dt\right\}$$
$$- E\left\{\int_0^T [-\psi_1(t)F(x^*(t),K^*(t)) + \psi_2(t)f(x^*(t),p^*(t),K^*(t))\right.$$
$$\left. - V_1(\pi_1^*(t)) - \lambda V_2(\pi_2^*(t))]dt\right\}$$

$$+ \int_0^T \theta(|r^*(t) - \tilde{r}(t)| + |R^*(t) - \tilde{R}(t)|$$
$$+ |K^*(t) - \tilde{K}(t)|)dt$$
$$< -\alpha \cdot mA(\alpha) + \theta(b \cdot mA(\alpha))$$

where $mA(\alpha)$ is the Lebesgue measure of the set $A(\alpha)$. Note that $\theta(b \cdot mA(\alpha))/b \cdot mA(\alpha) \to 0$ as $mA(\alpha) \to 0$. Therefore, by choosing $A(\alpha)$ with $mA(\alpha)$ sufficiently small, we can make

$$-\alpha \cdot mA(\alpha) + \theta(b \cdot mA(\alpha)) < 0$$

That is,

$$E\left\{\int_0^T V_1(\pi_1^*(t))dt\right\} - E\left\{\int_0^T V_1(\tilde{\pi}_1(t))dt\right\} < 0$$

which contradicts the fact that $(r^*(t), R^*(t), K^*(t))$ is an optimal control. Therefore the theorem is true.

§4. FIRST ORDER NECESSARY CONDITIONS

It is clear from Theorem 2 that if $(r^*(t), R^*(t), K^*(t))$ is an optimal control to problem (1)-(3) then the following three necessary conditions are satisfied for almost all t in [0,T].

$$E\{-\psi_1(t)F_K(x^*(t),K^*(t)) + \psi_2(t)f_K(x^*(t),p^*(t),K^*(t))$$
$$- V_1'(\pi_1^*(t))r^*(t)p^*(t)F_K(x^*(t),K^*(t))$$
$$- \lambda V_2'(\pi_2^*(t))[(1 - r^*(t))p^*(t)F_K(x^*(t),K^*(t))$$
$$- c]\} = 0 \tag{7}$$

$$E\{-V_1'(\pi_1^*(t))p^*(t)F(x^*(t),K^*(t))$$
$$+ \lambda V_2'(\pi_2^*(t))p^*(t)F(x^*(t),K^*(t))\} = 0 \tag{8}$$

and

$$E\{-V_1'(\pi_1^*(t) + \lambda V_2'(\pi_2^*(t)\} = 0 \tag{9}$$

Note that the left hand sides of (7), (8), and (9) are obtained by differentiating the left hand side of (5) with respect to K*, r*, and R*, respectively. From these three necessary conditions, we wish to draw some conclusions on the optimal royalty rate, r*(t), at time t.

First, $\lambda(t)$ can be obtained very easily from (9), where

$$\lambda = E\{V_1'(\pi_1^*(t))\}/E\{V_2'(\pi_2^*(t))\}$$

Substituting this λ into (8), we have the following auxiliary condition which (r*(t),R*(t),K*(t)) should satisfy:

$$-E\{V_1'(\pi_1^*(t))p^*(t)F(x^*(t),K^*(t))\}E\{V_2'(\pi_2^*(t))\}$$

$$+ E\{V_1'(\pi_1^*(t))\}E\{V_2'(\pi_2^*(t))p^*(t)F(x^*(t),K^*(t))\} = 0 \tag{10}$$

for almost all $t \in [0,T]$.

We claim that $r^*(t) \neq 0$ and $r^*(t) \neq 1$ for almost all $t \in [0,T]$. For if $r^*(t) = 0$ for all $t \in B$, where $mB > 0$, then $V_1'(\pi_1^*(t)|_{r^*=0}) > 0$ and the left hand side of (10) becomes

$$-V_1'(\pi_1^*(t)|_{r^*=0})[E\{p^*(t)F(x^*(t),K^*(t))\}$$

$$\cdot E\{V_2'(\pi_2^*(t)|_{r^*=0})\}$$

$$- E^3V_2'(\pi_2^*(t)|_{r^*=0})p^*(t)F(x^*(t),K^*(t))\}]$$

Now, if we let $\nu = p^*(t)F(x^*(t),K^*(t))$ and $\mu = \pi_2^*(t)|_{r^*=0}$, then the expression inside the bracket in the above becomes

$$E\{\nu\}E\{V_2'(\mu)\} - E\{V_2'(\mu)\nu\} = E\{\nu\}E\{V_2'(\mu)\} - E\{\nu\}V_2'(E\{\mu\})$$

$$+ E\{\nu\}V_2'(E\{\mu\}) - E\{V_2'(\mu)\nu\}$$

$$= -E\{(\mu - E\{\mu\})(V_2'(\mu) - V_2'(E\{\mu\}))\} > 0,$$

since $V_2'' < 0$. This implies that if $R^*(t) = 0$ for all t in a set of nonzero measure, then (10) is not satisfied. Similarly, we can show that if $r^*(t) = 1$ then (10) is also not satisfied. This conclusion coincides with the result obtained in the static case (see [3]).

§5. APPENDIX: PROOF OF LEMMA 1

In this section we shall prove Lemma 1 as follows: If $(r^*(t),R^*(t),K^*(t))$ is an optimal control with corresponding trajectory $(x^*(t),p^*(t))$ and $(r(t),R(t),K(t))$ is any admissible control with corresponding trajectory $(x(t),p(t))$, we have

$$E\left\{\int_0^T V_1(\pi_1^*(t))dt\right\} - E\left\{\int_0^T V_1(\pi_1(t))dt\right\}$$

$$= E\left\{\int_0^T V_1(\pi_1^*(t))dt\right\} - E\left\{\int_0^T V_1(\pi_1(t))dt\right\}$$

$$+ \lambda E\left\{\int_0^T V_2(\pi_2^*(t))dt - \int_0^T V_2(\pi_2(t))dt\right\}$$

$$= E\left\{\int_0^T V_1(\pi_1^*(t))dt - \int_0^T V_1(r(t)p^*(t)F(x^*(t),K(t)) + R(t))dt\right\}$$

$$+ E\left\{\int_0^T V_1(r(t)p^*(t)F(x^*(t),K(t)) + R(t))dt - \int_0^T V_1(\pi_1(t))dt\right\}$$

$$+ \lambda E\left\{\int_0^T V_2(\pi_2^*(t))dt - \int_0^T V_2((1 - r(t))p^*(t)F(x^*(t),K(t)) - cK(t) - R(t))dt\right\}$$

$$+ \lambda E\left\{\int_0^T V_2((1 - r(t))p^*(t)F(x^*(t),K(t)) - cK(t) - R(t))dt - \int_0^T V_2(\pi_2(t))dt\right\} \tag{11}$$

Now from assumption (A1) and Taylor's theorem, we have

$$E\left\{\int_0^T V_1(r(t)p^*(t)F(x^*(t),K(t)) + R(t))dt - \int_0^T V_1(\pi_1(t))dt\right\}$$

$$= E\left\{\int_0^T V_1(\pi_1(t))[r(t)F(x(t),K(t))(p^*(t) - p(t)) + r(t)p(t)F_x(x(t),K(t))(x^*(t) - x(t))]dt\right\}$$

$$+ E\left\{\int_0^T \theta(|p^*(t) - p(t)| + |x^*(t) - x(t)|)dt\right\}$$

$$= E\left\{\int_0^T V_1'(\pi_1^*(t))[r^*(t)F(x^*(t),K^*(t))(p^*(t) - p(t)) + r^*(t)p^*(t)F_x(x^*(t),K^*(t))(x^*(t) - x(t))]dt\right\}$$

$$+ Q + E\left\{\int_0^T \theta(|p^*(t) - p(t)| + |x^*(t) - x(t)|)dt\right\}$$

where

$$Q = E\left\{\int_0^T [V_1'(\pi_1(t))r(t)F(x(t),K(t)) - V_1'(\pi_1^*(t))r^*(t)F(x^*(t),K^*(t))] \cdot (p^*(t) - p(t))dt\right\}$$

$$+ \int_0^T [V_1'(\pi_1(t))r(t)p(t)F_x(x(t),K(t)] - V_1'(\pi_1^*(t))r^*(t)p^*(t)F_x(x^*(t),K^*(t))]$$

$$(x^*(t) - x(t))dt$$

and $\theta(\cdot)$ is a quantity such that $\lim_{h\to 0} \frac{\theta(h)}{h} = 0$.

Now from the fact that $V_1'(\pi_1)$ and $F_x(x,K)$ are continuous and (A3) it can be shown very easily that

$$Q = E\left\{\int_0^T \theta(|r^*(t) - r(t)| + |R^*(t) - R(t)| + |K^*(t) - K(t)|)dt\right\}$$

Thus

$$\begin{aligned} &E\left\{\int_0^T V_1(r(t)p^*(t)F(x^*(t),K(t)) + R(t))dt - \int_0^T V_1(\pi_1(t))dt\right\} \\ &= E\left\{\int_0^T V_1'(\pi_1^*(t))[r^*(t)F(x^*(t),K^*(t))(p^*(t) - p(t)) + r^*(t)p^*(t)F_x(x^*(t),K^*(t))(x^*(t) - x(t))]dt\right\} \\ &\quad + \int_0^T \theta(|r^*(t) - r(t)| + |R^*(t) - R(t)| + |K^*(t) - K(t)|)dt \end{aligned} \tag{12}$$

Similarly, one can also show that

$$\begin{aligned} &\lambda E\left\{\int_0^T V_2((1 - r(t))p^*(t)F(x^*(t),K(t)) - cK(t) - R(t))dt - \int_0^T V_2(\pi_2(t)dt\right\} \\ &= \lambda E\left\{\int_0^T V_2'(\pi_2^*(t))[(1 - r^*(t))F(x^*(t),K^*(t))(p^*(t) - p(t)) + (1 - r^*(t))p^*(t)F_x(x^*(t),K^*(t))(x^*(t) - x(t))]dt\right. \\ &\quad + \int_0^T \theta(|r^*(t) - r(t)| + |R^*(t) - R(t)| + |K^*(t) - K(t)|)dt \end{aligned} \tag{13}$$

Since $(\psi_1(t),\psi_2(t))$ is a stochastic process which satisfies (4), we have

$$E\left\{\int_0^T V_1(\pi_1^*(t))dt\right\} - E\left\{\int_0^T V_1(\pi_1(t))dt\right\} = E\left\{\int_0^T V_1(\pi_1^*(t))dt\right.$$

$$\left. - \int_0^T V_1(r(t)p^*(t)F(x^*(t),K(t)) + R(t))dt\right\}$$

$$+ \lambda E\left\{\int_0^T V_2(\pi_2^*(t))dt\right.$$

$$\left. - \int_0^T V_2((1 - r(t))p^*(t)F(x^*(t),K(t)) - cK(t) - R(t))dt\right\}$$

$$+ E\left\{\int_0^T [\dot{\psi}_1(t) - \psi_1(t)F_x(x^*(t),K^*(t))\right.$$

$$\left. + \psi_2(t)f_x(x^*(t),p^*(t),K^*(t))](x^*(t) - x(t))dt\right\}$$

$$+ E\left\{\int_0^T d\psi_2(t)(p^*(t) - p(t))\right.$$

$$\left. + \int_0^T \psi_2(t)f_p(x^*(t),p^*(t),K^*(t))(p^*(t) - p(t))dt\right\}$$

$$+ \int_0^T \psi_2(t)\sigma_p(p^*(t))(p^*(t) - p(t))dW_t$$

$$+ \int_0^T \theta(|r^*(t) - r(t)| + |R^*(t) - R(t)|$$

$$+ |K^*(t) - K(t)|)dt \tag{14}$$

Based on the same argument, we also have

$$\left|E\left\{\int_0^t [-F(x^*(s),K^*(s)) + F(x^*(s),K(s))]ds\right\}\right.$$

$$+ \int_0^t [f(x^*(s),p^*(s),K^*(s)) - f(x^*(s),p^*(s),K(s))]ds$$

$$- \int_0^t [-Fx(x^*(s),K^*(s)) + f_x(x^*(s),p^*(s),K^*(s))](x(s)$$

$$- x^*(s))ds$$

$$- \int_0^t f_p(x^*(s),p^*(s),K^*(s))(p(s) - p^*(s))ds$$

$$- \int_0^t \sigma_p(p^*(s))(p(s) - p^*(s))dW_s \Big|$$

$$= \int_0^t \theta(|r^*(s) - r(s)| + |R^*(s) - R(s)| + |K^*(s) - K(s)|)ds$$

Therefore, for the solution $(\psi_1(t), \psi_2(t))$ of (4), we have

$$E\left\{\int_0^T \dot{\psi}_1(t) \int_0^t [-F(x^*(s), K^*(s)) + F(x^*(s), K(s))]ds\, dt\right\}$$

$$+ E\left\{\int_0^T d\psi_2(t) \int_0^t [f(x^*(s), p^*(s), K^*(s))$$

$$- f(x^*(s), p^*(s), K(s))]ds\right\}$$

$$- E\left\{\int_0^T \dot{\psi}_1(t)(x(t) - x^*(t))dt + \int_0^T d\psi_2(t)(p(t) - p^*(t))\right\}$$

$$+ E\left\{\int_0^T \dot{\psi}_1(t) \int_0^t F_x(x^*(s), K^*(s))(x(s) - x^*(s))ds\, dt\right\}$$

$$- E\left\{\int_0^T d\psi_2(t) \int_0^t f_x(x^*(s), p^*(s), K^*(s))(x(s) - x^*(s))ds\right\}$$

$$- E\left\{\int_0^T d\psi_2(t) \int_0^t f_p(x^*(s), p^*(s), K^*(s))(p(s) - p^*(s))ds\right\}$$

$$= \int_0^T \sigma(|r^*(t) - r(t)| + |R^*(t) - R(t)|$$

$$+ |K^*(t) - K(t)|)dt \tag{15}$$

Integrating the integrals in (15) by parts and combining with (14), we finally have the following:

$$E\left\{\int_0^T V_1(\pi_1^*(t))dt\right\} - E\left\{\int_0^T V_1(\pi_1(t))dt\right\}$$

$$= E\left\{\int_0^T [-\psi_1(t)F(x^*(t), K(t)) + \psi_2(t)f(x^*(t), p^*(t), K(t))\right.$$

$$- V_1(r(t)p^*(t)F(x^*(t), K(t)) + R(t)$$

$$- \lambda V_2((1 - r(t))p^*(t)F(x^*(t),K(t)) - cK(t) - R(t))]dt\Big\}$$

$$- E\Big\{\int_0^T [-\psi_1(t)F(x^*(t),K^*(t)) + \psi_2(t)f(x^*(t),p^*(t),K^*(t))$$

$$- V_1(\pi_1^*(t)) - V_2(\pi_2^*(t))]\ dt\Big\}$$

$$+ \int_0^T \theta(|r^*(t) - r(t)| + |R^*(t) - R(t)|$$

$$+ |K^*(t) - K(t)|)dt$$

This comples the proof of Lemma 1.

ACKNOWLEDGMENT

Partial support from the National Science Foundation for this research is gratefully acknowledged. We also thank Pan-Tai Liu for his advice and assistance.

REFERENCES

1. Doob, J. L., Stochastic Processes, John Wiley and Sons, Inc., New York, 1953.

2. Fleming, W. H., Stochastic Lagrange Multipliers, Mathematical Theory of Control, edited by Neusdalt and Balakrishnan, Academic Press, New York, 1965.

3. Sutinen, J. G., The Rational Choice of Share Leasing and Implications for Efficiency, American Journal of Agricultural Economics, vol. 57, no. 4, November, 1975.

MULTIVARIABLE SELF-TUNING FILTERS

G. Ledwich and J. B. Moore

University of Newcastle
Newcastle, New South Wales, Australia

ABSTRACT

Self-tuning algorithms to achieve minimum variance performance are studied for multivariable filters in stochastic environments. Specialization of the algorithms including their adaption to achieve self-tuning predictors and self-tuning regulators have found applications in engineering situations. Sufficient conditions for convergence of the self-tuning process are found to consist of a very reasonable persistently exciting condition on the filter states and a passivity condition on a system derived from the optimal filter. Convergence for the nonstochastic case is studied using Lyapunov functions and almost sure convergence is studied for the stochastic case using the deterministic results together with martingale theory.

§1. INTRODUCTION

For filters, optimal design methods as in [1] require knowledge of signal statistics or knowledge of a signal model for the system dynamics and the disturbances. It is usually assumed that the parameters of an optimal filter are calculated or identified using off-line calculations on experimental data. However, for situations in which the optimal parameter may drift slowly about a nominal value, it is clearly preferable if a simple recursive on-line calculation for adaption of the filter parameters can be arranged. Of course, convergence to at least a neighborhood of the optimal parameters should be guaranteed. Algorithms which achieve this end but are not necessarily globally convergent or able to track rapid parameter changes are termed self-tuning rather than adaptive since the term adaptive carries the notion of convergence even for rapidly changing parameters of relatively large magnitudes.

Adaptive or self-tuning filter algorithms are reviewed in [2]. Several approaches for filtering with parameter uncertainties are investigated. In the adaptive filtering area the Bayesian, maximum likelihood and error residual approaches are considered. On the other hand, the closely related problem of adaptive control is discussed [3] and in particular a model reference approach is employed. Least squares identification to facilitate adaptive control is introduced [4]. More recently, an approach to self-tuning regulators using least squares estimation ideas and having the advantage of simplicity as well as good convergence properties is in [5]. Duals of these regulator results also lead to the self-tuning filter results of [6]. They have been successfully employed in various engineering situations [5,6].

In this paper we take the view that the work of [5,6,7] is a major contribution to adaptive filtering and control and should be explored more fully. Here an alternative approach to self-tuning filters is presented from which the algorithms of [5,6] can be derived as special cases. The specific advantages of the approach taken here are now listed.

1. The class of filters with unknown parameters for which self-tuning algorithms are designed are more general than those of [6]. The unknown parameters may be auto regressive and/or moving average parameters of a subsystem of the filter but are not so restricted as in [6]. The measurements z_t may be vector quantities in both the algorithms and convergence theory whereas in [5,7] to achieve insights into convergence results scalar measurements are assumed. They may be either continuous-time or discrete-time measurements in contrast to [7] where they are restricted to the discrete time case. The case when the optimal filter is time varying where the time variations are known is not excluded as in [7].

2. The selection of a particular optimal filter structure allows us to propose reasonable self-tuning algorithms more directly than using the approach of [5]. They are also more simply expressed than those of [5], although each can be derived from the other.

3. The convergence theory of [5,7] is in fact incomplete but does give "valuable insight" into the behavior of the self-tuning algorithms. The most useful results in [5,7] require assumptions of stationarity and knowledge of such quantities as $E[\hat{x}_k\hat{x}_k']$ where $\hat{x}_k$ is the state of the filter. There is a considerable research effort required to gain insights into the convergence properties. Here the convergence theory for the basic nonlinear stochastic difference or differential

equations gives conditions for almost sure convergence of the self-tuning equations to the optimal filter equations. It requires a test for passivity of a system readily derived from the optimal filter. There is also a very reasonable "persistently exciting" condition and a further condition excluding certain highly unstable closed loop systems. There is no assumption of ergodicity or stationarity.

4. One advantage of the approach taken in this paper is that the very close relationship between self-tuning filters and model reference adaptive identification algorithms [8,9] becomes evident and in fact is exploited in the derivation of convergence results. The same can be said for the relationship between self-tuning regulators and model reference adaptive controllers, although this point is not taken up in this paper.

This paper focusses almost exclusively on self-tuning filters since space does not permit application of the ideas to predictors and regulators. In Section 2, a class of self-tuning filters is proposed which includes the self-tuning filters of [6] as a special case. A crucial step is the selection of a suitable filter structure with unknown parameters which are updated recursively using least squares techniques. The algorithm properties are studied in Section 3 and the theory is presented going beyond the work of [6,7] in that sufficient conditions for convergence of the self-tuning process are derived.

§2. SELF-TUNING FILTER

In this section we consider in turn, a suitable structure for an optimal (minimum variance) filter expressed in terms of a known subsystem and unknown parameters, a suboptimal filter where parameter estimates are employed, and an adjustment law for updating the parameter estimates from the filter state estimates and residuals. A convergence theory for the resulting self-tuning filter is given in Section

Conditional optimal filter. Consider the special filter structure of Figure 2.1a with linear state equations

$$\hat{x}_{k+1/k,\theta} = F\hat{x}_{k/k-1,\theta} + G\hat{y}_{k/k-1,\theta} + K\nu_{k/\theta} \tag{2.1a}$$

$$\nu_{k/\theta} = z_k - \hat{y}_{k/k-1,\theta} \tag{2.1b}$$

$$\hat{y}_{k/k-1,\theta} = \theta'\hat{x}_{k/k-1,\theta} \quad \text{or} \quad \hat{y}_{k/k-1,\theta} = \hat{X}'_{k/k-1,\theta}\theta \tag{2.1c}$$

where F, G, and K are known matrices (possibly time varying) of the subsystem block denoted $\mathcal{W}$ in Figure 2.1a and θ is an unknown parameter matrix or vector. In the alternative expression $\hat{y}_{k/k-1,\theta} = \hat{X}'_{k/k-1,\theta}\theta$, $\hat{X}$ is a matrix with elements derived from $\hat{x}$ and θ is restricted to being a vector.

We further assume that the block $\mathcal{W}$ is asymptotically stable. Also, for finite initial times, the initial states $\hat{x}_{0/-1,\theta}$ are assumed known and for initial times in the infinitely remote past z_k is assumed stationary and the filter is assumed to be asymptotically stable.

The above filter is termed a conditional optimal filter in that it is the best linear filter in a minimum variance sense only if the parameter θ which parameterizes the measurements z_k is known and is employed in this filter. The estimate $\hat{x}_{k/k-1,\theta}$ is the minimum variance (or conditional mean) estimate of the state x_k of some signal model the details of which are unimportant to us at this stage in our development of the self-tuning filter. For the minimum variance filter, the residuals $\nu_{k/\theta}$ are white. (In fact, whiteness of residuals of such a filter implies that the filter is a minimum variance filter).

Rather than work directly with the conditional optimal filter, we could equally well have assumed an innovations representation in terms of an unknown θ with parameters F, G, K known. This approach is not explored in detail here.

We now claim that for a wide class of linear stochastic signal models with unknown parameters, the optimal filter can be organized as in (2.1) with θ as the unknown parameter matrix. If θ were known then the optimal filter can be constructed. This claim will be looked at more closely when specializing the self-tuning filter results below for AR, ARMA signal models with unknown AR, and ARMA parameters, respectively. The results are also specialized for the case when the usual Kalman state space signal model is known but the noise variances are unknown.

For the case when the parameter matrix θ is unknown, then some estimate of θ, denoted $\hat{\theta}_k$, can be employed to yield an approximation to the optimal filter as follows.

<u>Approximation to the optimal filter</u>. Consider the suboptimal filter of Figure 2.1b with equations

$$\hat{x}_{k+1} = F\hat{x}_k + G\hat{y}_k + K\hat{\nu}_k \tag{2.2a}$$

$$\hat{\nu}_k = z_k - \hat{y}_k \tag{2.2b}$$

$$\hat{y}_k = \hat{\theta}_k'\hat{x}_k \quad \text{or} \quad \hat{y}_k = \hat{X}_k'\hat{\theta}_k \tag{2.2c}$$

with obvious notational implications. The approximation to the conditional mean estimate $\hat{x}_{k/k-1,\theta}$ is denoted $\hat{x}_k$. Certainly as $\theta_k = (\tilde{\theta} - \hat{\theta}_k) \to 0$, this filter approaches the conditional optimal filter (2.1) and $\tilde{x}_k = (\hat{x}_{k/k-1,\theta} - \hat{x}_k) \to 0$ and $\hat{\nu}_k \to \nu_{k/\theta}$.

Adjustment law. The whole point of working with a filter structure with $z_k = \theta'\hat{x}_{k/k-1,\theta} + \nu_{k/\theta}$ is that least squares ideas can be employed to give recursive equations for updating an estimate of the unknown parameter matrix θ. Let us introduce the temporary assumption that $\hat{x}_{k/k-1,\theta}$ is known and since $\nu_{k/\theta}$ is white, than a least squares estimate of θ is

$$\hat{\theta}_{k+1} = P_{k+1} \sum_{i=0}^{k} \hat{x}_{i/i-1,\theta} z_i'$$

$$P_{k+1}^{-1} = \sum_{i=0}^{k} \hat{x}_{i/i-1}\, \hat{x}_{i/i-1}' \tag{2.3}$$

Such an estimate converges almost surely to the true parameter θ under certain stability and persistently exciting conditions in [8].

The idea of extended least squares algorithms is to replace the optimal estimate $\hat{x}_{i/i-1,\theta}$ in (2.3) by the suboptimal estimate $\hat{x}_i$ of (2.2) to yield

$$\hat{\theta}_{k+1} = \Lambda_{k+1} \sum_{i=0}^{k} \hat{x}_i z_i', \qquad \Lambda_{k+1}^{-1} = \sum_{i=0}^{k} \hat{x}_i \hat{x}_i' \tag{2.4}$$

which may be calculated recursively as

$$\hat{\theta}_{k+1} = \hat{\theta}_k + \Lambda_{k+1}\, \hat{x}_k \hat{\nu}_k' \tag{2.5a}$$

$$\Lambda_{k+1} = \Lambda_k - \Lambda_k \hat{x}_k (\hat{x}_k \Lambda_k \hat{x}_k + I)^{-1} \hat{x}_k' \Lambda_k \tag{2.5b}$$

Corresponding adjustment laws are readily conjectured for the case $z_k = \hat{X}_{k/k-1,\theta}\theta + \nu_{k/\theta}$ as

$$\hat{\theta}_{k+1} = \hat{\theta}_k + \Lambda_{k+1} \hat{X}_k \hat{\nu}_k \tag{2.6a}$$

$$\Lambda_{k+1} = \Lambda_k - \Lambda_k \hat{X}_k (\hat{X}_k' \Lambda_k \hat{X}_k + I)^{-1} \hat{X}_k' \Lambda_k \tag{2.6b}$$

For this case in fact, a weighted least squares approach can be employed where I in (2.6b) is replaced by an estimate of the covariance $E[\nu_{k/\theta}\nu_{k/\theta}']$ such as $\frac{1}{k}\sum_{i=0}^{k} \hat{\nu}_i \hat{\nu}_i'$. Further details on this latter notion will not be presented here.

Self-tuning filter. The self-tuning filter is simply the approximation to the optimal filter (2.2) coupled to the adjustment law equations (2.5) or (2.6); see Figure 2.2.

Before studying the properties of the self-tuning filter, we examine special classes of signal models and self-tuning filters for these.

Autoregressive signal model case. Consider the AR signal model

$$z_k + \alpha_1 z_{k-1} + \cdots \alpha_n z_{k-n} = \nu_k \tag{2.7}$$

where z_k is a p-vector full rank process* and the α_i are $p \times p$ matrices. We denote $\theta' = [-\alpha_1, -\alpha_2, \cdots, -\alpha_n]$ as the unknowns AR parameters and $x_k' = [z_{k-1}', z_{k-2}', \cdots, z_{k-n}']$ as the states of the model. This model (2.7) is readily reorganized as

---

*Here $E[\nu_k\nu_k'] = \gamma > 0$ assures that z_k is a full rank process.

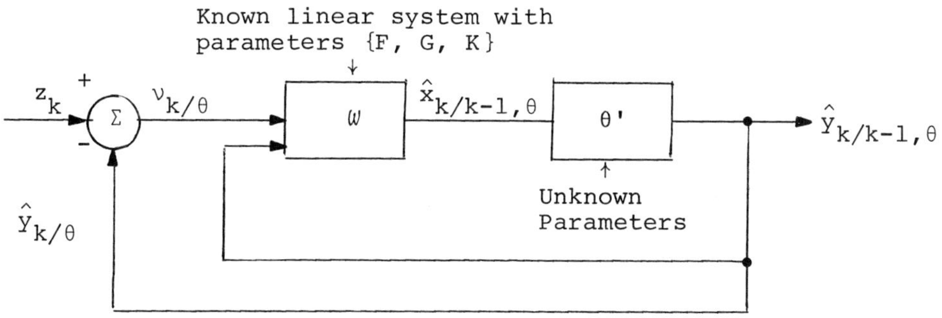

(a)

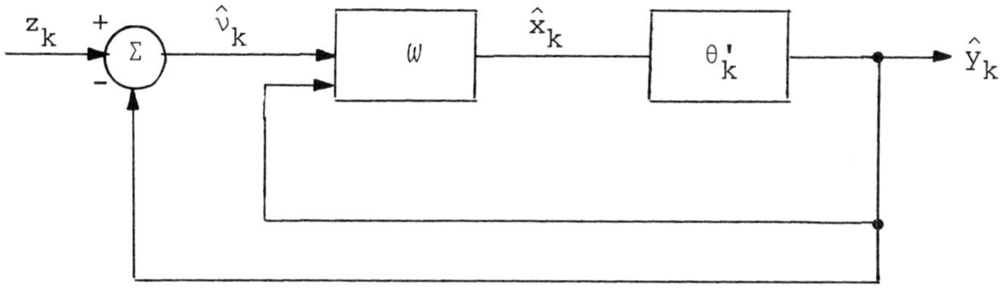

(b)

FIGURE 2.1. (a) Conditional optimal filter; (b) approximation to the optimal filters.

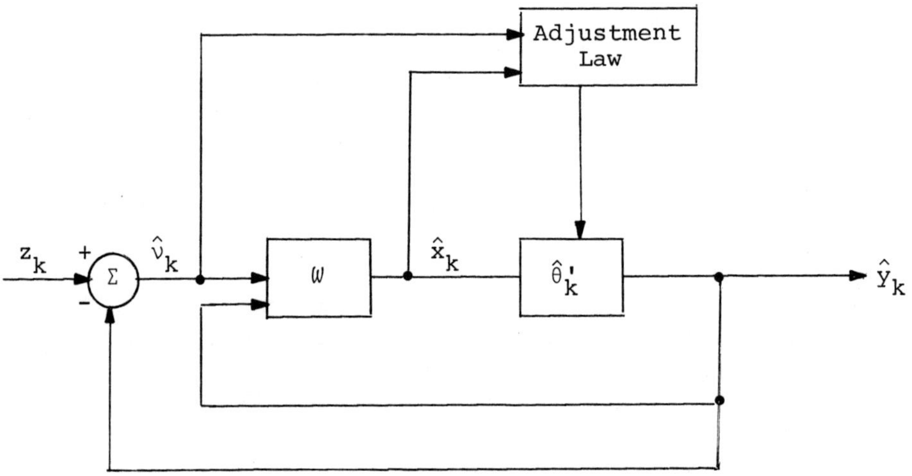

FIGURE 2.2. Self-tuning filter.

$$x_{k+1} = Fx_k + Gy_k + K\nu_k \tag{2.8a}$$

$$\nu_k = z_k - y_k \tag{2.8b}$$

$$y_k = \theta' x_k \tag{2.8c}$$

with

$$F = \begin{bmatrix} 0 & 0_p \\ I_{(n-1)p} & 0 \end{bmatrix}, \quad K = \begin{bmatrix} I_p \\ 0 \end{bmatrix}, \quad G = \begin{bmatrix} I_p \\ 0 \end{bmatrix} \tag{2.9}$$

when the initial time is in the infinitely remote past, z_k is stationary only if the AR model is asymptotically stable and only then is it an innovations representation.

The conditional optimal filter for the above innovations model (2.8) is given by (2.1) with matrices F, K, G, and θ as defined above. Of course, $\hat{x}_{k/k-1,\theta} = x_k$ and $\hat{y}_{k/k-1,\theta} = y_k$. The self-tuning filter for the AR model is derived as a specialization of the results earlier in the section as depicted in Figure (2.3a). Observe that the states $x_k = [z_{k-1}', \cdots,$

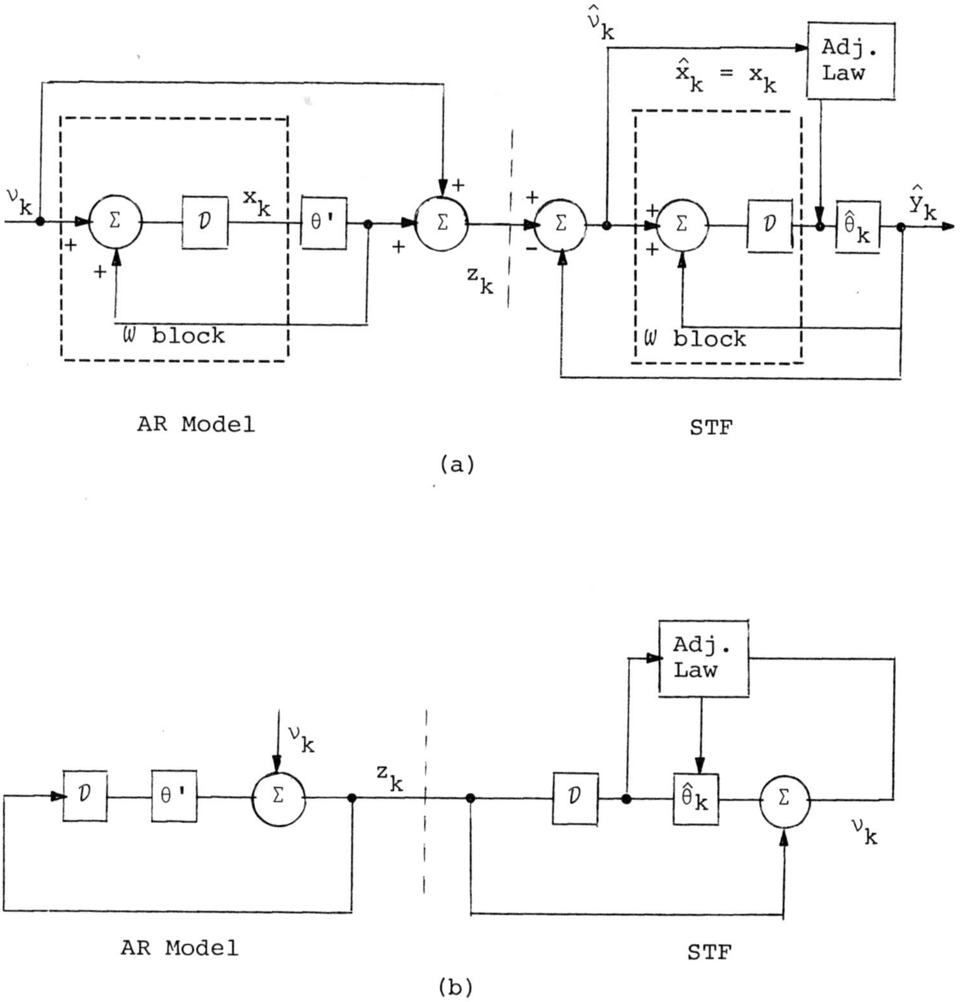

FIGURE 2.3. AR model and self-tuning filter (STF).

$z'_{k-n}]$ are derived from $z_k = y_k + \nu_k$ by passing z_k through an appropriate system of delays. Denoting this delay system by $\mathcal{D}$, then $x_k = \mathcal{D}z_k = \mathcal{D}(y_k + \nu_k) = \mathcal{D}(\hat{y}_k + \hat{\nu}_k) = \hat{x}_k$.

It is important here that $\hat{x}_k = \hat{x}_k = \hat{x}_{k/k-1,\theta}$ since then the estimation of θ is in fact nothing other than a least squares estimation process for which there is known almost sure convergence under very reasonable "persistently exciting" and "stability" conditions [8]. In fact the arrangement of

Figure (2.3a) simplifies as in Figure (2.3b) which is perhaps a less disguised version of a least squares AR estimator.

The application of the self-tuning filter ideas to the AR model case indicates that the algorithms reduce to least squares estimation when this is possible and are certainly no more sophisticated than least squares algorithms. For models such as ARMA models where the least squares theory can not be applied directly, the self-tuning filter algorithms now described are once again seen to be very simple.

<u>Autoregressive moving average signal model</u>. Consider the ARMA model for a p-vector full rank process z_k as

$$z_k = A_1 z_{k-1} + A_2 z_{k-2} + \cdots + A_n z_{k-n} = \nu_k$$
$$+ B_1 \nu_{k-1} + \cdots + B_n \nu_{k-n}$$

where A_i, B_i are $p \times p$ unknown AR and MA matrices. Its state space equations in the form of (2.8) can be worked out with the state $x_k' = [z_{k-1}', z_{k-2}', \cdots, z_{k-n}', \nu_{k-1}', \nu_{k-2}', \cdots, \nu'_{k-n}]$ and the unknown vector $\theta' = [-A_1, -A_2, \cdots, A_n, B_1, B_2, \cdots, B_n]$. The relevant matrices are

$$F = \begin{bmatrix} 0 & 0 & 0 & 0 \\ I_{(n-1)p} & 0 & 0 & 0 \\ 0 & 0 & 0 & 0_p \\ 0 & 0 & I_{(n-1)p} & \end{bmatrix}, \quad K = \begin{bmatrix} I_p \\ 0 \\ 0 \\ 0 \end{bmatrix}, \quad G = \begin{bmatrix} I_p \\ 0 \\ 0 \\ 0 \end{bmatrix} \qquad (2.10)$$

for the initial time at $-\infty$, the above model is an innovations model if and only if it is both asymptotically stable and minimum phase thus giving rise to an asymptotically stable inverse which can be viewed as an optimal filter.

The conditional optimal filter for the above innovations model (2.8) for the ARMA case is given by (2.1) with matrices F, K, G, and θ as defined above. Of course, $\hat{x}_{k/k-1,\theta} = x_k$ and $\hat{y}_{k/k-1,\theta} = y_k$ and the self-tuning filter for the ARMA model is derived as a specialization of the results earlier in the section. It takes the form of Figure (2.2) with the states of the $\mathcal{W}$ block being $\hat{x}'_k = [\hat{x}'_{1k}, \hat{x}'_{2k}]$ where $\hat{x}'_{1k} = [z'_{k-1}, z'_{k-2}, \cdots, z'_{k-n}]$, $x'_{2k} = [\hat{\nu}'_{k-1}, \hat{\nu}'_{k-2}, \cdots, \hat{\nu}'_{k-n}]$ which can be derived by passing z_k and $\hat{\nu}_k$ through an appropriate delay block such that

$$x_k = \bar{\mathcal{D}} \begin{bmatrix} z_k \\ \hat{\nu}_k \end{bmatrix}$$

Such a self-tuning filter specializes as depicted in Figure (2.4). Observe that $x_{1k} = x_{1k}$ is obtained directly from the measurements, but x_{2k} is obtained from the filter states.

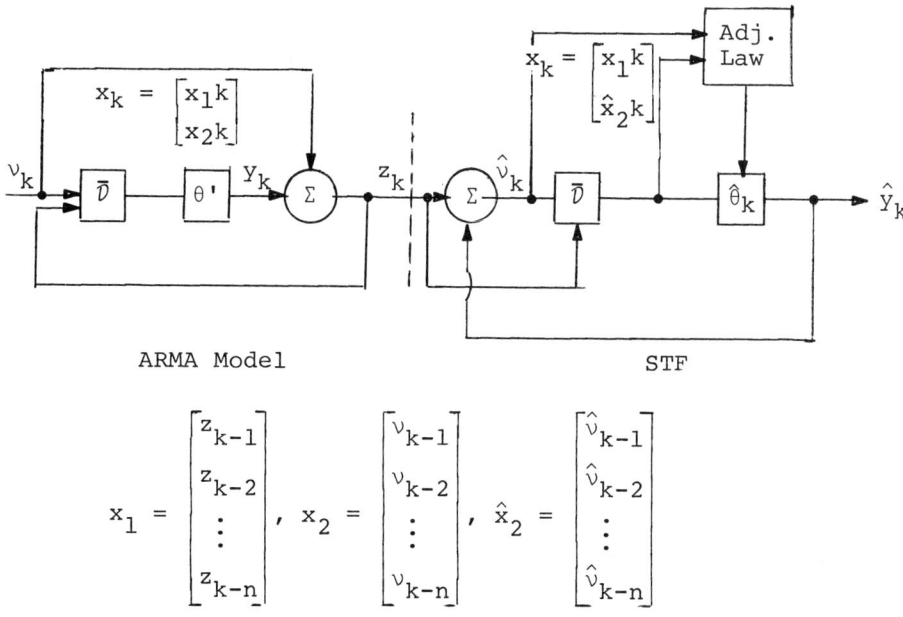

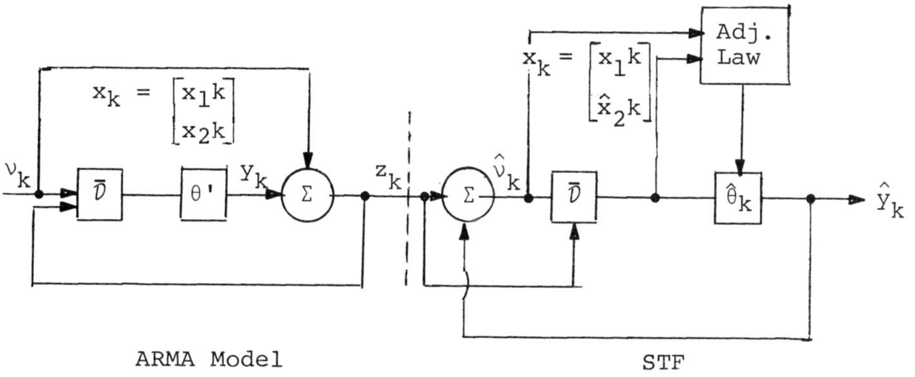

$$x_1 = \begin{bmatrix} z_{k-1} \\ z_{k-2} \\ \vdots \\ z_{k-n} \end{bmatrix}, \quad x_2 = \begin{bmatrix} \nu_{k-1} \\ \nu_{k-2} \\ \vdots \\ \nu_{k-n} \end{bmatrix}, \quad \hat{x}_2 = \begin{bmatrix} \hat{\nu}_{k-1} \\ \hat{\nu}_{k-2} \\ \vdots \\ \hat{\nu}_{k-n} \end{bmatrix}$$

FIGURE 2.4. Self-tuning filter for ARMA model - a simplified arrangement.

It may well be the case that the A_i in the ARMA model are restricted to being of the form a_iI for scalar a_i. This is because an arbitrary rational transfer function matrix V(z) can always be written as $V(z) = (\sum_{i=0}^{n} c_i z^{-i})^{-1}(\sum_{i=0}^{n} D_i z^{-i})$ for scalar c_i, the polynomial $\sum_{i=0}^{n} c_i z^{-i}$ being the lowest common denominator of all entries of V(z), expressed as ratios of polynomials in z^{-1}. In particular, then, a minimum phase ARMA model always has a representation with $A_i = a_iI$. It is clear that with this restriction, all entries of θ are not free--some are zero, others are equal--and thus θ is partially specified. Denote the matrix θ' as θ'_{ps}. In this case the term $y_k = \theta'_{ps}x_k$ can be reorganized as $y_k = X'_k\theta$ for a vector θ consisting of all the unknown elements of θ_{ps} and X'_k some matrix with elements either zero or the elements of x_k. For example, observe that

$$\begin{bmatrix} d_1 & 0 & d_2 \\ 0 & d_1 & d_3 \end{bmatrix} \begin{bmatrix} x_1 \\ x_2 \\ x_3 \end{bmatrix} = \begin{bmatrix} x_1 & x_3 & 0 \\ x_2 & 0 & x_3 \end{bmatrix} \begin{bmatrix} d_1 \\ d_2 \\ d_3 \end{bmatrix}$$

Clearly, for the case where ARMA models with A_i restricted as a_iI where a_i are scalars, the self-tuning filters can be organized in the form of (2.2) with $\hat{y}_k = \hat{X}'_k\hat{\theta}_k$ where the adjustment law (2.6) is employed.

<u>Known model dynamics, unknown noise covariance</u>. Frequently, in practice the signal model can be adequately represented by the state space model.

$$x_{k+1} = Ax_k + Bw_k$$
$$z_k = Cx_k + v_k$$

where A, B, C are known system matrices and the noise terms w_k and v_k are white and of mean zero but with the covariance

$$\begin{vmatrix} w_k w_k' & w_k v_k' \\ v_k w_k' & v_k v_k' \end{vmatrix} = \begin{vmatrix} Q & S \\ S & R \end{vmatrix} \quad \text{unknown}$$

The Kalman (minimum variance) filter for such a model is

$$\hat{x}_{k+1/k,K} = A\hat{x}_{k/k-1,K} + K\nu_{k/K} \tag{2.11a}$$

$$\nu_{k/K} = z_k - \hat{y}_{k/k-1,K},\ \hat{y}_{k/k-1,K} = C\hat{x}_{k/k-1,K} \tag{2.11b}$$

where the Kalman gain depends on the matrices Q, R, and S. Observe that this conditional optimal filter is immediately organized in terms of a known linear system block $\bar{W}$ and unknown parameters K as indicated in Figure (2.5a). For the *case of scalar measurements*, it is immediately clear that since $\bar{W}K = K'\bar{W}'$, then the filters may be reorganized as in Figure (2.5b) with $\theta' = K'$ and $W = \bar{W}'$. For the *case of vector measurements* the reorganization is not as direct as illustrated by the example

$$\bar{W} = \begin{bmatrix} \bar{W}_{11} & \bar{W}_{12} \\ \bar{W}_{21} & \bar{W}_{22} \end{bmatrix}, \qquad K = \begin{bmatrix} K_{11} & K_{12} \\ K_{21} & K_{22} \end{bmatrix}$$

for scalar $\bar{W}_{ij}$ and K_{ij}. Here it is seen that a selection of

$$W = \begin{bmatrix} \bar{W}_{11} & 0 \\ 0 & \bar{W}_{11} \\ \bar{W}_{12} & 0 \\ 0 & \bar{W}_{12} \\ \bar{W}_{21} & 0 \\ 0 & \bar{W}_{21} \\ \bar{W}_{22} & 0 \\ 0 & \bar{W}_{22} \end{bmatrix} \qquad \theta_{ps} = \begin{bmatrix} K_{11} & 0 \\ K_{12} & 0 \\ K_{21} & 0 \\ K_{22} & 0 \\ 0 & K_{11} \\ 0 & K_{12} \\ 0 & K_{21} \\ 0 & K_{22} \end{bmatrix}$$

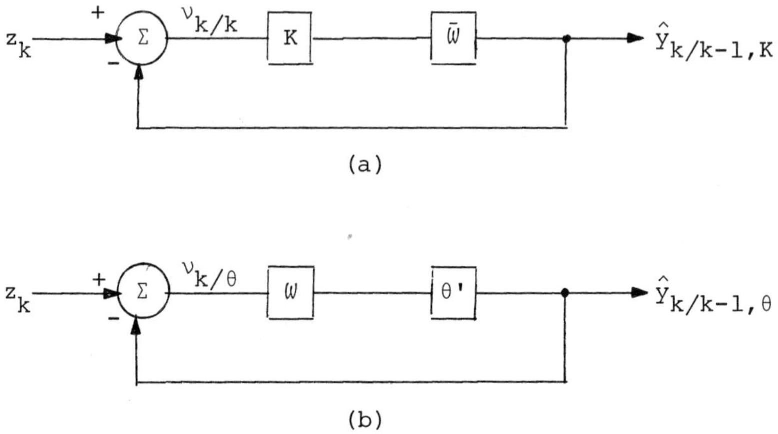

FIGURE 2.5. Conditional optimal filters for case of known model dynamics but unknown noise covariances.

ensures that $\bar{\omega}K = \theta'_{ps}\omega$. Again we are confronted with the task of identifying filters with θ_{ps} partially specified. For this case the equations (2.2) can be employed using the arrangement $\hat{y}_k = \hat{X}'_k\hat{\theta}_k$ for a vector $\hat{\theta}' = [\hat{K}_{11}\hat{K}_{12}\hat{K}_{21}\hat{K}_{22}]$ and the matrix $\ell_k = [\hat{X}_{1k} \quad \hat{X}_{2k}]$ where

$$\hat{X}_{1k} = \begin{bmatrix} \bar{\omega}_{11} & 0 \\ 0 & \bar{\omega}_{11} \\ \bar{\omega}_{21} & 0 \\ 0 & \bar{\omega}_{21} \end{bmatrix} \hat{\nu}_k, \qquad \hat{X}_{2k} = \begin{bmatrix} \bar{\omega}_{21} & 0 \\ 0 & \bar{\omega}_{21} \\ \bar{\omega}_{22} & 0 \\ 0 & \bar{\omega}_{22} \end{bmatrix} \hat{\nu}_k$$

Clearly, identification of $\theta = [K_{11}K_{12}K_{21}K_{22}]$ gives immediately the true Kalman gain K. Given an estimate $\hat{\theta}_k$, then a Kalman gain estimate R_k is available which can be employed in an approximate Kalman filter to achieve state estimates for the original signal model. The gain K in (2.11) is simply replaced by $\hat{K}_k$.

§3. CONVERGENCE RESULTS

For analysis purposes it is helpful to reformulate the self-tuning filter equations in terms of the state and parameter errors $\tilde{\theta}_k$ and $\tilde{x}_k$. We have

$$\tilde{\theta}_{k+1} = (I - \Lambda_{k+1}\hat{x}_k\hat{x}_k')\tilde{\theta}_k - \Lambda_{k+1}\hat{x}_k\tilde{x}_k'\theta' - (\Lambda_{k+1}\hat{x}_k\nu_k') \tag{3.1a}$$

$$\tilde{x}_{k+1} = (G - K)\hat{x}_k'\tilde{\theta}_k + [F + (G - K)\theta']\tilde{x}_k \tag{3.1b}$$

$$\tilde{y}_k = \theta_k'\hat{x}_k + \theta'\tilde{x}_k \tag{3.1c}$$

We seek conditions under which the states of (3.1) [the elements of $\tilde{x}$ and $\tilde{\theta}$] converge to zero as $k \to \infty$ since under such conditions the self-tuning filter converges to the optimal filter.

Observe that if $\tilde{\theta}_k$ in (3.1) approaches zero (the parameter estimate $\hat{\theta}_k$ approaches the true parameter θ), then (3.1) becomes

$$\tilde{x}_{k+1} = [F + (G - K)\theta']\tilde{x}_k, \quad \tilde{y}_k = \theta'\tilde{x}_k$$

as $k \to \infty$. If this system is asymptotically stable (the true filter is asymptotically stable), then $\tilde{x}_k \to 0$ as $k \to \infty$ and the state estimates $\tilde{x}_k$ approach the optimal estimates $\hat{x}_{k/k-1,\theta}$. Again, if $\tilde{x}_k$ in (3.1) approaches zero (as would certainly be the case for AR signal models), then (3.1) reduces to the least squares equations $\tilde{\theta}_{k+1} = [I - \Lambda_{k+1}\hat{x}_k\hat{x}_k']\tilde{\theta}_k - \Lambda_{k+1}\hat{x}_k\nu_k'$, $\tilde{y}_k = \hat{x}_k'\tilde{\theta}_k$. These equations give convergence of $\tilde{\theta}_k$ to zero as $k \to \infty$ under very reasonable "persistently exciting" and "stability" conditions [8].

For a stability analysis of the coupled equations we impose a restriction on the set of optimal filters (the set $\{\Theta\}$) in addition to the "persistently exciting" restrictions and "stability" restrictions imposed in least squares theory for stable signal models.

Two lemmas are now presented as preliminary results for the key convergence theorems concerning the convergence of the self-tuning filters or equivalently the convergence to zero of $\tilde{x}_k$ and $\tilde{\theta}_k$ as $k \to \infty$ in (3.1). The first lemma examines (3.1) as an unforced system by considering the special case when the driving term $\Lambda_{k+1}\tilde{x}_k\nu_k' = 0$. Unless (3.1) is asymptotically stable as an unforced system, then we could not expect that $\tilde{\theta}_k$, $\tilde{x}_k \to 0$ as $k \to \infty$ for the more interesting stochastic case where $\nu_k \neq 0$. The second lemma looks at the conditions under which the driving term $\Lambda_{k+1}\hat{x}_k\nu_k' \to 0$ as $k \to \infty$. The two lemmas together enable a convergence result to be obtained for the stochastic system equations (3.1). This is done in Theorem 3.1.

To achieve convergence of $\tilde{\theta}_k$ and $\tilde{x}_k$, certain stability assumptions are required. It turns out that the requirement that $\|\hat{x}_k\|$ be bounded is overly restrictive and there is only a need for a bound on $\hat{x}_k'\Lambda_k\hat{x}_k$, namely

$$\hat{x}_k'\Lambda_k\hat{x}_k + 1 \qquad (3.2)$$

for some scalar $\phi > 1$. This assumption eliminates certain classes of highly unstable self-tuning filter-signal model combinations in which the relevant eigenvalues are ever increasing. It in fact does not necessarily eliminate all unstable such combinations, although assumptions introduced at a later stage do. The condition (3.2) is termed here a "quasi-stability" condition.

Lemma 3.1 Consider the signal model and self-tuning filter equation (3.1) with F, G, K time invariant. Suppose that the "quasi-stability" condition (3.2) is satisfied. Then for the case when the equations are unforced in that $\Lambda_{k+1}\hat{x}_k\nu_k'$ is set to zero, a sufficient condition for $\tilde{x}_k$, $\tilde{y}_k$ (and thus also $\tilde{\theta}_k'\hat{x}_k$) to converge to zero as $k \to \infty$ is that the following transfer function matrix (derivable from the signal model parameters) is strictly positive real*:

$$W(z) = \frac{1}{2} I + \phi\theta'\{zI - [F + (G - k)\phi\theta']\}^{-1}(G - K) \quad (3.3)$$

with $|\lambda_i[F + (G - k)\phi\theta']| < 1$ and $|\lambda_i(F)| < 1$ for all i.

Proof Defining a function V_k as

$$V_k = tr(\tilde{\theta}_k'\Lambda_k^{-1}\tilde{\theta}_k) + \tilde{x}_k'P\tilde{x}_k \quad (3.4)$$

for some $P = P' > 0$, for the moment arbitrary, but later to be specified. Then tedious manipulations as in [9] show that

$$\Delta_k = V_{k+1} - V_k \leq \frac{1}{\phi^2} [\tilde{x}_k'(\tilde{\theta}_k'x_k)]D'RD \begin{bmatrix} \tilde{x}_k \\ \tilde{\theta}_k'x_k \end{bmatrix} \quad (3.5)$$

where

$$D = \begin{bmatrix} I & 0 \\ (\phi-1)\theta' & I \end{bmatrix}, \quad R = \begin{bmatrix} F_\#'PF_\# - P & F_\#'PG_\# - \phi\theta \\ G_\#'PF_\# - \phi\theta' & G_\#'PG_\# - I \end{bmatrix}$$

$$F_\# = F + (G - K)\phi\theta', \quad G_\# = (G - K)$$

---

*W(z) is strictly positive real here if (i) W(z) is real for real z; (ii) W(z) has no poles in $|z| \geq 1$; (iii) $W(e^{j\omega}) + W'(e^{-j\omega}) > 0$ for all real ω.

The fact that W(z) is positive real implies that there exists a particular choice of P ensuring that $R \leq 0$, see [10], provided that $[F_\#, G_\#]$ is completely reachable. It is actually not difficult to use the strict positive realness to conclude that simultaneously the condition $R \leq 0$ can be strengthened to $R < 0$ and the complete reachability can be relaxed to a stability requirement on $F_\#$. With such a choice of P, we have then

$$\Delta_k \leq - r^2 [\tilde{x}_k' \tilde{x}_k + (\hat{x}' \tilde{\theta}_k)(\tilde{\theta}_k' \hat{x}_k)] \tag{3.6}$$

for some $r \neq 0$. Since $V_k \geq 0$ for all k and $\Delta_k \leq 0$ for all k, we have $V_0 \geq V_0 - V_{k+1} = \sum_{j=1}^{k} (-\Delta_j) \geq 0$ for all k. Letting $k \to \infty$ shows that $\Delta_k \to 0$ and thus $\tilde{x}_k \to 0$, $\tilde{\theta}_k' \hat{x}_k \to 0$, and $\tilde{y}_k \to 0$.

Remarks

1. For fixed F, G, and K, the set $\{\phi\theta\}$ satisfying the positive real condition can be found if required but the computations are tedious [11]. Of course, when K = G and $|\lambda_i(F)| < 1$ as in asymptotically stable AR signal models, W(z) is strictly positive real for all $(\phi\theta)$. The important point for us here is that for many signal models, the positive real conditions for convergence of the signal estimate to the true signal will be satisfied in this unforced signal model case. The lower the bound ϕ is, the greater is the range of θ for which a self-tuning filter converges to the true filter.

2. The results can, at least in principle, be extended to time varying F, G, K. In the time-varying signal model case, the positive real condition is replaced by a passivity condition which is of course much harder to test.

3. The lemma can be extended to give conditions under which $\tilde{\theta}_k \to 0$ as $k \to \infty$, but such an extension is more conveniently embedded in a later theorem.

Lemma 3.2 The driving term $\Lambda_{k+1}\hat{x}_k\nu_k'$ of the filter equation (3.1) converges to zero almost surely.

Proof* First note that $X_m = \sum_{k=0}^{m} \Lambda_{k+1}\hat{x}_k\nu_k'$ is an F_m martingale (see Appendix A) where F_m is the σ-field generated by $\{\nu_1, \nu_2, \cdots, \nu_m\}$. To see this note that Λ_{k+1} and $\hat{x}_k$ are F_k measurable, ν_k is white, and so $E[\nu_, | F_{k-1}] = 0$, and thus

$$E[X_m | F_{m-1}] = E\left[\sum_{k=0}^{m-1} \Lambda_{k+1}\tilde{\ell}_, \nu_k' + \Lambda_{m+1}\hat{x}_m\nu_m' | F_{m-1}\right] = X_{m-1}$$

Moreover,

$$\begin{aligned} E[X_m X_m'] &= \sum_{j=0}^{m} E[X_j X_j' - X_{j-1}X_{j=1}'] \\ &= E\left\{\sum_{j=0}^{m} E[X_j X_j' = X_{j-1}X_{j-1}' | F_{j-1}]\right\} \\ &= E\left\{\sum_{j=0}^{m} E[(X_j - X_{j-1})(X_j' - X_{j-1}') | F_{j-1}]\right\} \\ &= E\left\{\sum_{j=0}^{m} \Lambda_{j+1}\left(\Lambda_{j+1}^{-1} - \Lambda_j^{-1}\right)\Lambda_{j+1}\right\} \end{aligned}$$

since $\Lambda_{j+1}^{-1} - \Lambda_j^{-1} = \hat{x}_j\hat{x}_j'$ and where without loss of generality we have taken $E[\nu_k\nu_k'] = I$. Observe now that

$$\begin{aligned} 0 \le (\Lambda_j - \Lambda_{j+1})\Lambda_j^{-1}(\Lambda_j - \Lambda_{j+1}) &= \Lambda_j - 2\Lambda_{j+1} \\ &+ \Lambda_{j+1}\Lambda_j^{-1}\Lambda_{j+1} \end{aligned}$$

Equivalently, $\Lambda_{j+1}(\Lambda_{j+1}^{-1} - \Lambda_j^{-1})\Lambda_{j+1} \le \Lambda_j - \Lambda_{j+1}$ which allows

*See also [8].

$$E[X_m X_m'] \le E\left\{\sum_{j=0}^{m} (\Lambda_j - \Lambda_{j+1})\right\} \le \Lambda_0$$

Thus X_m' is bounded in L^2. Standard martingale theorems [12] now yield that X_m is closed in L^2 and converges almost surely as m becomes infinite. The desired result follows.

The above two lemmas can now be employed along with further martingale theory to yield conditions for the almost sure convergence of x_k and $\tilde{\theta}_k$ to zero as $k \to \infty$. To achieve relevant convergence results, it appears that the "stability" condition (3.2) on $\hat{x}_k$ needs augmenting with the condition

$$\hat{x}_k' \Lambda_{k+1} \hat{x}_k < \beta/k^{\varepsilon} \text{ for some } \varepsilon > 0 \text{ and } \beta > 0 \tag{3.7}$$

which again falls short of requiring that $\|\hat{x}_k\|$ be bounded above. However, the most significant convergence results we achieve below appear to require a bound on $\|\hat{x}_k\|$. Also, the "persistently exciting" condition associated with "least squares" convergence which appears to be required is that

$$0 < \alpha_1 I \le \sum_{i=k+1}^{k+T+1} \hat{x}_i \hat{x}_i' = (\Lambda_{k+T}^{-1} - \Lambda_k^{-1}) \le \alpha_2 I \tag{3.8}$$

for some T, α_1, $\alpha_2 > 0$ and all k. This is a stronger condition than simply requiring that $\Lambda_{k+1} \to 0$ as $k \to \infty$ as in the conditions in [8]. This condition (3.8) constrains $\|x_k\|$ to be bounded above and gives the convergence rate of Λ_{k+1} to zero as $1/k$ convergence. Thus the "persistently exciting" condition (3.8) implies the "quasi-stability" conditions (3.2) and (3.7).

<u>Theorem 3.1</u> Suppose there is given the optimal filter (2.1) and self-tuning filter (2.2, 2.5) with associated

equations (3.1) where F, G, K are time-invariant, and the "stability" conditions (3.2) and (3.7) are satisfied. Then a sufficient condition for a subsequence of the signal estimation error y_k to converge to zero almost surely as $k \to \infty$ is $W(z) = \frac{1}{2} I + (\phi\theta)'\{zI - [F + (G - K)\phi\theta']\}^{-1}(G - K)$ be strictly positive real with $|\lambda_i[F + (G - K)\phi\theta']| < 1$ and $|\lambda_i[F]| < 1$ for all i. Moreover, if in addition the persistently exciting condition (3.8) is satisfied and $|\lambda_i[F + (G - K)\theta']| < 1$ for all i (i.e., the conditional mean filter is asymptotically stable), then $\tilde{\theta}_k$ and $\tilde{x}_k$ converge to zero almost surely as $k \to \infty$, or equivalently, the self-tuning filter converges almost surely to the conditional mean filter.

<u>Proof</u> For the function V_k of (3.4)

$$E[V_{k+1}|Z_k] = V_k + \Omega\hat{x}_k'\Lambda_{k+1}\hat{x}_k + \Delta_k \tag{3.9}$$

where $\Omega = \text{tr } E[\nu_k\nu_k']$ and $\Delta_k = V_{k+1} - V_k$ as defined in (3.5) but calculated with ν_k set to zero. Under the conditions of the theorem, it follows just as in the proof of Lemma 3.1 that $\Delta_k \leq 0$, and so

$$\frac{E[V_{k+1}|Z_k]}{k+1} \quad \frac{V_k}{k} + \frac{\beta\Omega}{k^{(1+\varepsilon)}}$$

Since $\sum_{i=0}^{\infty} i^{-(1+\varepsilon)} < \infty$ for $\varepsilon > 0$, a quantity $X_k = V_k/k + \beta\Omega \sum_{i=k}^{\beta} i^{-(1+\varepsilon)}$ can be defined with the property

$$E[X_{k+1}|Z_k] = \frac{E[V_{k+1}|Z_k]}{k+1} + \beta\Omega \sum_{i=k+1}^{\infty} i^{-(1+\varepsilon)}$$

$$\leq \frac{V_k}{k} + \beta\Omega \sum_{i=k}^{\infty} i^{-(1+\varepsilon)} = X_k$$

Taking expectations on this inequality yields that $E[X_{k+1}] \le E[X_k] \le E[X_0]$. As a consequence, the term $E[V_k/k]$ is bounded. This result and the inequality

$$\frac{E[V_{k+n}]}{k+n} \le \frac{E[V_k]}{k} + \beta\Omega \sum_{i=k}^{k+n} i^{-(1+\varepsilon)} + \sum_{i=k}^{k+n} \frac{E[\Delta_i]}{i}$$

ensure that $-\sum_{i=0}^{\infty} E[\Delta_i]/i < \infty$ and thus $E[\Delta_i] \to 0$ as $i \to \infty$. (Also $\Xi_i = \sum_{j=i}^{i+T} E[\Delta_j] \to 0$ as $i \to 0$.) Since $\Delta_i \le 0$, Δ_i converges in probability to zero as $i \to \infty$ and as a consequence, some subsequences of Δ_i (and $\Xi_i = \sum_{j=i}^{i+T} \Delta_j$) approach zero almost surely as $i \to \infty$. But as in Lemma 3.1, $\Delta_i \le r^2[\tilde{x}_k'\tilde{x}_k + \hat{x}_k'\tilde{\theta}_k\tilde{\theta}_k\hat{x}_k]$ for some $r \neq 0$ and thus a subsequence of $\tilde{x}_k$, $\tilde{\theta}_k'\hat{x}_k$ and $\tilde{y}_k$ converge to zero almost surely.

We now show the stronger result that with the additional persistently exciting condition (3.8), certain subsequences of $\tilde{\theta}_k \to 0$ almost surely as $k \to \infty$. We have already noted that there exist subsequences of

$$\Xi_i = \sum_{j=i}^{i+T} \Delta_i$$

which converge to zero almost surely. Therefore, so do subsequences of

$$\Phi_i = \sum_{j=i}^{i+T} \hat{x}_i'\tilde{\theta}_i\tilde{\theta}_i'\hat{x}_i$$

because $-\Xi_i \ge \Phi_i \ge 0$. Also, since $\tilde{\theta}_{k+1} = \tilde{\theta}_k - \Lambda_{k+1}\hat{x}_k\tilde{y}_k' - \Lambda_{k+1}\hat{x}_k\nu_k'$, we may use the fact that certain subsequences of $\tilde{y}_k$ converge to zero (the particular subsequences being determined by the subsequences of Ξ_i which converge to zero) and the result of Lemma 3.2, to conclude that subsequences of

$$\xi_i = \max_{j\in[i+1,i+T]} \|\tilde{\theta}_j - \tilde{\theta}_i\|$$

converge almost surely to zero. The indexing variables in the Φ_i and ξ_i subsequences are the same.

Now because of the ξ_i subsequence convergence,

$$\sum_{k=i}^{i+T} (\tilde{\theta}_k'\hat{x}_k\hat{x}_k'\tilde{\theta}_k) \to \tilde{\theta}_i' \sum_{k=i}^{i+T} \hat{x}_k\hat{x}_k'\tilde{\theta}_i$$

almost surely for a certain subsequence and because of the Φ_i subsequence convergence, we see that a subsequence of

$$\tilde{\theta}_i' \sum_{k=i}^{i+T} \hat{x}_k'\hat{x}_k'\tilde{\theta}_i$$

converges almost surely. The persistently exciting condition then shows that some subsequences of the $\tilde{\theta}_k$ sequence converges to zero almost surely.

Now we shall show that $\tilde{\theta}_k \to 0$ and $\tilde{x}_k \to 0$ almost surely. Observe that X_k above is in fact a positive supermartingale, and thus converges almost surely [12]. In turn V_k/k converges almost surely. But now (3.1) implies that $\Lambda_k^{-1} \leq kI\gamma$ for some γ, and therefore

$$0 \leq \frac{V_k}{k} = \frac{1}{k}\left[\tilde{x}_k'\bar{P}x_k + \tilde{\theta}_k'\Lambda_k^{-1}\tilde{\theta}_k\right] \leq \frac{\tilde{x}_k'P\tilde{x}_k}{k} + \gamma\tilde{\theta}_k'\tilde{\theta}_k$$

Subsequences of the upper bound on V_k/k converge to zero almost surely and so V_k/k converges almost surely to zero. Using (3.8) to give a lower bound for V_k/k, we see that $\tilde{\theta}_k \to 0$ almost surely. Finally, because [see (3.1)]

$$\tilde{x}_{k+1} = [F + (G - k)\theta']\tilde{x}_k + (G - K)\tilde{\theta}_k'\hat{x}_k$$

the eigenvalue constraint on $[F + (G - K)\theta']$, the boundedness of $\hat{x}_k$ and the convergence of $\tilde{\theta}_k$ show that $\tilde{x}_k \to 0$ almost surely.

Remarks

1. The case when the assumed signal model dimension is too low results in biased convergence and is not considered further here.

2. The time invariant constraint on F, K, and G can be relaxed by generalizing the positive real condition to a passivity condition. Also results for the case when $y_k = X_k'\theta_k$ for some matrix X_k and vector θ_k are readily derived.

3. The above theorem tells us that under certain conditions, the self-tuning algorithm will converge, but it is unlikely that all the conditions can be tested realistically in practice before actually constructing a self-tuning filter. In practice it is more likely that one would build a self-tuning filter and see if it worked. But then one needs a mechanism to decide whether it it working or not. For this, the following two theorems are useful.

Theorem 3.2 For a measurement sequence $\{z_n\}$ where the optimal filter is (2.1) with time-invariant parameters, suppose that the parameter adjustment of the self-tuning filter (2.2) and (2.5) converges in that $\|\tilde{\theta}_i - \delta\| \le \alpha_i^{-\varepsilon}$ for some α, $\varepsilon > 0$ and some δ, and $\hat{x}_k$ is bounded. Then

$$\lim_{k\to\infty} \frac{1}{k} \sum_{0}^{k} \hat{x}_i \hat{\nu}_i' = 0 \qquad (3.10)$$

If in addition it is known that the self-tuning filter converges to a time invariant filter (for asymptotically stationary measurements), then asymptotic ergodicity yields that

$$\lim_{k\to\infty} E[\hat{x}_k \hat{\nu}_k'] = 0 \qquad (3.11)$$

Proof Premultiplying (2.4) by $(\sum_{i=0}^{k} \hat{x}_i\hat{x}_i')$ and replacing z_k by $(\hat{\theta}'\hat{x}_k + \hat{\nu}_k')$ yields

$$\sum_{i=0}^{k} \hat{x}_i\hat{x}_i'\hat{\theta}_{k+1} = \sum_{i=0}^{k} \hat{x}_i\hat{\nu}_i' + \sum_{i=0}^{k} \hat{x}_i\hat{x}_i'\hat{\theta}_i$$

from which we have that

$$\frac{1}{k}\sum_{0}^{k} \hat{x}_i\hat{\nu}_i' = \frac{1}{k}\sum_{0}^{k} \hat{x}_i\hat{x}_i'(\hat{\theta}_{k+1} - \hat{\theta}_i)$$

The Schwarz inequality now yields

$$\|\frac{1}{k}\sum_{0}^{k} \hat{x}_i\hat{\nu}_i'\| \le \frac{1}{k}\sum_{0}^{k} \|\hat{x}_i\|^2\|\hat{\theta}_{k+1} - \hat{\theta}_i\|$$

Now with $\|\hat{x}_i\|$ bounded by $\beta < \infty$, then for $\delta = \lim_{k\to\infty} \theta_k$

$$\lim_{k\to\infty} \|\frac{1}{k}\sum_{0}^{k} \hat{x}_i\hat{\nu}_i'\| \le \lim_{k\to\infty} \frac{\beta}{k}\sum_{0}^{k} \|\tilde{\theta}_i - \delta\|$$

Classical analysis now yields that if $\|\tilde{\theta}_i - \delta\| \le \alpha_i^{-\varepsilon}$ for some $\alpha, \varepsilon > 0$, then the right hand side of the above inequality is zero and the first result (3.10) is established. That (3.11) follows from (3.10) under asymptotical ergodicity assumptions is immediate.

Theorem 3.3 Consider the self-tuning filter of Figure 2.2 and the block $\mathcal{W}$ a linear causal time-invariant completely reachable subsystem. Consider also that the parameter adjustment converges and $\lim E[\hat{x}_k\hat{\nu}_k'] = 0$ (as in Theorem 3.2). Then if it is known that the r-vector $\hat{\nu}_k$ has innovations representation with reachability index n or equivalently with an ARMA representation

$$\hat{\nu}_k = \sum_{i=0}^{n} A_{ik}\hat{\nu}_{k-1} + \sum_{i=0}^{n} B_{ik}\nu_{i-1},\ B_{0k} = I \tag{3.12}$$

for some matrices A_{ik} and B_{ik} and that x_k is an nr-vector, then

$$\lim_{k\to\infty} E[\hat{\nu}_k \hat{\nu}'_{k-\ell}] = 0 \text{ for all } \ell \neq 0 \tag{3.13}$$

and thus the self-tuning filter converges to the conditional mean filter with $\tilde{y}_k = (\hat{y}_{k/k-1} - \hat{y}_k) \to 0$ as $k \to \infty$.

Proof The reachability assumption on $\mathcal{W}$ and the assumption on the convergence of $\tilde{\theta}$ ensures that the subsystem of the filter with inputs $\{\hat{\nu}_k\}$ and states $\{\hat{x}_k\}$ is completely reachable and asymptotically time invariant. This implies that as $k \to \infty$ there exists a full rank transformation T such that $T\hat{x}_k = [s'_{k-1}, s'_{k-2}, \cdots, s'_{k-n}]'$ for some r vector sequence $\{s_k\}$ derived from $\hat{\nu}_k$ by the AR equations

$$s_{k+1} = - \sum_{i=1}^{n} \alpha_i s_{k-i+1} + \hat{\nu}_{k+1} \tag{3.14}$$

with α_i some matrix AR parameters. Now $E[\hat{x}_k \hat{\nu}'_k] = 0$, $E[T\hat{x}_k \hat{\nu}'_k] = 0$, $E[s_{k-i} \hat{\nu}'_k] = 0$ for all $0 < i \leq n$. Post multiplying (3.12) by $s'_{k-n-1}, s'_{k-n-2}, \cdots$, in turn, it follows that

$$E[s_{k-i} \hat{\nu}'_k] = 0 \text{ for all } i > 0 \tag{3.15}$$

Returning to (3.14), post multiplying by $\hat{\nu}_{k+i}$, taking expectations and applying (3.15) leads to the condition (3.13). To show the filter converges to the optimal filter, note the equations for the Kalman filter for z_k may be written

$$\hat{x}_{k+1/k} = F\hat{x}_{k/k-1} + K\nu_k, \quad \nu_k = z_k - H'\hat{x}_{k/k-1} \tag{3.16}$$

where the innovation ν_k is a white noise sequence. From (3.16) it is clear that z_k may be written

$$y_k = \nu_k + H'K\nu_{k-1} + H'FK\nu_{k-2} + H'F^2K\nu_{k-3} \cdots$$

The estimate $\hat{y}_k$ being driven by y_k is also expressible as

$$\hat{y}_k = A_1\nu_{k-1} + A_2\nu_{k-2} + A_3\nu_{k-3} \cdots \text{ for some } A_i$$

So

$$\hat{\nu}_k = \nu_k + (H'K - A_1)\nu_{k-1} + (H'FK - A_2)\nu_{k-2}$$
$$+ (H'F^2K - A_3)\nu_{k-3} \cdots$$

But by (3.17) $\hat{\nu}_k$ converges to moving average of white noise of length one and hence $\hat{\nu}_k$ converges to ν_k. But since $\nu_k = z_k - \hat{y}_{k/k-1}$ and $\hat{\nu}_k = z_k - \hat{y}_k$ this implies the desired result $\lim_{k\to\infty} (\hat{y}_{k/k-1} - y_k) = 0$.

Remarks

1. The Theorems 3.2 and 3.3 are a generalization of results quoted in [5], to the class of signal models considered here. The nature of the assumed convergences of $\tilde{\theta}_i$ to some constant value is less restrictive than in [5].

2. It might be questioned whether or not under the conditions of the Theorem 3.2, $\tilde{\theta}_k$ can converge to other than zero if $\tilde{y}_k$ converges to zero. Following earlier arguments it can be shown that if $\tilde{y}_k$ converges to zero and the persistently exciting condition $\Lambda_{k+1} \to 0$ as $k \to 0$ is satisfied, then both $\tilde{\theta}_k$ and $\tilde{x}_k$ must also approach zero. In this context it is also helpful to view the persistently exciting condition as an observability condition.

3. The conditions in Theorem 3.3 concerning the dimensionality of an innovations model for $\hat{\nu}_k$ are always satisfied

when $\mathcal{W}$ and θ are the models used in [9] for AR and ARMA innovations representations.

§4. CONCLUSIONS

Using an innovations representation with unknown parameters organized in a special form with the measurement vector as $z_k = \theta' x_k + \nu_k$ where θ is the unknown parameter, least squares ideas suggest a class of self-tuning filter algorithms which lead to least squares estimation of θ in the event that x_k is known and lead to the conditional mean filter in the event that θ is known. In the event that estimates of both θ and x_k must be used, the equations for the estimation error $\tilde{x}_k$ are coupled to the equations for the parameter estimation error $\tilde{\theta}_k$.

Under certain "pseudo-stability" constraints for the self-tuning filter scheme, and "persistently exciting" conditions, Lyapunov theory and martingale convergence theorem yield almost sure convergence of the self-tuning filter to the conditional mean filter for a restricted class of signal models (or optimal filters) constrained by a positive real condition.

For the case when an actual self-tuning algorithm parameter estimate $\hat{\theta}_k$ converges, then appropriate asymptotic ergodicity, filter reachability, and signal model dimensionality assumptions ensure that the filter tunes to the conditional mean filter.

The cases when the algorithms of this paper do not converge require further study.

REFERENCES

1. Astrom, K. J., Introduction to Stochastic Control Theory, Academic Press, New York (1970).

2. Leondes, C. T. and J. O. Pearson, Kalman Filtering of Systems with Parameter Uncertainties - A Survey, Int. J. Control, vol. 17, no. 4, 1973, pp. 785-801.

3. Landau, I. D. and B. Courtoil, Design of Multivariable Adaptive Model Following Control Systems, Automatica, vol. 10, no. 5, pp . 483-494.

4. Peterka, V., Adaptive Digital Regulation of Noisy Systems, 2nd Prague IFAC Symposium on Identification and Process Parameter Estimation (1970).

5. Astron, K. S. and B. Wittenmark, On Self-Tuning Regulators, Automatica, vol. 9 (1973), pp. 185-199.

6. Wittenmark, B., A Self-Tuning Predictor, IEEE Trans. on Auto Control, vol. AC-19, no. 6, December, 1974, pp. 848-851.

7. Ljung, L. and B. Wittenmark, Asymptotic Properties of Self-Tuning Regulators, Lund Institute Report 7404, February, 1974.

8. Moore, J. B., Martin C. Clark, and B. D. Anderson, On Martingales and Least Squares Linear System Identification, IFAC Symposium of Identification and System Parameter Estimation, USSR (Tblissi), 1976.

9. Ledwich, G. and J. B. Moore, On Multivariable Linear System Identification, to appear.

10. Anderson, B. D. O., K. L. Hitz, and N. D. Diem, Recursive Algorithm for Spectral Factorization, IEEE Transactions on Circuits and Systems, No. 6, Nov., 1974, pp. 742-750.

11. Jury, E., Inners and Stability of Dynamic Systems, Wiley, New York, 1974.

12. Meyer, P., Martingales and Stochastic Integrals, Lecture Notes on Mathematics Series, No. 284, Springer-Verlag, New York, 1972.

STOCHASTIC DIFFERENTIAL GAMES WITH STOPPING TIMES

A. Bensoussan and J. L. Lions

University of Paris IX and
LABORIA, Paris, France

and

Collège de France and
LABORIA, Paris, France

§1. INTRODUCTION

In a previous paper [2], we have noticed that the solution of a variational inequality (VI) can be interpreted as the pay-off of a stochastic control problem, where the control variable is a stopping time. More precisely, let the evolution of the system be governed by the Ito equation

$$\begin{aligned} dy &= g(y)dt + \sigma(y)dw(t) \\ y(0) &= x \end{aligned} \tag{1.1}$$

Define next the pay-off

$$J_x(\theta) = E\left[\int_0^{\theta\wedge\tau_x} f(y(s))ds + \psi(y(\theta))\chi_{\theta<\tau_x}\right] \tag{1.2}$$

where θ is a stopping time, and τ_x is the first exit time of the trajectory, from a region $\mathcal{O}$ (bounded for simplicity). If one sets

$$u(x) = \operatorname*{Inf}_{\theta} \tau_x(\theta) \tag{1.3}$$

then under mild assumptions, the function u(x) can be

characterized as the solution of a VI, which broadly speaking, can be written as follows

$$\begin{aligned} &Au \le f \\ &u \le \psi \\ &(u-\psi)(Au-f) = 0 \\ &u/\Gamma = 0 \end{aligned} \tag{1.4}$$

where $\Gamma = \partial\mathcal{O}$ is the boundary of $\mathcal{O}$. The writing (1.4) requires regularity assumptions on the function ψ, in order to obtain regularity results on u, such that Au have a meaning (A is the 2nd order differential operator connected with the Ito equation (1.1)).

In [1] we have given a review of the various cases which can be envisaged. We have in mind a similar program here for two-sided variational inequalities, which can be interpreted as the pay-off of a differential game with stopping times. The analog of (1.4) is

$$\begin{aligned} &(Au - f)(v - u) \ge 0 \qquad \forall \ \psi_1 \le v \le \psi_2 \\ &\psi_1 \le u \le \psi_2 \end{aligned} \tag{1.5}$$

The writing (1.5) requires regularity assumptions on ψ_1, ψ_2. Under those regularity assumptions, the study of (1.5) and its interpretation as the pay-off of a differential game with stopping times has been done by Krylov [5] for the stationary case and by Friedman [4] for the nonstationary case. We will here for both the stationary and nonstationary cases review their results and compare them to what can be expected in nonregular situations. Some of the proofs will be omitted. All the details can be found in a forthcoming book [3].

§2. STATIONARY CASE

2.1 Notation - Setting of the Problem

Let $\mathcal{O}$ be a bounded open subset of R^n, whose boundary is C^2. Let $a_{ij}(x)$, $g_i(x)$ be functions defined on $\bar{\mathcal{O}}$, such that

$$a_{ij} = a_{ji}, \quad \Sigma a_{ij}\xi_i\xi_j \geq \beta\Sigma\xi_i^2, \ \beta > 0,$$
$$\forall\ \xi_1, \cdots, \xi_n \in R; \qquad a_{ij} \in C^0(\bar{\mathcal{O}}) \tag{2.1}$$

$$g_i(x) \in L^\infty(\mathcal{O}) \tag{2.2}$$

We assume that $a \equiv a_{ij}$ and $g \equiv g_i$ are conveniently extended to R^n, in order that properties (2.1) and (2.2) be preserved. We set

$$a = \frac{\sigma^2}{2} \tag{2.3}$$

and define the second order differential operator

$$A = -\sum_{ij} a_{ij}\frac{\partial^2}{\partial x_i \partial x_j} - \sum_i g_i \frac{\partial}{\partial x_i} \tag{2.4}$$

Let f, ψ_1, ψ_2 be functions defined on $\bar{\mathcal{O}}$ such that

$$f \in L^p(\mathcal{O}), \quad \psi_i \in C^0(\bar{\mathcal{O}}), \quad A\psi_i \in L^p(\mathcal{O}),\ p > \frac{n}{2}$$
$$\psi_1 \leq \psi_2 \text{ on } \bar{\mathcal{O}}, \quad \psi_1/\Gamma \leq 0 \leq \psi_2/\Gamma$$
$$f - A\psi_2 \leq C, \quad f - A\psi_1 \geq -C,\ C \geq 0 \tag{2.5}$$

We consider the following problem: find a function u(x) such that

$$u \in W^{2,p}(\mathcal{O})$$
$$\psi_1 \leq u \leq \psi_2 \text{ on } \bar{\mathcal{O}}, \quad u/\Gamma = 0$$
$$Au \leq f \quad \text{a.e. on } \mathcal{O} \quad \text{if } \psi_1 < u < \psi_2$$
$$Au \geq f \quad \text{a.e. on } \mathcal{O} \quad \text{if } u = \psi_1$$
$$Au \leq f \quad \text{a.e. on } \mathcal{O} \quad \text{if } u = \psi_2 \tag{2.6}$$

We shall show the existence and uniqueness of the solution u of (2.6) and interpret it as the value of a differential game with stopping times. This result is essentially in Krylov [5], or Friedman [4]. Our proof differs from their's as far as existence is concerned. It is based on the extensive use of the Maximum Principle instead of probabilistic methods[1]. This approach follows a suggestion of T. Kurz [6]. Before giving the proof, we describe the probabilistic set up that we need for the interpretation of the solution. Let

$$\Omega = C\left([0,\infty);\ R^n\right),\quad \omega \equiv \omega(s) \text{ and } F^t = \sigma(\omega(s)),\ 0 \le s \le t$$

We set

$$y(t;\ \omega) = \omega(t)$$

For any $x \in R^n$, there exists one and only one measure P^x on $(\Omega,\ F^\infty)$ such that there exists a standard Wiener process w(t), which is an F^t martingale, and

$$\begin{aligned} dy &= g(y)dt + \sigma(y)dw(t) \\ y(0) &= x \end{aligned} \tag{2.7}$$

This result is due to Stroock-Varadhan [8]. Next, let $v_1(t)$, $v_2(t)$ be two nonanticipative processes such that

$$\text{a.s.} \quad v_i(t) \in [0,1], \quad \forall t \tag{2.8}$$

Let τ be the exit time of y(t) from $\mathcal{O}$. We set

$$J^x_\varepsilon(v_1,v_2) = E^x\left[\int_0^\tau dt\left(f + \frac{1}{\varepsilon}v_1\psi_1 + \frac{1}{\varepsilon}v_2\psi_2\right)(y(t)) \cdot \left(\exp - \frac{1}{\varepsilon}\int_0^t (v_1 + v_2)ds\right)\right] \tag{2.9}$$

---

[1] It is therefore not a fundamentally different one.

Next, let θ_1, θ_2 be two stopping times. We set

$$J^x(\theta_1,\theta_2) = E^x \int_0^{\tau\wedge\theta_1\wedge\theta_2} f(y(t))dt + \psi_1(y(\theta_1))\chi_{\theta_1\le\theta_2,\theta_1<\tau} + \psi_2(y(\theta_2))\chi_{\theta_2<\theta_1\wedge\tau} \tag{2.10}$$

We introduce the penalized problem. It is defined as follows:

$$\begin{aligned} &Au_\varepsilon + \frac{1}{\varepsilon}(u_\varepsilon - \psi_2)^+ - \frac{1}{\varepsilon}(u_\varepsilon - \psi_1)^- = f \text{ in } \mathcal{O} \\ &u_\varepsilon/\Gamma = 0 \\ &u_\varepsilon \in W^{2,p}(\mathcal{O}) \end{aligned} \tag{2.11}$$

We have the following theorems.

Theorem 2.1 Under the assumptions (2.1, 2.2, and 2.5), there exists one and only one solution of (2.11). Moreover, one has

$$u_\varepsilon(x) = \min_{v_2}\max_{v_1} J^x_\varepsilon(v_1,v_2) = \max_{v_1}\min_{v_2} J^x_\varepsilon(v_1,v_2) \tag{2.12}$$

Theorem 2.2 Under the assumptions (2.1, 2.2, and 2.5), there exists one and only one solution of (2.6). Moreover, one has

$$u(x) = \min_{\theta_2}\max_{\theta_1} J^x(\theta_1,\theta_2) = \max_{\theta_1}\min_{\theta_2} J^x(\theta_1,\theta_2) \tag{2.13}$$

2.2 Proof of Theorem 2.1

Let $\alpha < 0$. For $\phi \in C^0(\bar{\mathcal{O}})$, we define

$$z = S_\alpha(\phi)$$

as the unique solution of

$$Az + \alpha z + \frac{2}{\varepsilon} z = f - \frac{1}{\varepsilon}(\phi - \psi_2)^+ + \frac{1}{\varepsilon}(\phi - \psi_1)^- + \frac{2}{\varepsilon}\phi$$
$$z/\Gamma = 0$$
$$z \in W^{2,p}(\mathcal{O}) \tag{2.14}$$

By Sobolev imbedding theorems, $S_\alpha(\phi) \in C^0(\bar{\mathcal{O}})$. We shall show that $S_\alpha(\phi)$ has a unique fixed point, which is equivalent to showing that the equation

$$Aw_\varepsilon + \alpha w_\varepsilon + \frac{1}{\varepsilon}(w_\varepsilon - \psi_2)^+ - \frac{1}{\varepsilon}(w_\varepsilon - \psi_2)^- = f$$
$$w_\varepsilon/\Gamma = 0$$
$$w_\varepsilon \in W^{2,p}(\mathcal{O}) \tag{2.15}$$

has one and only one solution. To prove the existence and uniqueness of a fixed point of $S_\alpha(\phi)$, we shall show that $S_\alpha(\phi)$ is a contraction mapping on $C^0(\bar{\mathcal{O}})$. Take indeed $\phi_1, \phi_2 \in C^0(\mathcal{O})$ and let z_1, z_2 be the corresponding solutions of (2.14). We have

$$A(z_1 - z_2) + \alpha(z_1 - z_2) + \frac{2}{\varepsilon}(z_1 - z_2) = X$$
$$z_1 - z_2/\Gamma = 0 \tag{2.16}$$

where

$$\begin{aligned} X = {} & + \frac{2}{\varepsilon}(\phi_1 - \phi_2) - \frac{1}{\varepsilon}(\phi_1 - \psi_2)^+ + \frac{1}{\varepsilon}(\phi_2 - \psi_2)^+ \\ & + \frac{1}{\varepsilon}(\phi_1 - \psi_1)^- - \frac{1}{\varepsilon}(\phi_2 - \psi_1)^- \\ = {} & - \frac{1}{\varepsilon}(\phi_1 - \psi_2)^- + \frac{1}{\varepsilon}(\phi_2 - \psi_2)^- \\ & + \frac{1}{\varepsilon}(\phi_1 - \psi_1)^+ - \frac{1}{\varepsilon}(\phi_2 - \psi_1)^+ \end{aligned} \tag{2.17}$$

hence

$$|X| \le \frac{2}{\varepsilon}\|\phi_1 - \phi_2\|_{C^0(\bar{\mathcal{O}})} \tag{2.18}$$

By the Maximum principle, it follows from (2.16) that

$$\|z_1 - z_2\|_{C^0(\bar{\mathcal{O}})} \leq \frac{1}{\alpha + \frac{2}{\varepsilon}} \frac{2}{\varepsilon} \|\phi_1 - \phi_2\|_{C^0(\bar{\mathcal{O}})} \tag{2.19}$$

hence the contraction property.

Now, let α go to 0, in order to prove the existence of a solution of (2.11). Let ζ_ε be the solution of

$$A\zeta_\varepsilon = |f| + \frac{1}{\varepsilon}|\psi_1| + \frac{1}{\varepsilon}|\psi_2|$$

$$\zeta_\varepsilon/\Gamma = 0 \tag{2.20}$$

We have

$$\|\zeta_\varepsilon\|_{C^0(\bar{\mathcal{O}})} < \infty, \qquad \zeta_\varepsilon \geq 0 \tag{2.21}$$

But from (2.15) and (2.20) it follows that

$$\begin{aligned} A(w_\varepsilon - \zeta_\varepsilon) + \alpha(w_\varepsilon - \zeta_\varepsilon) &= f - |f| - \alpha\zeta_\varepsilon \\ &- \frac{1}{\varepsilon}(w_\varepsilon - \psi_2)^+ - \frac{1}{\varepsilon}|\psi_2| + \frac{1}{\varepsilon}(w_\varepsilon - \psi_1)^- \\ &- \frac{1}{\varepsilon}|\psi_2| \end{aligned}$$

or

$$\begin{aligned} A(w_\varepsilon - \zeta_\varepsilon) + \alpha(w_\varepsilon - \zeta_\varepsilon) + \frac{1}{\varepsilon}(w_\varepsilon - \zeta_\varepsilon) &= f - |f| \\ &- \left(\alpha + \frac{1}{\varepsilon}\right)\zeta_\varepsilon + \frac{1}{\varepsilon}\left(\psi_1 - |\psi_1|\right) - \frac{1}{\varepsilon}|\psi_2| \\ &+ \frac{1}{\varepsilon}(w_\varepsilon - \psi_1)^+ - \frac{1}{\varepsilon}(w_\varepsilon - \psi_2)^+ \\ &\leq f - |f| - \left(\alpha + \frac{1}{\varepsilon}\right)\zeta_\varepsilon + \frac{1}{\varepsilon}\left(\psi_1 - |\psi_2|\right) \\ &- \frac{1}{\varepsilon}|\psi_2| + \frac{1}{\varepsilon}(\psi_2 - \psi_1) \\ &\leq 0 \end{aligned}$$

Since $w_\varepsilon - \zeta_\varepsilon = 0$ on Γ, it follows from the Maximum principle that

$$w_\varepsilon - \zeta_\varepsilon \le 0 \tag{2.22}$$

Hence $w_\varepsilon = w_{\varepsilon\alpha}$ is bounded above by a constant independent of α. A similar proof holds true for a bound below. Therefore

$$\|w_{\varepsilon\alpha}\|_{C^0(\bar{\mathcal{O}})} \le C \qquad \text{independent of } \alpha$$

From this and the equation, it follows that

$$\|w_{\varepsilon\alpha}\|_{W^{2,p}(\mathcal{O})} \le C \qquad \text{independent of } \alpha$$

Letting α tend to 0, one easily shows the existence of a solution of (2.11).

The uniqueness of the solution of (2.11) is a consequence of the probabilistic interpretation. Indeed, if $v_1(t)$, $v_2(t)$ is any pair of nonanticipative processes satisfying (2.8), a straightforward application of Ito's formula yields

$$\begin{aligned} u_\varepsilon(x) = {} & J_\varepsilon^x(v_1,v_2) + \frac{1}{\varepsilon} E^x \int_0^\tau \left[\exp - \frac{1}{\varepsilon}\int_0^t (v_1 + v_2)ds\right] \\ & \cdot \Big\{v_1(t)\left[u_\varepsilon(y(t)) - \psi_1(y(t))\right] + \left[u_\varepsilon(y(t))\right. \\ & \left. - \psi_1(y(t))\right]^- + v_2(t)\left[u_\varepsilon(y(t)) - \psi_2(y(t))\right] \\ & - \left[u_\varepsilon(y(t)) - \psi_2(y(t))\right]^+\Big\}dt \end{aligned} \tag{2.23}$$

from which one easily proves (2.12) and the existence of a saddle point.

2.3 <u>Proof of Theorem 2.2</u>

From (2.11) we have

$$A(u_\varepsilon - \psi_2) + \frac{1}{\varepsilon}(u_\varepsilon - \psi_2)^+ - \frac{1}{\varepsilon}(u_\varepsilon - \psi_2)^- = f - A\psi_2$$

But since $\psi_1 \le \psi_2$, one has

$$- (u_\varepsilon - \psi_1)^- \geq - (u_\varepsilon - \psi_2)^-$$

hence

$$A(u_\varepsilon - \psi_2) + \frac{1}{\varepsilon} (u_\varepsilon - \psi_2)^+ - \frac{1}{\varepsilon} (u_\varepsilon - \psi_2)^- \leq f - A\psi_2 \tag{2.24}$$

or

$$\varepsilon A(u_\varepsilon - \psi_2) + u_\varepsilon - \psi_2 \leq \varepsilon C$$

$$(u_\varepsilon - \psi_2)|_\Gamma \leq 0$$

and the Maximum principle yields

$$u_\varepsilon - \psi_2 \leq \varepsilon C$$

Therefore

$$\frac{(u_\varepsilon - \psi_2)^+}{\varepsilon} \leq C \tag{2.25}$$

A similar proof yields

$$\frac{(u_\varepsilon - \psi_2)^-}{\varepsilon} \leq C \tag{2.26}$$

Those estimates and (2.11) imply that

$$\|u_\varepsilon\|_{W^{2,p}(\mathcal{O})} \leq C \tag{2.27}$$

We can extract a subsequence, still denoted by u_ε such that

$$u_\varepsilon \to u \text{ in } W^{2,p}(\mathcal{O}) \text{ weakly and } C^0(\bar{\mathcal{O}}) \text{ strongly}$$

From (2.25) and (2.26) it follows that

$$(u - \psi_2)^+ = (u - \psi_1)^- = 0$$

hence

$$\psi_1 \le u \le \psi_2 \tag{2.28}$$

Let now v(x) be a measurable function such that

$$\psi_1 \le v \le \psi_2 \qquad \text{a.e. on } \mathcal{O}$$

One has

$$\begin{aligned} Au_\varepsilon(v - u_\varepsilon) &= f(v - u_\varepsilon) - \frac{1}{\varepsilon}(u_\varepsilon - \psi_2)^+(v - u_\varepsilon) \\ &\quad + \frac{1}{\varepsilon}(u_\varepsilon - \psi_2)^-(v - u_\varepsilon) \\ &\ge f(v - u_\varepsilon) + \frac{1}{\varepsilon}(u_\varepsilon - \psi_2)^{+2} + \frac{1}{\varepsilon}(u_\varepsilon - \psi_1)^{-2} \\ &\ge f(v - u_\varepsilon) \end{aligned}$$

Hence

$$\int_{\mathcal{O}} (Au - f)(v - u)dx \ge 0$$

from which it easily follows that

$$(Au - f)(v - u) \ge 0 \qquad \forall\, v \text{ real such that}$$

$$\psi_1(x) \le v \le \psi_2(x) \tag{2.29}$$

We thus have proved the existence of a solution of (2.6).

The uniqueness is again a consequence of the probabilistic interpretation. Let indeed θ_1, θ_2 be two stopping times. By Ito's formula, one obtains

$$u(x) = E^x\left[\int_0^{\tau\wedge\theta_1\wedge\theta_2} Au\big(y(t)\big)dt + u\big(y(\tau\wedge\theta_1\wedge\theta_2)\big)\right] \tag{2.30}$$

Let next $\hat{\theta}_1$, $\hat{\theta}_2$ be defined by

$$\hat{\theta}_1 = \inf\left\{t \geq 0 \mid u\big(y(t)\big) = \psi_1\big(y(t)\big)\right\}$$

$$\hat{\theta}_2 = \inf\left\{t \geq 0 \mid u\big(y(t)\big) = \psi_2\big(y(t)\big)\right\} \tag{2.31}$$

It is easy to check that $\hat{\theta}_1$, $\hat{\theta}_2$ is a saddle point of $J^x(\theta_1, \theta_2)$. The proof of Theorem 2.2 is complete.

2.4 Estimation of the Penalization Error

We have shown in Theorem 2.2 that

$$u_\varepsilon \to u \text{ in } C^0(\bar{\mathcal{O}})$$

An interesting question concerns the rate of convergence. In [3], the following theorem is proved.

Theorem 2.3 Under the assumptions (2.1, 2.2, and 2.5), one has the following estimate

$$\|u_\varepsilon - u\|_{C^0(\bar{\mathcal{O}})} \leq C\left[\rho_1(\varepsilon) + \rho_2(h) + \exp - \frac{h}{\varepsilon}\right]$$

for any $h \geq 0$, and ρ_1, ρ_2 being positive functions on R^+ tending to 0 when the argument tends to 0.

The proof is rather technical and is based on probabilistic arguments. Actually the method consists in comparing the functions

$$\min_{v_2} \max_{v_1} J^x_\varepsilon(v_1, v_2) \quad \text{and} \quad \min_{\theta_2} \max_{\theta_1} J^x(\theta_1, \theta_2)$$

There is an interesting open problem here. Can one obtain an estimate as (2.32) without the smoothness assumptions on ψ_1, ψ_2, say assuming only ψ_1, $\psi_2 \in C^0(\bar{\mathcal{O}})$? This may be expected since the derivatives of ψ_1, ψ_2 never enter into the pay-off functions. In the case of control, the answer is positive.

We now are going to give another estimate involving not only the functions u_ε and u, but also the derivatives. We need some additional assumptions and notation. We assume

$$\frac{\partial a_{ij}}{\partial x_k} \in L^\infty(\mathcal{O}) \tag{2.33}$$

We can then rewrite the operator A in the divergence form

$$A = - \sum_{i,j} \frac{\partial}{\partial x_i}\left(a_{ij} \frac{\partial}{\partial x_j}\right) - \sum_i \left(g_i - \sum_j \frac{\partial a_{ij}}{\partial x_j}\right)\frac{\partial}{\partial x_i} \tag{2.34}$$

We will consider A + λ instead of A, in order to make some coerciveness assumption. Actually, let $V = H_0^1(\mathcal{O})$ be the Sobolev space of order 1 of functions which are 0 on the boundary. We associate to A a bilinear continuous form on V defined by

$$a(u,v) = \sum_{ij}\int_{\mathcal{O}} a_{ij} \frac{\partial u}{\partial x_j} \frac{\partial v}{\partial x_i} dx + \sum_i \int_{\mathcal{O}} a_i \frac{\partial u}{\partial x_i} v \, dx \tag{2.35}$$

where

$$a_i = - g_i + \sum_j \frac{\partial a_{ij}}{\partial x_j}$$

For λ large enough, one has

$$a(v,v) + \lambda |v|^2 \geq \beta \|v\|^2 \qquad \beta > 0 \qquad \forall \ v \in V \tag{2.36}$$

where $|\ |$ denotes the L^2 norm, and $\|\ \|$ denotes the H_0^1 norm. To save notation we will call u the solution of (2.6), and u_ε the solution of (2.11) corresponding to A + λ, instead of A. It is easy to check that u is the solution of the VI.

$$a(u,v-u) + \lambda(u,v-u) \geq (f,v-u)$$

$$\forall \ v \in K = \left\{v \in H_0^1(\mathcal{O}) \mid \ \psi_1 \leq v \leq \psi_2 \text{ a.e.}\right\}$$

$$u \in K \tag{2.37}$$

The equivalence of (2.6) (with A changed in $A + \lambda$) and (2.37) is easy when u is regular [i.e., $u \in W^{2,p}(\mathcal{O})$]. We then have the following.

<u>Theorem 2.6</u> Under the assumptions of Theorem 2.3 (2.33), (2.36) and $\psi_i \in H^1(\mathcal{O})$, then one has

$$\|u_\varepsilon - u\|_{H^1(\mathcal{O})} \leq C\varepsilon^{1/2} \tag{2.38}$$

<u>Proof</u> We note that

$$\left((v - \psi_2)^+, (v - \psi_2)^-\right) = 0 \tag{2.39}$$

We multiply (2.11)[1] by $(u_\varepsilon - \psi_2)^+$. We obtain, by virtue of (2.39)

$$a\left(u_\varepsilon - \psi_2, (u_\varepsilon - \psi_2)^+\right) + \lambda\left(u_\varepsilon - \psi_2, (u_\varepsilon - \psi_2)^+\right) + \frac{1}{\varepsilon}\,|(u_\varepsilon - \psi_2)^+|^2 = \left(f - A\psi_2, (u_\varepsilon - \psi_2)^+\right) \tag{2.40}$$

Hence

$$|(u_\varepsilon - \psi_2)^+| \leq C\varepsilon \;^{2} \tag{2.41}$$

and

$$\|(u_\varepsilon - \psi_2)^+\| \leq C\sqrt{\varepsilon} \tag{2.42}$$

Similarly

$$\|(u_\varepsilon - \psi_1)^-\| \leq C\sqrt{\varepsilon} \tag{2.43}$$

Let us next define

---

[1] We recall that A is changed into $A + \lambda$.

[2] which was already known [see (2.25)]

$$r_\varepsilon = u_\varepsilon - (u_\varepsilon - \psi_2)^+ + (u_\varepsilon - \psi_1)^-$$

which belongs to K. We shall prove that

$$\|u - r_\varepsilon\| \le C\sqrt{\varepsilon}$$

which, by virtue of (2.42, 2.43, 2.44) is sufficient to ensure (2.38). We take $v = r_\varepsilon$ in (2.37) and multiply (2.11) scalarly by $-(r_\varepsilon - u)$. One obtains, by addition

$$\begin{aligned} &a(u_\varepsilon-u,\ r_\varepsilon-u) + \lambda(u_\varepsilon-u,\ r_\varepsilon-u) + \frac{1}{\varepsilon} X_2 + \frac{1}{\varepsilon} X_1 \le 0, \\ &X_1 = -\big((u_\varepsilon - \psi_2)^-,\ r_\varepsilon - u)\big) \\ &X_2 = \big((u_\varepsilon - \psi_2)^+,\ r_\varepsilon - u)\big) \end{aligned} \tag{2.46}$$

But

$$\begin{aligned} X_1 &= -\big((u_\varepsilon - \psi_1)^-,\ r_\varepsilon - \psi_1 - (u_\varepsilon - \psi_2)\big) \\ &= -\big((u_\varepsilon - \psi_1)^-,\ (u_\varepsilon - \psi_1)^+ - (u_\varepsilon - \psi_2)^+\big) \\ &\quad + \big((u_\varepsilon - \psi_1)^-,\ u - \psi_1\big) \\ &= \big((u_\varepsilon - \psi_1)^-,\ u - \psi_1\big) \ge 0 \end{aligned}$$

Similary

$$\begin{aligned} X_2 &= \big((u_\varepsilon - \psi_2)^+,\ r_\varepsilon - \psi_2 - (u - \psi_2)\big) \\ &= \big((u_\varepsilon - \psi_2)^+,\ -(u_\varepsilon - \psi_2)^- + (u_\varepsilon - \psi_1)^-\big) \\ &\quad + \big((u_\varepsilon - \psi_2)^+,\ -(u - \psi_2)\big) \\ &= \big((u_\varepsilon - \psi_2)^+,\ -(u - \psi_2)\big) \ge 0 \end{aligned}$$

Therefore (2.46) yields

$$a(u_\varepsilon - u,\ r_\varepsilon - u) + \lambda(u_\varepsilon - u,\ r_\varepsilon - u) \le 0$$

hence

$$a(r_\varepsilon-u,\ r_\varepsilon-u) + \lambda(r_\varepsilon-u,\ r_\varepsilon-u) \le a(r_\varepsilon-u_\varepsilon,\ r_\varepsilon-u)$$
$$+ \lambda(r_\varepsilon-u\ ,\ r_\varepsilon-u)$$

and

$$\|r_\varepsilon - u\| \le C\|r_\varepsilon - u_\varepsilon\| \le C\sqrt{\varepsilon}$$

2.5 Interpretation of the Solution of a VI

The VI (2.37) has a solution (which is unique) under much less stringent smoothness assumptions than (2.5). Similarly, the pay-off of the differential game $J^x(\theta_1,\theta_2)$ does not need such a smoothness to make sense, either. What requires smoothness is the writing (2.6). Therefore, it is reasonable to ask whether the solution of (2.37) can be interpreted as the value of a differential game when both problems make sense independently. We will assume (2.1, 2.2, 2.33, 2.36) and

$$(p-1)\sum_{ij} a_{ij}\xi_i\xi_j + \sum_i a_i\xi_i\xi_o + \lambda\xi_o^2 \ge 0$$
$$\forall\ \xi_i,\ \xi_o \in R \qquad (p > n/2) \tag{2.47}$$

$$f \in L^p(\mathcal{O}), \qquad p > n/2, \qquad \psi_i \in C^0(\bar{\mathcal{O}}),$$
$$\psi_1 \le \psi_2 \text{ on } \bar{\mathcal{O}}, \qquad \psi_1|_\Gamma \le 0 \le \psi_2|_\Gamma \tag{2.48}$$

$$K \text{ defined in (2.37) is not empty} \tag{2.49}$$

Under assumptions (2.1, 2.2, 2.33, 2.36, and 2.49) the problem (2.37) has one and only one solution (see Lions-Stampacchia [7]). On the other hand, the payoff

$$J_x(\theta_1,\theta_2) = E^x\left[\int_0^{\tau\wedge\theta_1\wedge\theta_2} e^{-\lambda t} f(y(t))dt + \psi_1(y(\theta_1))\chi_{\theta_1\le\theta_2,\theta_1<\tau} e^{-\lambda\theta_1} + \psi_2(y(\theta_2))\chi_{\theta_2<\theta_1\wedge\tau} e^{-\lambda\theta_2}\right] \tag{2.50}$$

has a meaning under (2.1, 2.2, 2.48). The only difference with the regular situation is that we cannot write (2.6) any more. We then have the following.

Theorem 2.5 Under the assumptions (2.1, 2.2, 2.33, 2.36) and (2.47, 2.48, 2.49) the solution u of (2.37) belongs to $C^0(\bar{\mathcal{O}})$ and one has

$$u(x) = \min_{\theta_2}\max_{\theta_1} J^x(\theta_1,\theta_2) = \max_{\theta_1}\min_{\theta_2} J^x(\theta_1,\theta_2) \tag{2.51}$$

Moreover, there exists a saddle point of $J^x(\theta_1,\theta_2)$ defined by

$$\hat{\theta}_1 = \inf\{t \ge 0 \mid u(y(t)) = \psi_1(y(t))\}$$
$$\hat{\theta}_2 = \inf\{t \ge 0 \mid u(y(t)) = \psi_2(y(t))\} \tag{2.52}$$

Proof Let ψ_1^N, ψ_2^N be a sequence of functions such that

$$\psi_i^N \in \mathcal{D}(\bar{\mathcal{O}}), \quad \psi_1^N \le \psi_2^N \text{ on } \bar{\mathcal{O}}, \quad \psi_1^N \le 0 \le \psi_2^N \text{ on } \Gamma$$
$$\psi_i^N \to \psi_i \text{ in } C^0(\bar{\mathcal{O}}) \tag{2.53}$$

Then the assumptions of (2.5) are verified for ψ_i^N, except maybe for the last ones (unless $f \in L^\infty$). However, by virtue of (2.47), one can use L^p estimates instead of the Maximum principle, to obtain the same results as in Theorem 2.2 (for details see [3]). We note u^N, u_ε^N the solutions of the VI and penalized problem, corresponding to ψ_i^N. Let also $J_N^x(\theta_1,\theta_2)$, $J_{N\varepsilon}^x(v_1,v_2)$ be the pay-offs associated with ψ_i^N. It is easy to check that

$$|J_N^x(\theta_1,\theta_2) - J^x(\theta_1,\theta_2)| \le \max\Big(\sup_{x\in\bar{\mathcal{O}}}|\psi_1(x) - \psi_1^N(x)|,$$
$$\sup_{x\in\bar{\mathcal{O}}}|\psi_2(x) - \psi_2^N(x)|\Big) \tag{2.54}$$

$$|J_{N\varepsilon}^x(v_1,v_2) - J_\varepsilon^x(v_1,v_2)| \le \max\Big(\sup_{x\in\bar{\mathcal{O}}}|\psi_1(x) - \psi_1^N(x)|,$$
$$\sup_{x\in\bar{\mathcal{O}}}|\psi_2(x) - \psi_2^N(x)|\Big) \tag{2.55}$$

Since ψ_i^N are smooth, we have

$$u_\varepsilon^N(x) = \inf_{v_2}\sup_{v_1} J_N^x\,(v_1,v_2) = \sup_{v_1}\inf_{v_2} J_{N\varepsilon}^x(v_1,v_2)$$

$$u^N(x) = \inf_{\theta_2}\sup_{\theta_1} J_N^x(\theta_1,\theta_2) = \sup_{\theta_1}\inf_{\theta_2} J_N^x(\theta_1,\theta_2) \tag{2.56}$$

From (2.54, 2.55), we obtain

$$|u^N(x) - \inf_{\theta_2}\sup_{\theta_1} J^x(\theta_1,\theta_2)| \le \max\big(\|\psi_1-\psi_1^N\|,\ \|\psi_2-\psi^N\|\big)$$

$$|u^N(x) - \sup_{\theta_1}\inf_{\theta_2} J^x(\theta_1,\theta_2)| \le \max\big(\|\psi_1-\psi_1^N\|,\ \|\psi_2-\psi_2^N\|\big) \tag{2.57}$$

$$|u_\varepsilon^N(x) - \inf_{v_2}\sup_{v_1} J^x(v_1,v_2)| \le \max\big(\|\psi_1-\psi_1^N\|,\ \|\psi_2-\psi_2^N\|\big)$$

$$|u_\varepsilon^N(x) - \sup_{v_1}\inf_{v_2} J^x(v_1,v_2)| \le \max\big(\|\psi_1-\psi_1^N\|,\ \|\psi_2-\psi_2^N\|\big) \tag{2.58}$$

We also have

$$u_\varepsilon(x) = \inf_{v_2}\sup_{v_1} J_\varepsilon^x(v_1,v_2) = \sup_{v_1}\inf_{v_2} J_\varepsilon^x(v_1,v_2) \tag{2.59}$$

This follows from the fact that Theorem 2.1 holds true under the assumptions of the present theorem, as it is easily checked by the proof. By (2.57) we necessarily have

$$\inf_{\theta_2} \sup_{\theta_1} J^x(\theta_1,\theta_2) = \sup_{\theta_1} \inf_{\theta_2} J^x(\theta_1,\theta_2) \tag{2.60}$$

Using the preceding relationships and the fact that for N fixed, one has

$$\sup_x |u_\varepsilon^N(x) - u^N(x)| \to 0, \text{ as } \varepsilon \to 0 \tag{2.61}$$

one easily checks that

$$\lim_{\varepsilon\to 0}\left(\sup_x \left|u_\varepsilon(x) - \inf_{\theta_2} \sup_{\theta_1} J^x(\theta_1,\theta_2)\right|\right) = 0 \tag{2.62}$$

But

$$u_\varepsilon \to u$$

in say $L^2(\mathcal{O})$. Hence necessarily, from (2.62)

$$u(x) = \inf_{\theta_2} \sup_{\theta_1} J^x(\theta_1,\theta_2) \in C^0(\bar{\mathcal{O}}) \tag{2.63}$$

The uniform convergence (2.62) plays a fundamental role in showing the existence of a saddle point. We omit the proof (see [3]).

Remark 2.1 In the proof of Theorem 2.5, we have shown (indirectly) that

$$u_\varepsilon \to u \text{ in } C^0(\bar{\mathcal{O}})$$

It would be interesting to give a direct proof of that result. This is possible in the case of Control Theory (see [3]).

§3. NONSTATIONARY CASE

We only state the results (details can be found in [3]).

3.1 Operators Not in Divergence Form

Let A(t) be a family of differential operators

$$A(t) = \sum_{ij} a_{ij}(x,t) \frac{\partial^2}{\partial x_i \partial x_j} - \sum_i g_i(x,t) \frac{\partial}{\partial x_i} \tag{3.1}$$

where

$$(Q = \mathcal{O}\times]0,T[,\ \Gamma = \partial\mathcal{O} \text{ of class } C^2,\ \Sigma = \Gamma\times]0,T[)$$

$$a_{ij} = a_{ji} \in C^0(\bar{Q}),\ \Sigma a_{ij}\xi_i\xi_j \geq \beta\Sigma\xi_i^2,\quad \beta > 0,$$
$$\forall\ \xi_1, \cdots, \xi_n \in R;\ g_i \in L^\infty(Q) \tag{3.2}$$

We extend those functions to $R^n \times [0,T)$, respecting their properties. Next, let f, ψ_1, ψ_2, $\bar{u}$ such that

$$f \in L^p(Q),\ \psi_i \in W^{2,1,p}(Q)\ ^1,\quad \bar{u} \in W^{2,p}(\mathcal{O}) \cap W_o^{1,p}(\mathcal{O}),$$
$$p > \frac{n}{2} + 1;\quad \psi_1|_\Sigma \leq 0 \leq \psi_2|_\Sigma,\quad \psi_1 \leq \psi_2 \text{ on } \bar{Q},$$
$$\psi_1(x,T) \leq \bar{u}(x) \leq \psi_2(x,T);\quad f + \frac{\partial\psi_2}{\partial t} - A(t)\psi_2 \leq C;$$
$$f + \frac{\partial\psi_1}{\partial t} - A(t)\psi_1 \geq -C \tag{3.3}$$

We note $a = \sigma^2/2$. Let $\Omega = C^0([o,\infty[;\ R^n)$, $F_t^s = \sigma$ algebra generated by $\omega(\lambda)$, $\lambda\varepsilon[t,s]$, and P^{xt} the unique probability measure on (Ω,F_t) such that if $y(s;\omega) = \omega(s)$, there exists a standard Wiener process, which is an F_t^s martingale, and

$$dy = g(y,s)ds + \sigma(y,s)dw(s)$$
$$y(t) = x \tag{3.4}$$

If $v_1(s)$, $v_2(s)$ are two scalar nonanticipative processes such that

$$\text{a.s.}\quad v_i(s) \in [0,1]\ \forall\ s \tag{3.5}$$

and θ_1, θ_2 are two F_t^s stopping times such that $\theta_i \in [t,T]$, one defines

---

[1] Space of functions $z \in L^p(Q)$, such that $\frac{\partial z}{\partial x_i}$, $\frac{\partial^2 z}{\partial x_i\partial x_j}$, $\frac{\partial z}{\partial t} \in L^p(Q)$.

$$J^{xt}(v_1,v_2) = E^{xt}\Bigg[\int_t^{\tau\wedge T}(f + \frac{1}{\varepsilon} v_1\psi_1 + \frac{1}{\varepsilon} v_2\psi_2)(y(s),s)$$

$$\cdot \left(\exp - \frac{1}{\varepsilon}\int_t^s (v_1 + v_2)\right) + \chi_{T<\tau}\bar{u}(y(T))$$

$$\cdot \exp - \frac{1}{\varepsilon}\int_t^T (v_1 + v_2)\,ds\Bigg] \tag{3.6}$$

$$J^{xt}(\theta_1,\theta_2) = E^{xt}\Bigg[\int_t^{\tau\wedge\theta_1\wedge\theta_2} f(y(s),s)\,ds$$

$$+ \psi_1(y(\theta_1),\theta_1)\chi_{\theta_1\le\theta_2,\,\theta_1<\tau\wedge T}$$

$$+ \psi_2(y(\theta_2),\theta_2)\chi_{\theta_2<\theta_1\wedge\tau\wedge T}$$

$$+ \bar{u}(y(T))\chi_{T=\theta_1\wedge\theta_2\le\tau}\Bigg] \tag{3.7}$$

One defines the following problem: to find a function $u(x,t)$ such that

$$u \in W^{2,s,p}(Q)$$

$$\psi_1 \le u \le \psi_2 \text{ on } \bar{Q}, \quad u|_\Sigma = 0, \quad u(x,T) = \bar{u}(x) \text{ on } \bar{\mathcal{O}}$$

$$-\frac{\partial u}{\partial t} + A(t)u = f \quad \text{if} \quad \psi_1 < u < \psi_2$$

$$-\frac{\partial u}{\partial t} + A(t)u \ge f \quad \text{if} \quad u = \psi_1$$

$$-\frac{\partial u}{\partial t} + A(t)u \le f \quad \text{if} \quad u = \psi_2 \tag{3.8}$$

One then has

Theorem 3.1 Under the assumptions (3.2, 3.3), there exists one and only one solution of (3.8). Moreover, one has

$$u(x,t) = \min_{\theta_2}\max_{\theta_1} J^{xt}(\theta_1,\theta_2) = \max_{\theta_1}\min_{\theta_2} J^{xt}(\theta_1,\theta_2) \tag{3.9}$$

If one sets

$$\hat{\theta}_1 = \inf\{s \in [t,T] \mid u(y(s),s) = \psi_1(y(s),s)\}$$

$$\hat{\theta}_2 = \inf\{s \in [t,T] \mid u(y(s),s) = \psi_2(y(s),s)\} \tag{3.10}$$

then $\hat{\theta}_1$, $\hat{\theta}_2$ is a saddle point of $J^{xt}(\theta_1,\theta_2)$.

3.2 Operators in Divergence Form

We now make the following set of assumptions

$$a_{ij} = a_{ji},\ \frac{\partial a_{ij}}{\partial x_k} \in L^\infty(Q),\ a_i = -g_i + \sum_j \frac{\partial a_{ij}}{\partial x_j} \in L^\infty(Q)$$

$$\Sigma a_{ij}\xi_i\xi_j \geq \beta\Sigma\xi_i^2,\ \beta > 0,\ \forall\ \xi_1, \cdots, \xi_n \in R \tag{3.11}$$

$$f \in L^p(Q),\ p > \frac{n}{2} + 1,\ \psi_i \in C^0(\bar{Q})$$

$$\psi_1 \leq \psi_2 \text{ on } \bar{Q},\quad \psi_1|_\Sigma \leq 0 \leq \psi_2|_\Sigma \tag{3.12}$$

$$\psi_i,\ \frac{\partial\psi_i}{\partial t} \in L^2(o,T;H_0^1(\mathcal{O})),\quad \frac{\partial^2\psi_i}{\partial t^2} \in L^2(Q)$$

$$\frac{\partial\psi_i}{\partial t}\Big|_\Sigma = 0 \tag{3.13}$$

$$-\frac{\partial}{\partial t}(\psi_2-\psi_1) \geq 0,\quad -\frac{\partial^2}{\partial t^2}(\psi_2-\psi_1) \geq 0 \tag{3.14}$$

$$\bar{u} \in H_0^1(\mathcal{O}),\quad \psi_1(x,T) \leq \bar{u}(x) \leq \psi_2(x,T);\quad A(T)\bar{u} \in L^2(\mathcal{O})$$

$$\bar{u} \in C^0(\bar{\mathcal{O}}) \tag{3.15}$$

The differential operators A(t) are rewritten as follows

$$A(t) = -\sum_{ij}\frac{\partial}{\partial x_i}\left(a_{ij}(x,t)\frac{\partial}{\partial x_j}\right) + \sum_i a_i(x,t)\frac{\partial}{\partial x_i} \tag{3.16}$$

One next sets for $u,v \in H^1(\mathcal{O})$

$$a(t;u,v) = \sum_{ij}\int_{\mathcal{O}} a_{ij}(x,t)\frac{\partial u}{\partial x_j}\frac{\partial v}{\partial x_i}\,dx + \sum_i\int_{\mathcal{O}} a_i\frac{\partial u}{\partial x_i}\,v\,dx \tag{3.17}$$

One then has the following.

Theorem 3.2 Under the assumptions (3.11, 3.12, 3.13, 3.14, and 3.15) there exists one and only one function u such that

$$u \in L^2(o,T;H_0^1(\mathcal{O})) \cap L^\infty(0,T;L^2(\mathcal{O})) \cap C^0(\bar{Q})$$

$$\frac{\partial u}{\partial t} \in L^2(o,T;H_0^1(\mathcal{O})) \cap L^\infty(0,T;L^2(\mathcal{O})) \tag{3.18}$$

$$\psi_1 \le u \le \psi_2 \quad \text{a.e. in } Q \tag{3.19}$$

$$-\left(\frac{\partial u}{\partial t}, v-u\right) + a(t;u,v-u) - (f,v-u) \ge 0$$

$$\forall\ v \in H_0^1(\mathcal{O}),\quad \psi_1 \le v \le \psi_2 \tag{3.20}$$

$$u(x,T) = \bar{u}(x) \quad \text{a.e.} \tag{3.21}$$

One has again (3.9) and (3.10).

BIBLIOGRAPHY

1. Bensoussan, A., Variational Inequalities and Optimal Stopping Time Problems, Proceedings of the Conference on Calculus of Variations and Control Theory, Madison, 1975, to appear.
2. Bensoussan, A. and J. L. Lions, Problemes de temps d'arrêt optimal et Inéquations variationnelles paraboliques, Applicable Analysis, vol. 3 (1973), pp. 267-294.
3. Bensoussan, A. and J. L. Lions, Temps d'arrêt optimal et Inéquations variationnelles, in preparation.
4. Friedman, A., Stochastic Games and Variational Inequalities, Arch. Rat. Mech. Anal. 51 (1973), pp. 321, 346.
5. Krylov, N. V., On a Problem with Two Free Boundaries for an Elliptic Equation and Optimal Stopping of a Markov Process, Dokl. Akad. Nauk SSSR Tom 194 (1970), no. 6.

6. Kurz, T., personal communication.

7. Lions, J. L. and G. Stamzacchia, Variational Inequalities, _Comm. Pure Applied Math._ _XX_ (1967), 493-519.

8. Stroock, D. and S. Varadhan, Diffusion Processes with Continuous Coefficients, I, II, _Comm. Pure Applied Math._ _XXII_, (1969, 345-400, 479, 530.

A DIFFERENTIAL GAME ON POINT PROCESSES

A. Ephremides

University of Maryland
College Park, Maryland

§1. INTRODUCTION

A large number of classical queueing problems, such as traffic control, inventory control, resource allocation in service systems, as well as problems in the design of photon detectors in optical communications or in the modeling of high error-rate channels, and problems in bio-modeling involve, or aim at, the optimization of a suitable criterion of performance. Furthermore, they involve random occurrences of events, be they arrivals of customers, photons, errors, or neural spikes.

The queueing problems have been traditionally analyzed via heuristic techniques, descriptions of which abound in the operations research literature. These techniques have used the statistical theory of point processes [1-3], which has had a narrow mathematical basis. On the other hand, the strong mathematical theory of optimization has been traditionally used in the analysis of deterministic control problems or in estimation and stochastic control of models involving ordinary random processes (Markov dynamics and additive perturbations or control). In conflict situations or in large scale systems, although the theory of games (plain, multistage, differential) has been considerably advanced [4-8], very little has been accomplished in obtaining implementable solutions to even quite simple problems (particularly in their stochastic versions).

However, recently, the work of Beutler [9,10], Bar-David [11], Snyder [12,13], Rubin [14,15] followed by the recently completed work by Kailath [16], Segall [17], Bremaud [18], Wong [19], and Variaya [20] has placed the "counting" process associated with a point process, or more generally, a jump process in its natural framework of martingale theory, where the decomposition theorem [21,22] allows a generalization of the innovations method to apply to the estimation of a Poisson signal and to the derivation of likelihood ratio formulas.

Thus, at this point it appears possible (and certainly desirable) to tackle stochastic control problems and differential games involving jump processes by using a similar approach. In the sequel, an alternate approach to such problems is developed where the state does not coincide with the jump process governed by a stochastic differential equation, but is rather a probability distribution function with an associated observation equation where the observation is the jump process.

More generally, applying the approach to a Markov process, the diffusion equation becomes the observation equation and the Fokker-Planck equation becomes the state equation. The virtue of this approach lies in the fact that standard separation theorem [23,24] methods may be attempted for the solution, fitting the problem in the usual control theoretic framework of models and thus combining control (and game) theory of bilinear systems [25,26] with the theory of point processes. As argued in the sequel, it is often sensible for a number of queueing problems to consider the probability function as the "state."

In addition to the theoretical and practical aspects of this approach that need investigation, it is proposed to study related problems in channel modeling, optical communications, and biological information transmission by combining the theory of point processes with a corresponding optimization problem.

The proposed approach can best be described through an example that possesses many of the features of more complex problems, yet has a simple formulation and an intuitively appealing associated terminology, which can be easily modified to apply to cooperative and other contexts.

§2. THE BORDER VIOLATION PROBLEM - AN EXAMPLE

This problem, in its simplest form, is worked out to a substantial detail level. In the process a variety of its interesting features are revealed.

Description A border can be penetrated through several corridors each of which is protected by a number of sentries. In a typical corridor it is assumed that attackers are dispatched randomly in time with

$$p\{\text{one arrival in } (t, t + dt]\} = \alpha_t \, dt$$

The quantity α_t is the intensity rate of the offensive point process which is assumed to be Poisson. Of course, one may consider only a regular process by defining $\alpha_t dt$ as the conditional probability given the σ-field of the past. A single sentry is available to protect the corridor. It is assumed that he engages the attackers for a period of time γ_s, starting at $t = s$. At the end of the engagement (fight) period the attacker is neutralized. The "service" time γ_s is a random variable, independent of the attack process, and distributed in an exponential-like fashion. Namely, at any instant t > s

$$p\{\gamma_s \leq t - s + dt/\gamma_s > t - s\} = \beta_t dt$$

Thus β_t can be thought of as the defense capability; generally, the larger its value, the smaller the average service time. Realistically, one may interpret the liquidation of a sentry as followed by an instant replacement, hence his invincibility. Therefore, an intruder may penetrate only if at the time of his arrival the sentry is busy.

The objective of the attack strategist, termed player 1, is to maximize the average number of penetrators over a fixed period of time T, whereas the objective of the defender, termed player 2, is to minimize the same quantity. Player 1 is free to choose α_t, $0 \leq t \leq T$, and player 2 selects β_t, $0 \leq t \leq T$. Thus they may be called the α-player and β-player, respectively. As seen later, one may approximate a real conflict by allowing player 1 to choose the actual instants of attack, rather than their density and player 2 to control more than just his "average" service time.

Of course, some cost weight must be placed upon the choices of α_t, and β_t. Thus, if M_t is the counting process of penetrators, the cost function of this zero-sum, two-player game is defined by

$$K = E[M_T] + c\int_0^T (\beta_t - \alpha_t)dt \tag{1}$$

Under the previous assumptions it can be easily shown that

$$E[M_T] = \int_0^T \alpha_t p_t \, dt$$

where p_t is the probability that the sentry is engaged at time t, and that

$$\dot{p}_t = -(\alpha_t + \beta_t)p_t + \alpha_t \tag{2}$$

with the choice of initial condition given by

$$p_0 = 0$$

By regarding p_t as the "state" of the system, Eq. (2) is the system equation. The observation equation is given by

$$dy_t = dN_t \tag{2a}$$

where N_t is the attack counting process (Poisson) or by

$$dy_t = dM_t \tag{2b}$$

where M_t is the penetrator counting process (Poisson too, under some conditions). For a first analysis one may consider

$$y_t = p_t$$

that is, perfect information on the state.

Comments

1. In view of the fact that the system is linear with a linear cost, solutions on the boundary are expected. Hence, a reasonable set of constraints for the control functions α_t, β_t is that they be bounded. Thus

$$0 \le \alpha_t \le \alpha_M, \qquad 0 \le t \le T$$

$$0 \le \beta_t \le \beta_M, \qquad 0 \le t \le T$$

Furthermore, we will look for piece-wise continuous rather than measurable functions.

2. The necessary constraint for the state is that

$$0 \le p_t \le 1, \qquad 0 \le t \le T$$

which defines the "playing space" $[0,1] \times [0,T]$. This is automatically ensured by the form of Eq. (2) and by the chosen initial condition.

3. The general form of the chosen cost function allowing for randomized strategies or strategies that are functions of the random observation process y_t (or its entire past) is given by

$$K(\alpha_t, \beta_t, p_t) = E\left\{\int_0^T \alpha_t p_t dt + c \int_0^T (\beta_t - \alpha_t) dt\right\}$$

where

$$\alpha_t = \alpha'(y_s, s \le t) = \alpha(p_s, s \le t; \omega)$$

$$\beta_t = \beta'(y_s, s \le t) = \beta(p_s, s \le t; \omega) \qquad (2c)$$

4. Since the system is linear, a quadratic cost function might facilitate the analysis. With the system described above it is not easy to formulate a quadratic criterion. For example, an attempt to consider as cost the variance of N_T given that $E[N_T] <$ threshold, yields an even more complicated criterion function. However, possibilities for other criteria include requiring $p_t \gtrless$ threshold. The bilinear structure of the system and the availability of results [26] in the controllability of bilinear systems suggests such an approach. That might be an appropriate criterion in a surveillance radar system context or in other communication-detection versions of

point process games.

Perfect Information Case In this case it is assumed that both players can observe the state p_t at every $0 \le t \le T$. First, one verifies that the system satisfies trivially various sets of sufficient conditions for the existence of a saddle point, as for example the ones in Theorem 2.5.1 of [6], namely the existence of a unique solution to the state equation for any given controls α_t, β_t, the latter's continuity in the controls, and the linearity of the pay-off in α_t, β_t. That the saddle point consists of pure strategies will be concluded a posteriori, after finding and exhibiting such pure strategies. To find them, Pontryagin's maximum (minimum) principle will be used [27] in the standard control theoretic fashion by finding the $\max_\alpha[\min_\beta(K)]$, which, by the existence of value for the game, will be equal to $\min_\beta[\max_\alpha(K)]$. This approach has the advantage of supplying the solution to the control problem of designing the defense β_t for known rate α_t (typical in simple models of air traffic control systems), by specifying the function $\beta_t(\alpha_t)$ in the intermediate stage of the maxmin procedure.

Of course, the analysis can be carried out in terms of cost-to-go from an arbitrary starting position (π, τ) in the playing space $[0,1] \times [0,T]$. Thus prescriptions for optimal strategies will be supplied to each player in case the opponent deviates from optimal play and consequently forces the state out of the optimal trajectory. Furthermore, arbitrary boundary conditions for p at the beginning of the play must be allowed.

<u>The Min Part</u> The system is described by

$$\dot{p}_t = -(\alpha_t + \beta_t)p_t + \alpha_t$$

$$\dot{q}_t = (\alpha_t + \beta_t)q_t - \alpha_t$$

where q_t is the co-state, and with boundary conditions

$$p_\tau = \pi, \; q_T = 0$$

The Hamiltonian is given by

$$H(p_t, q_t; \alpha_t, \beta_t) = (c - p_t q_t)\beta_t + (p_t + q_t - p_t q_t - c)\alpha_t$$

The H-minimal control will be

$$\beta_t = \begin{cases} 0, & p_t q_t < c \\ \text{indeterminate}, & p_t q_t = c \\ \beta_M, & p_t q_t > c \end{cases}$$

Already it should be noted that for $c \geq 1$ the overall game assumes the trivial solution $\alpha_t \equiv \beta_t \equiv 0$, since it is then more expensive to both players to play the game. So it will be assumed that $0 < c < 1$.

The solution is illustrated in Figures 1 and 2. As it turns out the relative magnitude of β_M and α_t is critical to the form of the solution. In Figure 1 the case of the so-called *weak* defense is depicted, that is when

$$\beta_M \leq \min_{0 \leq t \leq t_1} \alpha_t \frac{1 - \sqrt{c}}{\sqrt{c}}$$

In Figure 2 we have the case of *strong* defense, that is

$$\beta_M \geq \max_{0 \leq t \leq t_1} \alpha_t \frac{1 - \sqrt{c}}{\sqrt{c}}$$

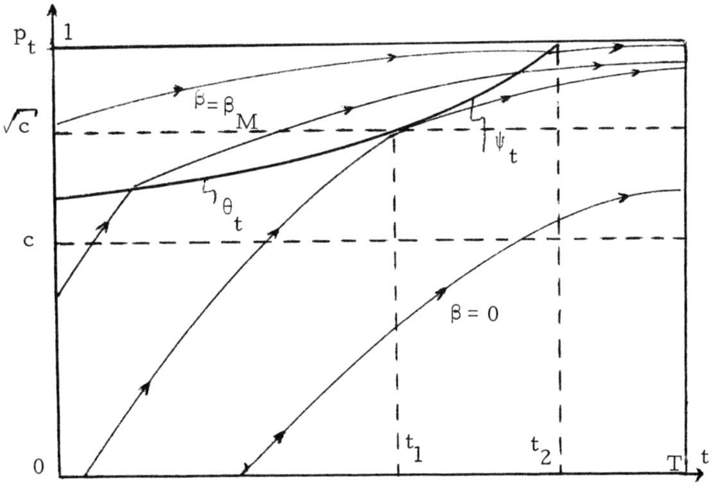

FIG. 1. H-minimal β_t for given α_t and for <u>weak</u> defense.

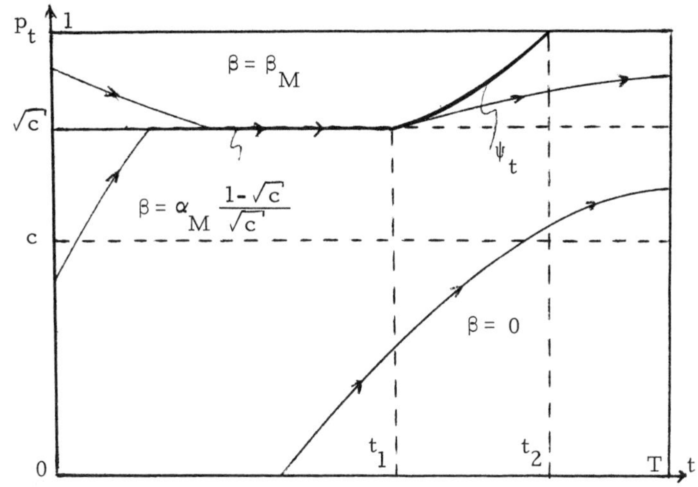

FIG. 2. H-minimal β_t for given α_t and for <u>strong</u> defense.

For the case that

$$\min_{0\le t\le t_1} \alpha_t \frac{1-\sqrt{c}}{\sqrt{c}} < \beta_M < \max_{0\le t\le t_1} \alpha_t \frac{1-\sqrt{c}}{\sqrt{c}}$$

which represents an intermediate situation we get the answer in all but a portion of the playing space, as will be explained later. However, in the maximization portion of the maxmin process this difficulty will be circumvented. The quantities that appear in the figures that need explanation are the following. The two critical instants, t_1 and t_2, are defined by

$$\int_{t_1}^{T} \alpha_t dt = \ell n \frac{1}{1-\sqrt{c}} \tag{3}$$

$$\int_{t_2}^{T} \alpha_t dt = \ell n \frac{1}{1-c} \tag{4}$$

Equation (3) implies that $q_{t_1} = \sqrt{c}$, and Eq. (4) that $q_{t_2} = c$ while the function ψ_t denotes the locus of the points at which $p_t q_t = c$, with $\psi_t = p_t$.

In Figure 1 the function θ_t is defined parametrically by

$$c = \left\{ \psi_s - \int_t^s \alpha_\xi \exp\left[- \int_\xi^s (\alpha_\eta + \beta_M) d\eta\right] d\xi \right\}$$

$$\cdot \left\{ \frac{c}{\psi_s} + \int_t^s \alpha_\xi \exp\left[\int_\xi^s (\alpha_\eta + \beta_M) d\eta\right] d\xi \right\} \tag{5}$$

and

$$\theta_t = \psi_s - \int_t^s \alpha_\xi \exp\left[-\int_\xi^s (\alpha_\eta + \beta_M) d\eta\right] d\xi, \quad t_1 \le s \le t_2,$$

$$t \le t_1$$

In Figure 2 this function does not show because in that case, the above equations do not have solutions. The fact that we do not know the conditions for existence of solutions

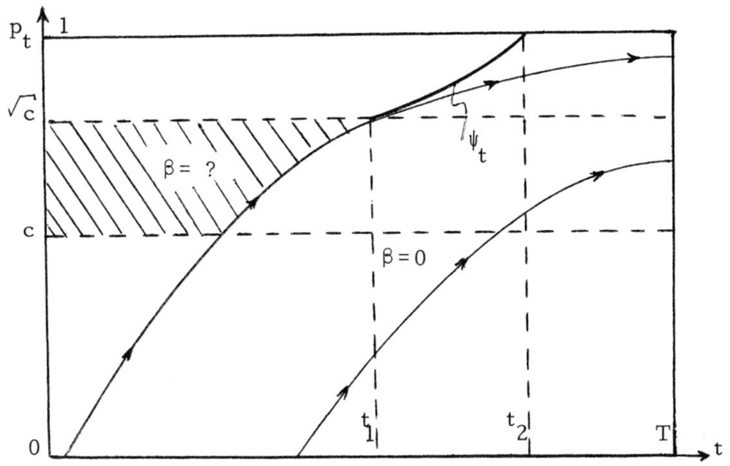

FIG. 3. H-minimal β_t for a given α_t and for intermediate defense.

in the intermediate case accounts for the incomplete specification of the solution as shown in Figure 3. The function θ_t represents the locus of the points at which $p_t \cdot q_t = c$ with $\theta_t = p_t$.

<u>The Max Part</u> There are three regimes of solutions for β_t in the min part, namely, $\beta = 0$, $\beta =$ indeterminate, and $\beta = \beta_M$. For each regime Pontryagin's principle is used to obtain that the H-maximal control is given by

$$\alpha_t = \begin{cases} 0, & p_t + q_t - p_t q_t < c \\ \text{indeterminate}, & p_t + q_t - p_t q_t = c \\ \alpha_M, & p_t + q_t - p_t q_t > c \end{cases}$$

for the regimes corresponding to $\beta = 0$ and $\beta = \beta_M$, and by

$$\alpha_t = \begin{cases} 0, & p_t + q_t - (1/\sqrt{c})p_tq_t < c \\ \text{indeterminate}, & p_t + q_t - (1/\sqrt{c})p_tq_t = c \\ \alpha, & p_t + q_t - (1/\sqrt{c})p_tq_t > c \end{cases}$$

for the regime β = indeterminate. In all cases it is argued that α_t cannot be indeterminate. Therefore always α_t is either α_M or 0. The solution is illustrated in Figures 4 and 5, for the cases of weak and strong defenses, respectively. The intermediate case does not arise, since in the region of possible ambiguity we have $\alpha_t = \alpha_M$, so that the defense can be determined to be either weak or strong. The function δ_t, which is a dispersal surface, as one can easily show by computing the pay-offs, and below which we have the only region where $\alpha_t = 0$, is defined by

$$\delta_t = 1 - \frac{\delta_M(1 - c)(T - t)}{1 - \exp[-\alpha_M(T - t)]}$$

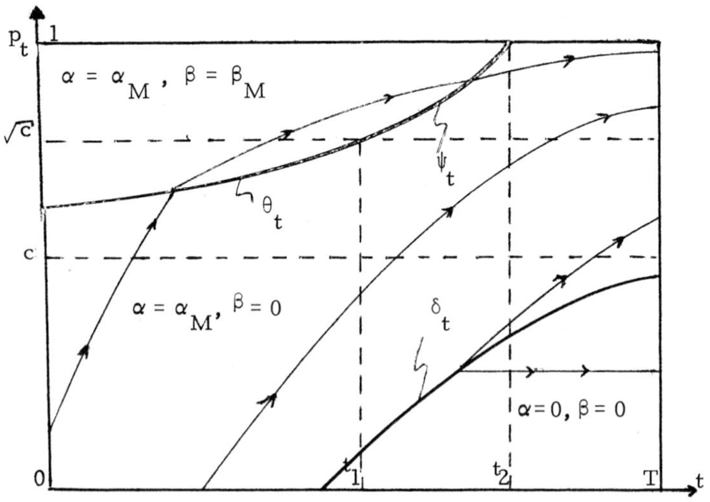

FIG. 4. Saddle point solution regimes for <u>weak</u> defense.

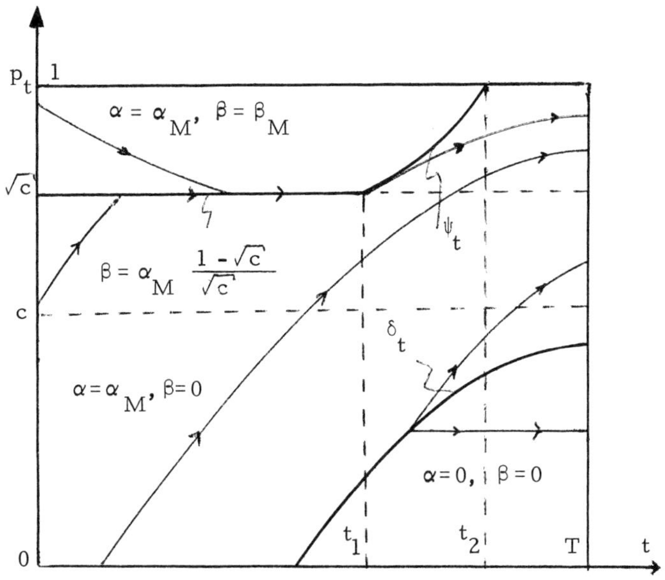

FIG. 5. Saddle point solution regimes for strong defense.

Thus the closed-loop, feedback strategies are completely specified for both players, in the perfect information case. The solution equation (5) is given by

$$t = s - \frac{1}{\alpha_M + \beta_M} \ln \frac{\dfrac{\alpha_M}{\alpha_M + \beta_M} - \dfrac{c}{\psi_s}}{\dfrac{\alpha_M}{\alpha_M + \beta_M} - \psi_s}, \quad t_1 \le s \le t_c \le t_2$$

when

$$\beta_M < \alpha_M \frac{1 - \sqrt{c}}{\sqrt{c}}$$

Imperfect Information Case The quantity p_t that was considered to be the "state" in the formulation of the problem is clearly unobservable and one can only compute it from the differential equation (2) if α_t and β_t are completely known. This not being the case, it is clear that the observation equation (2a) or (2b) has to be considered. Since a solution for the perfect information case has been obtained, one would hope that a separation theorem [23,24] here might hold; that is, that Eqs. (2c) might take the form

$$\alpha_t = \alpha(\hat{p}_t), \qquad \beta_t = \beta(\hat{p}_t)$$

where $\alpha(\cdot)$, $\beta(\cdot)$ are the saddle point solution derived before, and $\hat{p}_t$ the max likelihood estimate of p_t given y_s, $s \leq t$.

Showing a separation theorem has not been accomplished. Thus, as a suboptimal attempt, one may try the min part solution assuming such a separation. Reasonable approximate solutions would then be obtained for resource allocation problems (like airport facilities control). Towards this end it can be shown that $\hat{p}_t$ can be obtained by just solving Eq. (2) casually, with $\hat{\alpha}_t$ substituted for α_t, and where $\hat{\alpha}_t$ is the max likelihood estimate of α_t given y_s, $s \leq t$. This result can be proven rigorously for $\alpha_t \equiv$ unknown constant and heuristically at this point for arbitrary α_t. Methods for obtaining $\hat{\alpha}_t$ are numerous [13,15,17] and since Eq. (2) is relatively simple, the method can be tried in the computer by simulation. Of course, the cost function then assumes the form supplied in Eq. (2c). Another realistic observation equation is

$$y_t = I_t$$

where I_t is an indicator function of engagement of the sentry, given by

$$I_t = \begin{cases} 1, & \text{if sentry busy at time } t \\ 0, & \text{if sentry idle at time } t \end{cases}$$

Clearly, $E[I_t] = p_t$. Therefore, one can attempt to use a separation procedure using I_t as $\hat{p}_t$. Clearly, I_t is a jump process.

The imperfect information case has only been investigated imperfectly, but indications are that it is amenable to further analysis.

Generalizations

Dimension By considering many sentries one arrives at a matrix equation with the state vector being $\bar{p}(t)$ where $p_i(t)$ is the probability that i sentries are engaged at time t. The method of analysis generalizes straightforwardly, although analytical solutions in closed form are hard to obtain. Similar extensions can be accomplished when more than one corridor is considered. The form of the system is then nontrivially bilinear and recent results [25], in studying such systems through their equivalent simple, almost nondynamical, linear systems can be used to infer properties of the trajectories. The generalization becomes more interesting when switching of sentries from corridor to corridor is allowed. Then one gets an improved model for resources allocation problems.

Assignment When switching is allowed, the model can be constructed as follows. Consider 2 corridors and 1 sentry. Let

$$p\left[\frac{\text{of sentry being transferred to corridor 2 in } (t, t+dt)}{\text{given he is in corridor 1 and idle at } t}\right] = \delta_t^1 \, dt$$

and let the probability of the opposite switch be δ_t^2 dt. The attack rates for the two corridors are α_t^1, α_t^2 and the parameter of the sentry's capability is β_t as before. Let p_t^0, p_t^1, p_t^2, p_t^3 represent the probabilities of the four possibilities that the sentry is in corridor 1 and idle, or in 1 and busy, or in 2 and idle, or in 2 and busy at time t, respectively. Then allowing for switches only during idle periods and denoting by $\bar{p}_t$ the vector of the previous probabilities, the system is described by

$$\dot{\bar{p}}_t = \begin{bmatrix} -(\alpha_t^1 + \delta_t^1) & \beta_t & \delta_t^2 & 0 \\ \alpha_t^1 & -\beta_t & 0 & 0 \\ \delta_t^1 & 0 & -(\alpha_t^2 + \delta_t^2) & \beta_t \\ 0 & 0 & \alpha_t^2 & -\beta_t \end{bmatrix} \bar{p}_t$$

with pay-off (cost) function given by

$$K = \int_0^T c\left(\beta_t + \delta_t^1 + \delta_t^2 - \alpha_t^1 - \alpha_t^2\right)dt + E[M_T]$$

It is of course assumed that the two attack processes are mutually independent and that the switching intensities are independent of the service times.

The above system is nontrivially bilinear, and similar analysis as before is still applicable. More elementary treatments of restricted cases have been already done [28]. They involve simplifying the assignment strategy to allow only one switch during the course of the game. Then the optimum delta rule is to find a max likelihood decision time t_c and then make a MAP decision on the two attack processes.

Of course, the general problem that involves many corridors, several sentries and includes assignment functions has a large dimensionality and therefore, although properties of the solutions can be inferred analytically, the actual solutions would be rather obtained computationally.

Finally, the possibility of sentry liquidation can be incorporated in the previous model by defining an additional intensity h_t such that $h_t dt$ represents the probability of elimination in (t, t + dt] conditioned appropriately on the sentry's state at time t.

§3. AREAS OF INVESTIGATION

It is easy to see from the explicit outline of the previous example that:

a. a number of similar queueing problems can be modeled in a similar fashion. For example, a birth-death process with rate α_t for the births and β_t for the deaths and

$$\hat{p}_t(k) = \text{prob}[\text{at } t, \text{ population} = k]$$

can be visualized as a population control problem or as a cooperative or competitive game with, say, a criterion function related to the size of the population and the "fuel" costs α_t, β_t of the players. The quantity p_t satisfies a differential equation

$$\dot{p}_t(k) = -(\alpha_t + \beta_t)p_t(k) + \alpha_t p_t(k-1) + \beta_t p_t(k+1)$$

similar to Eq. (2), only slightly more complex. Therefore such problems are also amenable to solutions via the proposed method.

b. there are fundamental questions of theoretical nature that have to be looked into, namely

i) Does a separation theorem hold, that would allow decomposition of the problem into a state-estimation one, and a deterministic bilinear optimal control problem?

ii) Since the system with state p_t is bilinear, how can one apply recent results in the theory of bilinear systems to answer the questions of existence of saddle-points in more general cases, or even to analyze the system by using the bijective map between Lie Algebras and Lie Groups obtained by Lo [25].

iii) Since the information patterns in such problems are imbedded in random point processes, the use of useful representations of these processes in the related problems of prediction, control, filtering, etc., is important. Already the martingale approach [16-20] and Meyer's [21] decomposition, as well as related theoretical work [29], provide such representations which are more powerful than the ones used previously [9-15] which were given in terms of local intensity functions. Of interest to this author is the general form of the stochastic integral representation of a martingale [20] which requires the <u>sum</u> of individual integrals of the form

$$\int^t \phi_i(\tau)\, dq_i(\tau)$$

where the q_i's are the "basis" martingales. This general decomposition is reminiscent of the multiplicity representation of a second-order random process [29-33] in terms of integrals with respect to "basis" wide-sense martingales (i.e., processes with orthogonal increments). In view of the fact that the multiplicity representation provides the general framework for "wide-sense" or linear causal estimation, and that the martingale approach solves or generalizes causal estimation to classes of nonlinear problems, in extension of the innovations results [34-35], the similarity of the two representations is worth being looked at. Previous work done by the author in the multiplicity area makes this comparison an attractive side-question.

c. Finally, adjustments of these models can be made to fit problems of random channel modeling. Recent work by Kanal, et al. [36], as well as by the author [37] has focused on the jump process model for the error occurrences. Similarly, if time and luck permit, one may examine the usefulness of such models in problems of information transmission in the peripheral nervous system for muscle movement control messages.

REFERENCES

1. Cox, D. R. and P. A. W. Lewis, Statistical Analysis of Series of Events, Methuen, London, 1966.
2. Saaty, T. L., Elements of Queueing Theory with Applications, McGraw-Hill, New York, 1961.
3. Lewis, P. A. W., Stochastic Point Processes, Wiley-Interscience, 1972.
4. Dresher, M., L. S. Shapley, and A. W. Tucker (editors), Advances in Game Theory, Princeton University Press, 1964.
5. Friedman, A., Differential Games, Wiley-Interscience, 1971.
6. Danskin, J. M., Jr., Value in Differential Games, to be published.
7. Ho, Y. C. and K. C. Chu, Information Structure in Dynamic Multi-Person Control Problems, Technical Report 642, Division of Engineering and Applied Physics, Harvard University, May 1973.
8. Rhodes, J. B. and D. G. Luenberger, Nondeterministic Differential Games with Constrained State Estimators, IEEE Trans. AC, vol. 14, October 1969, pp. 476-481.
9. Beutler, F. J. and O. A. Z. Lenerman, The Spectral Analysis of Impulse Processes, Information and Control, vol. 12, no. 3, March 1968.
10. Beutler, F. J., On the Statistics of Random Pulse Processes, Information and Control, vol. 18, no. 4, May 1971.
11. Bar-David, I., Communication Under the Poisson Regime, IEEE Trans. IT, vol. 15, no. 1, January, 1969.
12. Snyder, D. L., Smoothing for Doubly Stochastic Poisson Processes, IEEE Trans. IT, vol. 18, September 1972, pp. 558-562.
13. Snyder, D. L., Filtering and Detection for Doubly Stochastic Poisson Processes, IEEE Trans. IT, vol. 18, January 1972.

14. Rubin, I., Regular Point Processes and their Detection, IEEE Trans. IT, September 1972, pp. 547-557.

15. Rubin, I., Information Rates for Poisson Sequences, IEEE Trans. IT, May 1973, pp. 283-294.

16. Segall, A., A Martingale Approach to Modeling, Estimation, and Detection of Jump Processes, Ph.D dissertation, Stanford University, August 1973.

17. Kailath, T., Signal Detection and Estimation by Martingale Methods, IEEE International Symposium on Information Theory, Ashkelon, Israel, June 1973.

18. Bremaud, P. M., A Martingale Approach to Point Processes, Memorandum No. ERL-M345, Electronics Research Laboratory, University of California, Berkeley, August 1972.

19. Boel, R., P. Variaya, and E. Wong, Martingales on Jump Processes I, II, Memoranda No. ERL-M407, ERL-M409, Electronics Research Laboratory, University of California, Berkeley, September and December 1973.

20. Variaya, P., The Martingale Theory of Jump Processes, Proceedings of IEEE Conference on Decision and Control, San Diego, December 1973, pp. 48-57.

21. Meyer, P. A., Probability and Potentials, Blaisdell, Waltham, Mass., 1966.

22. Kunita, H. and S. Watanabe, On Square-integrable Martingales, Nagoya Mathematical Journal, vol. 30, 1967, pp. 209-245.

23. Snyder, D. L. and I. B. Rhodes, Filtering and Control Performance Bounds with Implications on Asymptotic Separation, Automatica, vol. 8, 1972, pp. 747-753.

24. Witsenhausen, H. S., Separation of Estimation and Control for Discrete-Time Systems, IEEE Proceedings, 1971.

25. Lo, J. T., Representation of Continuous Curves on 3-Dimensional Rotation Group, to be published.

26. Brockett, R. W., Systems Theory on Group Manifolds and Coset Spaces, SIAM J. on Control, vol. 10, May 1972, p. 265.

27. Athans, M. and P. Falb, Optimal Control, McGraw-Hill, 1966.

28. Ephremides, A. and R. J. Corn, Differential Point Process Games, Proceedings of IEEE Conference on Decision and Control, San Diego, December 1973, p. 301.

29. Leadbetter, M. R., On Basic Results of Point Process Theory, Proc. 6th Berkeley Symposium on Math. Stat. and Prob., 1972, pp. 449-462.

30. Cramer, H., On the Structure of Purely Nondeterministic Stochastic Processes, Arkiv för Matematik, vol. 4, 1961, pp. 249-266.

31. Hida, T., Canonical Representations of Gaussian Processes and their Applications, Memoirs of the College of Science, University of Kyoto, Ser. A, vol. 23, Math. no. 1, 1960.

32. Ephremides, A. and J. B. Thomas, On the Multiplicity of a Class of Multivariate Random Processes, Annals of Mathematical Statistics, vol. 43, no. 6, December 1972, pp. 1-7.

33. Ephremides, A. and L. H. Brandenburg, On the Reconstruction Error of Sampled Data Estimates, IEEE Trans IT, vol. 19, no. 3, pp. 365-367, May 1973.

34. Kailath, T., Likelihood Ratios for Gaussian Processes, IEEE Trans. IT, vol. 16, May 1970, pp. 276-288.

35. Kailath, T., The Innovations Approach to Detection and Estimation Theory, Proceedings of IEEE, vol. 58, no. 5, 1970, pp. 680-695.

36. Adoul, J. P. A., B. D. Fritchman, and L. N. Kanal, A Critical Statistic for Channels with Memory, IEEE Trans. IT, vol. 18, January 1972, pp. 133-141.

37. Ephremides, A. and R. O. Snyder, Modeling of High Error Rate Binary Communication Channels, submitted to IEEE Trans. on Communications.

PIPELINE, PARALLEL AND SERIAL REALIZATION OF PHASE DEMODULATORS

R. S. Bucy, K. D. Senne, and H. Youssef

University of Southern California
Los Angeles, California

M.I.T. Lincoln Laboratory
Lexington, Massachusetts

and

Lockheed Aircraft Company
Burbank, California

§1. INTRODUCTION

A theory of designing the "best" processor for estimating the state of a Markov process or signal observed indirectly as a zero memory nonlinear function of the state, corrupted by additive white noise, was developed in the early and middle 1960s. "Best" here is to mean the estimate that minimizes the expected loss, with the loss a function of the estimation error. This theory is a generalization of the Kalman-Bucy linear filtering theory. However, unlike the Kalman-Bucy theory, no blueprint for the black box which accepts as inputs the observations and produces as outputs the optimal estimate, the minimum loss estimate, can be deduced from the theory. In fact, in general, the black box, nonlinear filter corresponding to the optimal estimate is infinite dimensional in that the state of the nonlinear filter is not finite dimensional. For important technological problems, the signal process estimators generated by linearization and application of the linear Kalman-Bucy design often exhibit poor performance, sometimes even divergence. Further, even when the linear design seems effective, one is interested in what estimator has the best possible performance and what this performance is, in order to know whether further effort on this optimal design is justified. Finally, study of the optimal estimator allows one to generate effective suboptimal designs.

For the above reasons, it became clear that building nonlinear filters was an important task. Now it is quite easy to see, for example see [10], that the black box, determining the optimal filter, is specified by the maps with domain at time n the observation process sample path up to and including time n and range, the conditional probability of the signal at time n given the observation process sample path up to and including n, $J_n(\cdot/\underline{z}_o, \ldots, \underline{z}_n)$, with $\underline{z}_i$ the observations. Our problem then has two phases, one the replacement of $J_n(\cdot/\underline{z}_o, \ldots, \underline{z}_n)$ by a finite vector $\underline{J}_n$, the representation problem, and second, the realization problem which is the means used to generate $\underline{J}_{n+1}$ from $\underline{J}_n$ and $\underline{z}_{n+1}$.

The realization tools which have been used have been digital and hybrid systems. The problems considered have been the cubic sensor, passive receiver and the one, two and three state dimensional phase demodulation problems. These problems have been chosen for their importance, but also because the state dimension was low, a necessity if sufficiently precise estimates of the error performance of the optimal filter are to be generated. This is because error performance can so far only be evaluated by Monte Carlo simulation, N. B. This is also true for suboptimal filters. The problems which we have studied the most are the cubic sensor problem and the two-state dimensional phase demodulation problem; see [4,11,12, 13].

The representations considered for the two-dimensional phase demodulation problem have been: point mass, Gauss-Hermite polynomials, Fourier Series, and Cubic Splines under tension.

On the basis of the structure of the equations to be solved in order to build an optimal nonlinear filter, it became clear early in our synthesis research effort that parallel or associative digital computers offered considerable speed-up in estimate production over standard serial machines of third generation; see [1,4] where these effects are discussed. Because of our lack of access to machines of the parallel or associative type in the early 1970s, a parallel processor was constructed using a contemporary hybrid system. In [2] the feasibility was considered, while in [3,5] the construction of the actual system was reported along with subsequent Monte Carlo runs, all for a one-dimensional signal process cubic sensor problem. The results were encouraging; in particular the hybrid realization proved to be 16 times faster than an all-digital realization, using the digital portion of the hybrid system. Our ultimate aim, however, was to study the effect of using CDC Star 100 and Illiac IV as synthesis tools for the nonlinear filter. This paper is devoted to detailing the impact of these machines on the nonlinear filtering problem of phase demodulation.

In the Summer of 1975, we obtained access to the Illiac IV and made plans to access the CDC Star 100, when it was delivered to NASA Langley. It was decided that in order to effectively be able to judge the improvement, we should initially realize a problem, the two-dimensional phase demodulation problem where we had extensive numerical results and experience on the serial CDC 6600. Further, for this problem we could generate estimate and signal sequences with an existing Serial 6600 program. We were initially convinced that this serial program was close to the fastest possible, and the fact that this was not the case was an interesting byproduct of producing an effective Star program. While studying this two-dimensional phase demodulation problem on the Illiac and the Star gave us good data on speed-up factors possible with pipeline and parallel machines, our real objective was to do a problem which was beyond the capabilities of third generation serial machines. Consequently, we developed a generalized three-state dimensional demodulation problem where phase, phase rate, and amplitude all must be estimated. Beginning in the fall of 1975, Ken Senne developed a two-dimension optimal phase demodulator program for the Illiac IV using the Glypnir language and the Arpa net. In the month of June, 1976, the authors were visiting scientists at ICASE, NASA Langley. During this time, the corresponding program for the Star 100 was developed. This latter program not only was very effective for Star, a large percentage of operations were of long vector type with over 4,000 component vectors; consequently, the Star was doing long streaming. This Star program was used as a prototype of an extremely fast serial program which was over 2-1/3 times faster on the 6600 than our original program reported in [4]; also see the listing of the cyclic point mass program. Finally, a Star program for a three-dimensional signal process, combining amplitude and phase estimation, was developed and time comparison between it and the corresponding serial version obtained.

Methods of obtaining extremely fast serial realization of the optimal filter are being investigated using optical and surface wave devices. It is curious that the mathematical mapping which permits a time correlator surface wave device to calculate multi-dimensional convolutions, is intrinsic in the Star program--compare [9]. We will discuss later in this paper the possible application of array processors coupled with minicomputers to this problem.

In [10], a discussion is given comparing the speed-up possibilities inherent in realization method versus density representation. This paper is concerned only with machine realizations and only deals with the point mass representation. Other representations can be used, but in general they must be accuracy calibrated against the point mass method, and the effect of pipeline and parallel realization on estimate speed-up is unclear.

§2. COMPUTER REALIZATION SURVEY

2.1 The Physical Problem

It has been observed in previous studies [4] that nonlinear filters can be efficiently implemented on parallel computers. Since every so-called vector machine has unique limitations, however, it is necessary to adapt the algorithm to each particular architecture. Candidate architectures for the present study include the array processor (Illiac IV), the pipeline processor (CDC-Star 100), and the look-ahead processors (CDC 6600 and 7600). Benchmarks on fast serial machines (IBM 370-168 and PDP 11-70) have also been included for reference.

An interesting application for optimal nonlinear estimation was introduced by Mallinckrodt, Bucy, and Cheng [7], who considered the problem of tracking a first-order phase process based on measurements of a modulated signal in noise of the form

$$ds(t) = A \cos[\omega_o t + x_1(t)]dt + dv(t) \tag{2.1}$$

where A is a known amplitude, ω_o is a known carrier frequency, and $x_1(t)$ is the message process being tracked. The measurement noise is assumed white. Using a voltage-controlled oscillator, the known carrier may be removed by heterodyning down to base band, producing both in-line and quadrature components and resulting in an equivalent two-dimensional measurement process of the form

$$\begin{bmatrix} dz_1(t) \\ dz_2(t) \end{bmatrix} = \begin{bmatrix} \cos x_1(t) \\ \sin x_1(t) \end{bmatrix} dt + \begin{bmatrix} dv_1(t) \\ dv_2(t) \end{bmatrix} \tag{2.2}$$

where A has been taken as unity without loss of generality, and the noise has been replaced by a vector of mutually independent quantities.

The first-order phase process studied in [7] consisted of Brownian motion with increment of length h having variance qh. In this paper, we describe a study of a second order phase process involving the integral of Brownian motion, expressed as

$$\begin{bmatrix} dx_1 \\ dx_2 \end{bmatrix} = \begin{bmatrix} 0 & 1 \\ 0 & 0 \end{bmatrix} \begin{bmatrix} x_1 \\ x_2 \end{bmatrix} dt + \begin{bmatrix} 0 \\ 1 \end{bmatrix} d\beta_t \tag{2.3}$$

We will retain the same measurement model (2.2) and let the noises v_1 and v_2 be independent Brownian motions with path of length h having variance rh.

The optimal filter is specified by

$$J_{n+1|n+1}\begin{pmatrix} y_1 \\ y_2 \end{pmatrix}$$

the conditional density of the phase and phase rate at time $(n + 1)\Delta$, given discrete observations up to this time at sampling interval Δ. The following equations determine this density:

$$J_{n+1\ n+1}\begin{pmatrix} y_1 \\ y_2 \end{pmatrix} = D_{n+1}(y_1)\int_{-\infty}^{\infty} \exp\left\{-\frac{(y_2 - \mu)^2}{2q\Delta}\right\} J_{n|n}\begin{pmatrix} y_1 - \mu\Delta \\ \mu \end{pmatrix} d\mu \tag{2.4}$$

where

$$D_{n+1}(y_1) \triangleq c_o \exp\left\{\frac{z_1(n+1)\cos y_1 + z_2(n+1)\sin y_1}{r/\Delta}\right\} \tag{2.5}$$

It can be shown (see [4]) that a modulated density $\tilde{J}_{n|n}$ defined as

$$\tilde{J}_{n|n}\begin{pmatrix} \sigma \\ \tau \end{pmatrix} \triangleq \sum_{k=-\infty}^{\infty} \sum_{l=-\infty}^{\infty} J_{n|n}\begin{pmatrix} \sigma + 2\pi k \\ \tau + \frac{2\pi l}{\Delta} \end{pmatrix} \tag{2.6}$$

with $-\pi \le \sigma < \pi$, and $-\pi/\Delta \le \tau < \pi/\Delta$, carries all the information necessary for nonlinear filtering when the cyclic loss function $\frac{1}{2}(1 - \cos \varepsilon_1)$ with ε_1 the phase error, is used. The modulated density satisfies

$$\tilde{J}_{n+1|n+1} = D_{n+1}(\sigma) \int_{-\pi/\Delta}^{\pi/\Delta} a(\tau - \xi)\, \tilde{J}_{n|n}\begin{pmatrix} \sigma - \xi\Delta \\ \xi \end{pmatrix} d\xi \tag{2.7}$$

where

$$a(\tau - \xi) \triangleq \sum_{i=-\infty}^{\infty} \exp\left\{- \frac{(\tau - \varepsilon + 2\pi i/\Delta)^2}{2q\Delta}\right\} \tag{2.8}$$

We recall that the cyclic estimate, the one that minimizes the cyclic loss, is given by

$$x_n^* = \arg E\, e^{ix_1(n)} \Big| z_i^1,\ z_i^2 \quad i \le n) \tag{2.9}$$

2.2 Point Mass Placement

The most attractive formulation for numerical solution of the filtering problem on parallel machines involves the use of point-mass representation of the densities. A fixed rectangular grid will consist of m points in the phase variable and n points in the phase rate, with coordinates (i, j) corresponding to (σ, ξ) by the formulae

$$\sigma(i) = -\pi + 2\pi\left(\frac{i}{m}\right) + \pi\left(\frac{1}{m}\right) = \pi\left(\frac{1 + 2i}{m} - 1\right)$$

$$i = 0, \cdots, m-1 \tag{2.10}$$

$$\xi(j) = -\pi/\Delta + 2\pi/\Delta\left(\frac{j}{n}\right) + \pi/\Delta\left(\frac{1}{n}\right) = \pi/\Delta\left(\frac{1 + 2j}{n} - 1\right)$$

$$j = 0, \cdots, n-1 \tag{2.11}$$

Note that the phase rate variables may be used directly in the convolution (2.7) but the phase variables must be interpolated in general, so that

$$\eta(k) = \sigma(i) - \Delta\xi(j) \Rightarrow k = i + \frac{m}{2} - \frac{m}{n}\left(\frac{1}{2} + j\right) \tag{2.12}$$

If m is taken as an even integer, such that n/m is an integer, then (2.12) may be decomposed into a subdominant integer [k] and a fractional part Δk, as follows:

$$[k(i)]^{*} = i + \frac{m}{2} - \left(\frac{1}{2} + j\right) \text{ DIV } \left(\frac{n}{m}\right) - 1 \tag{2.13}$$

where DIV denotes integer division. Note that the addition of 1/2 in (2.13) will not change the result since n/m is an integer. Next, we observe that the remainder after the division in (2.13) is the fractional part Δk:

$$\Delta k = \left(\frac{1}{2} + j\right) \text{ MOD } \left(\frac{n}{m}\right), \qquad j = 0, \cdots, n-1 \tag{2.14}$$

or

$$1 - \Delta k = 1 - \left(\frac{1}{2} + j\right) \text{ MOD } \left(\frac{n}{m}\right)$$

The interpolated result desired will be between [k(i)] and [k(i)] + 1 (evaluated modulo m), with weighting Δk on the former and $1 - \Delta k$ on the latter. If the convolution is formed for fixed phase (i), then

$$[k(i + 1)] = [k(i)] + 1 \quad \text{(modulo m)} \tag{2.15}$$

The relation (2.15) suggests the recursion which will be used on the Illiac. (2.15) will be used to iterate, based on an initial condition

$$[k(0)] = \left(\frac{m}{2} - 1 - j \text{ DIV } \frac{n}{m}\right) \text{ MOD } m \tag{2.16}$$

During such iteration, i, Δk will be used to weight the term evaluated at [k(i)], while $1 - \Delta k$ will be used to weight the term evaluated at [k(1 + i)].

---

*Note that [k] should be interpreted as modulo m.

$$J^{i}_{n|n}\begin{pmatrix} n(k) \\ \xi(j) \end{pmatrix} \cong (1-\Delta k)\, J_{n|n}\begin{pmatrix} \eta([k(i+1)]) \\ \xi(j) \end{pmatrix} + \Delta k\, J_{n|n}\begin{pmatrix} \eta([k(i)]) \\ \xi(j) \end{pmatrix} \tag{2.17}$$

2.3 Evaluating the Filtered Data Term

The term (2.8) consists of an infinite sum. The value which Γ may take on in the grid coordinates consist of integer multiples of $2\pi/n\Delta$, resulting in

$$a(\Gamma(p)) = \sum_{\ell=-\infty}^{\infty} \exp\left[-\frac{2}{q\Delta}\left(\frac{\pi}{\Delta}\right)^2\left(\frac{p}{n}+\ell\right)^2\right]$$

$$-2(n-1) \le p < 2(n-1) \tag{2.18}$$

It may be seen from (2.18) that $a(\cdot)$ is an even function which is cyclic, modulo n. Further, for reasonable values of q the contribution of all terms except $\ell = 0$, is negligible. Thus, we compute

$$a(\Gamma(p)) = \exp\left[-\frac{2}{q\Delta}\left(\frac{\pi}{\Delta}\right)^2\left(\frac{|p|}{n}\right)^2\right] \quad 0 \le |p| \le n-1 \tag{2.19}$$

In the examples described below, five terms of $a(\cdot)$ were taken to be nonzero for the computing convolutions.

2.4 Assessment of Required Computations

The computations required to implement the point mass filter are summarized in Table 2.1, as a function of m and n. The sensor terms are the only ones which require math functions (exponentials). Since only m exponentials are required, this computation is generally insignificant compared with the overall filter update, so no special effort has been expended to optimize the required computations.

TABLE 2.1

Arithmetic Operations for Filter

Function	Number of Operations			
	Multiples	Adds	Divisions	Exponentials
Sensor Terms	2m	m	0	m
Interpolation	mn	2mn	0	0
Convolution	5mn	10mn	0	0
Row Sums	0	mn-m	0	0
Estimates	3m	3m-3	1	0
Normalization	mn+m	0	0	0
Total	7mn+6m	13mn+3m-3	1	m
Example m = 32 n = 128	28864	53341	1	32

2.5 The Illiac IV Algorithm

The primary considerations for programming the Illiac IV array are the proper utilization of all of the 64 Processor Elements (PEs) and minimization of data routing between PEs. In order to accomplish the efficient use of the PEs, the values of m = 32 and n = 128 were selected and utilized for all machine examples. On the Illiac, two rows of PE storage were used for each value of the phase samples in JN. The interpolation and convolution were accomplished in a totally parallel fashion, using all 64 PEs, except for the operation depicted in Figure 2.1. It is necessary to perform a cyclic rotation of each phase row to accumulate the terms for the

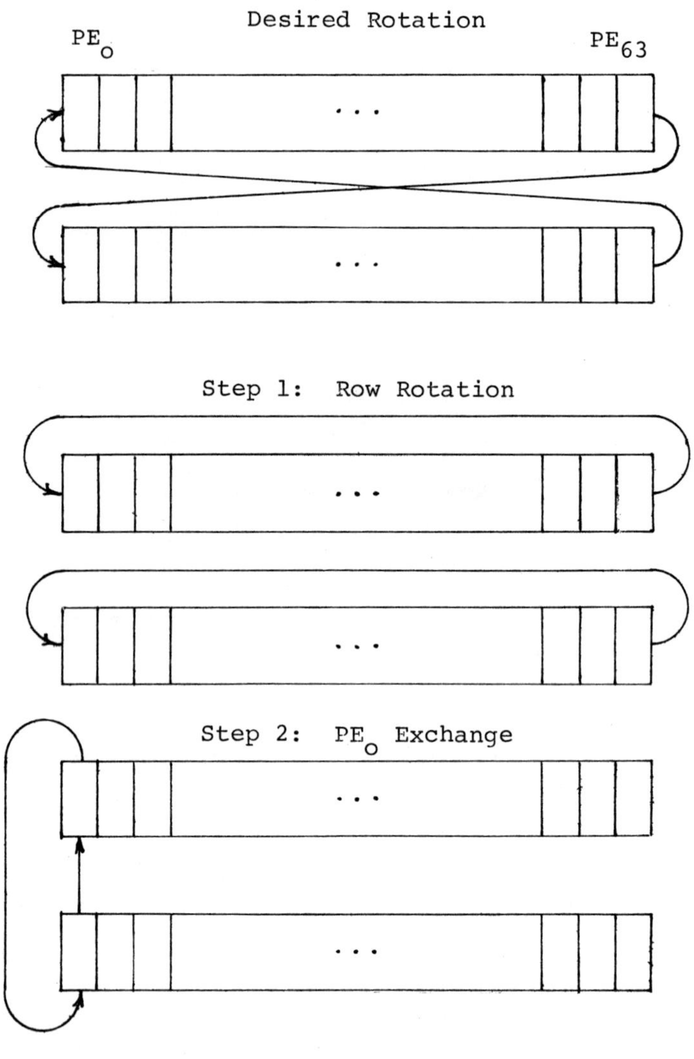

FIG. 2.1. Cyclic Rotation Modulo 128 on Illiac

convolution (a cyclic rotation to the right is combined with a cyclic rotation to the left at each step). Since a cyclic routing on Illiac IV involves only one row at a time, it was necessary to form the two-row rotation by two single-row rotations, followed by an end element switch (involving three transfers with only one PE enabled).

The overall effectiveness of the Illiac IV algorithm is evident from the fact that so few operations are required which involve fewer than 64 PEs enabled. The sensor terms are computed with only 32 PEs enabled, but this operation only involves about 5% of the estimate update. The single PE transfers and the convolution are also only responsible for about 5% of the computation time. Finally, the row sums are done logarithmically with a PE utilization efficiency of 16.7%, and they account for another 5% of the estimated computation time. Thus, the net efficiency of this algorithm is 87.5%, or the equivalent of 56 times faster than a single PE program.

The Illiac program was coded in the Glypnir language, utilizing assembly language listings to reduce the overhead computations. The resulting program was timed in the non-overlap mode, wherein the Control Unit (CU) and the PEs operate serially and considerable overhead cycles are required to allow error conditions to ring out between instructions. Thus, although we would expect the Illiac program to run at least twice as fast as the equivalent program for the CDC-STAR, in fact it ran five times slower. It will be reported in [17] how the program operates in overlap mode. It is also important to indicate that the Illiac IV Clock is running at 80 nsec as compared with the design goal of 50 nsec.

2.6 The CDC-STAR Algorithm

The pipeline architecture is unconstrained by small fixed resources (i.e., 64 processors). On the other hand, efficient utilization of the pipeline requires detailed attention to prearrangement of vectors to allow for streaming from consecutive memory locations. This consideration is particularly important for STAR, which has a relatively slow memory cycle time. Since the nonlinear filter is recursive, it is necessary to include the vector rearrangement as part of the filter update and therefore the rearrangement constitutes the major overhead of the STAR program. The operations on the matrix JN to precondition it for the convolution are shown in Figures 2.2 and 2.3. First, the column-ordered JN matrix is column-shuffled with itself to produce a matrix which has two copies of every phase variable in each column. Then, a scrambled JN can be formed which has the property that each row in the final convolved matrix can be generated by operating on a suitable interpolation between the two adjacent rows of the scrambled JN. The interpolation which does this is depicted in Figure 2.3.* This interpolation may be done by vector operations of length (m-1)n, or 4224. The row corresponding to m+1 in the result may be compressed out of the final result to reduce the subsequent calculations.

The cyclic convolution is shown in Figure 2.4. First, the end columns of the interpolated JN are cyclically copied to produce a matrix from which each of the terms of the convolution may be obtained as m x n matrices.

---

*The Figures 2.3-2.4 are shown with m = 4 and n = 16 for illustrative purposes only.

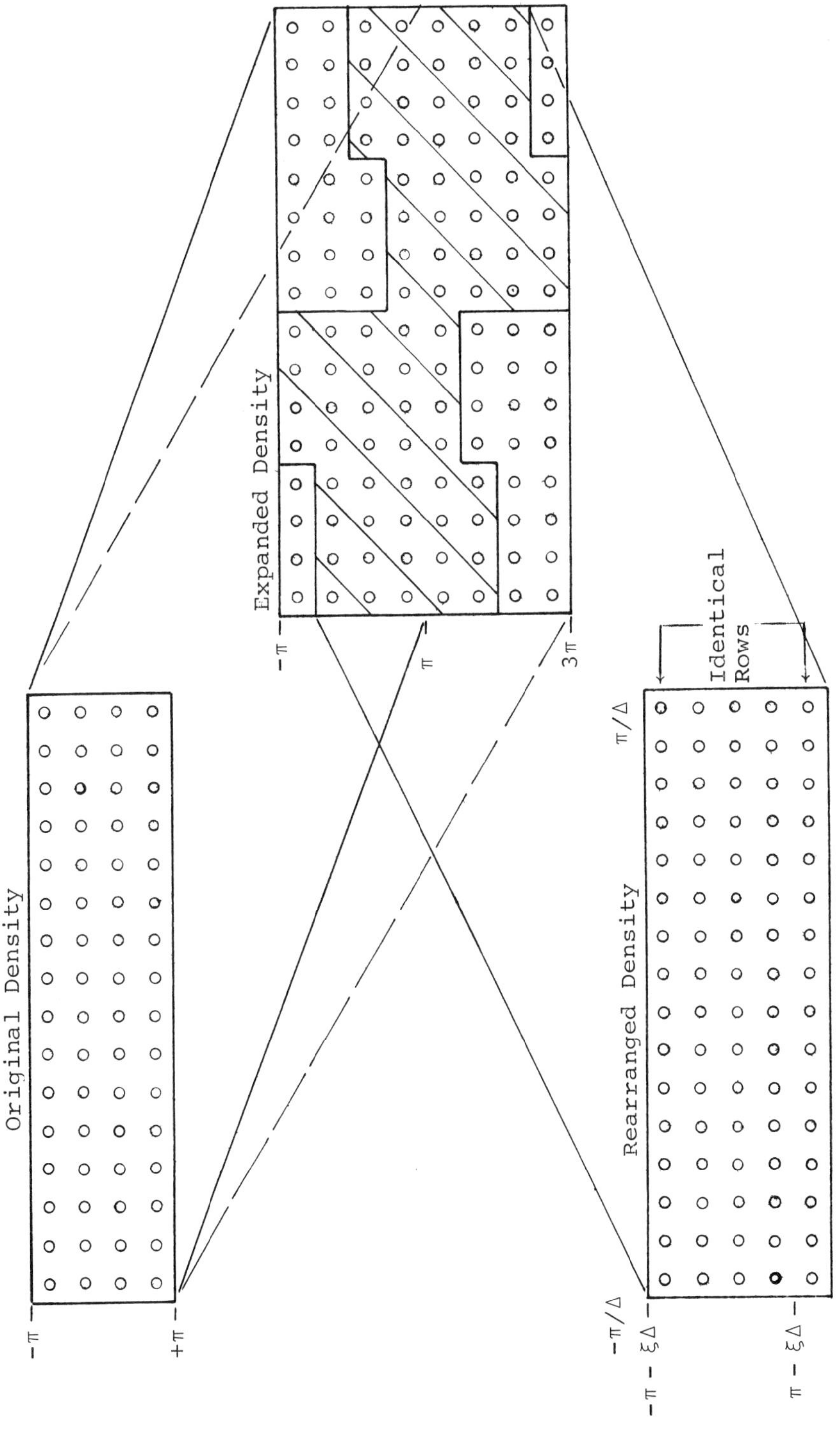

FIG. 2.2. Scrambling of Phase Variables Modulo 2π

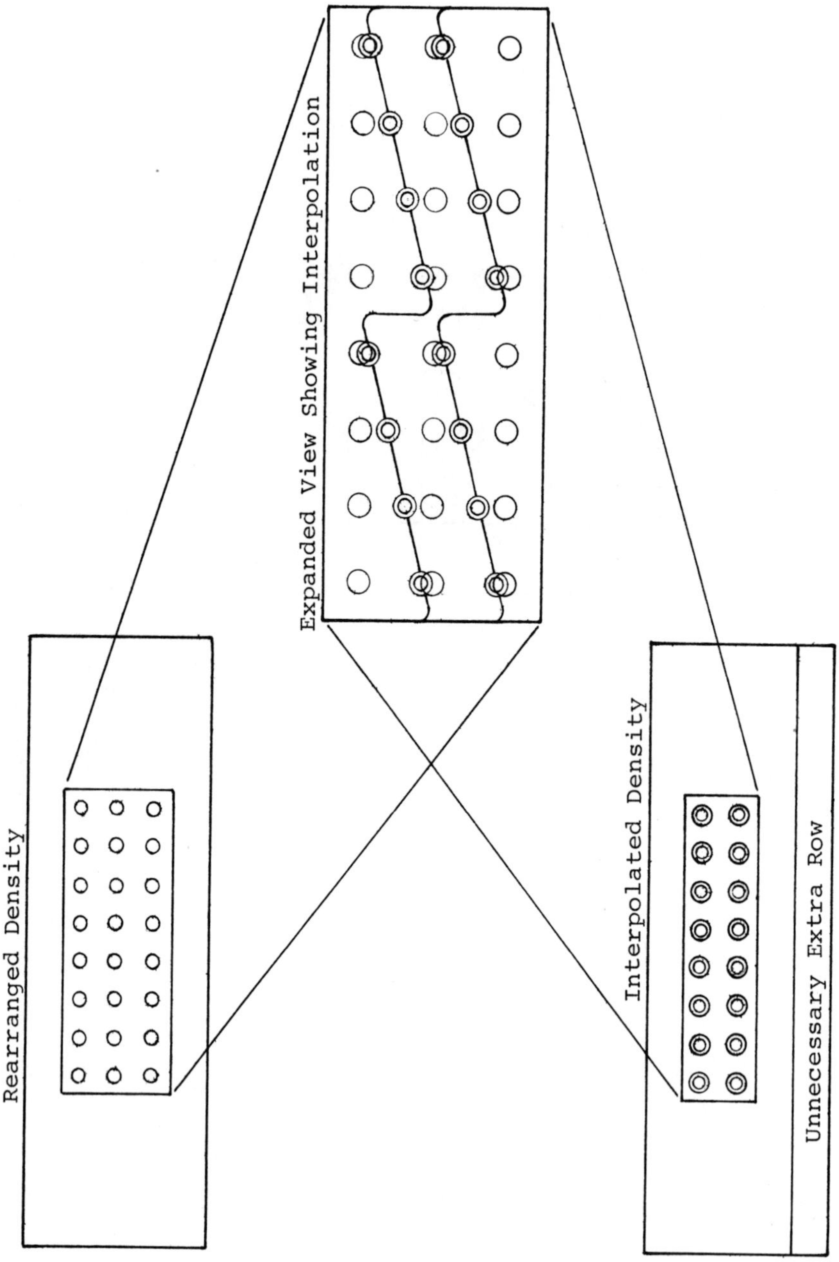

FIG. 2.3. Interpolation of Scrambled Matrix

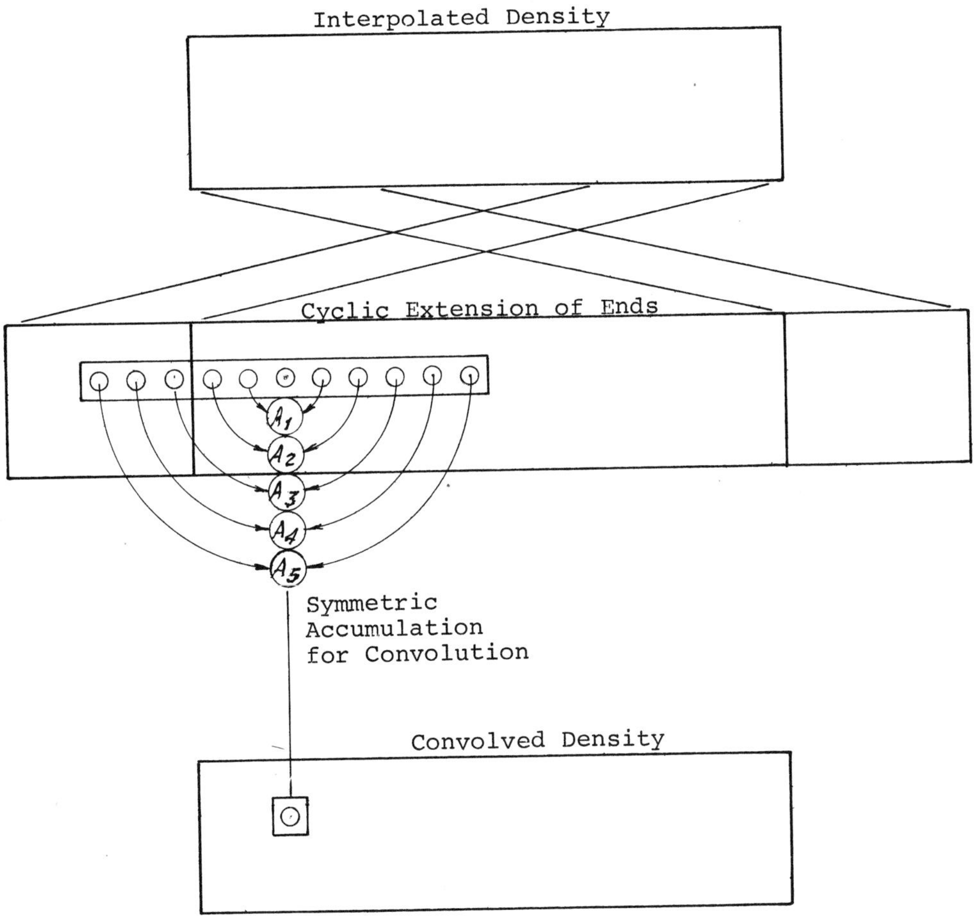

FIG. 2.4. Convolution Operation

The production of a 5-term symmetric convolution is done in parallel for all 4096 points by sequence of 10 vector adds and 5 vector multiplies, all of length 4096.

The only computations which are done on the STAR that are less than 100% efficient are the vector sums and the determination of the estimates. This is reflected in Table 2.2, which gives the breakdown of the various functions in the STAR program.

TABLE 2.2

STAR Program Breakdown

Operations	% of Total	Cycles Start-ups	Cycles Compute	Cycles Required
Vector Arithmetic	63.4	5174	76793	78.5%
12 Multiplies	33.3	1908	41152	70.1%
46 Adds	26.4	3266	30861	86.4%
1 Exponential	3.7	--	4780	100.0%
Vector Rearrangement	31.3	1233	39292	83.7%
1 Vector Transfer	5.8	1001	6464	68.0%
2 Indexed Transfers (Block lengths 32 & 33)	22.1	144	28480	100.0%
1 Vector Compress	3.4	88	4348	0%
Scalar Arithmetic	--	--	79	100.0%
1 Divide			46	
3 Adds			33	
Subroutine Overhead 3 Calls	4.1	--	5142	0%
Miscellaneous Memory Conflicts, etc.	1.2	--	1537	100.0%
TOTAL		6407	122843 ⟹ 5.17 msec	
		5%	95%	
Minimum Achievable		6224	111296 ⟹ 4.70 msec	

2.7 CDC 6600 Program

The CDC 6600 (and 7600) has instruction look-ahead and multiple arithmetic processor units which provide a partial overlap parallelism. The most efficient 6600 programs contain many tight loops instead of complicated computations within large loops. By recoding the vectorized STAR program in analogous FORTRAN for the CDC 6600, we were able to achieve near optimal utilization of the available resources with functional loops which are for the most part contained within the 8-word instruction stack.

The basic data flow of the CDC 6600 is illustrated in Figure 2.5. Reads from memory are accomplished by setting the A Registers A1-A5 with the appropriate address; writes are obtained by loading A6-A7. The B Registers are used for incrementing and address computation.

The breakdown of the computations for the CDC 6600 is shown in Table 2.3. Note that the major overhead is for reading and writing and miscellaneous waiting. It is interesting to note also that the ratio of multiply rates between STAR and 6600 is 25 to 1, while the add rate ratio is 17.5 (including normalization on 6600). Thus, for arithmetic alone, STAR would be expected to be 21 times faster than the 6600 on this problem. The achieved speed-up of 29 is explained by the slightly lower overhead on STAR (36.6% versus 52.9% for 6600).

2.8 Implications for Future Synthesis of Higher Dimensional Filters

The main purpose of this study was to measure the speed-up achievable by parallel and pipeline machines, in order to determine the feasibility of realizing 3 and 4 state dimensional

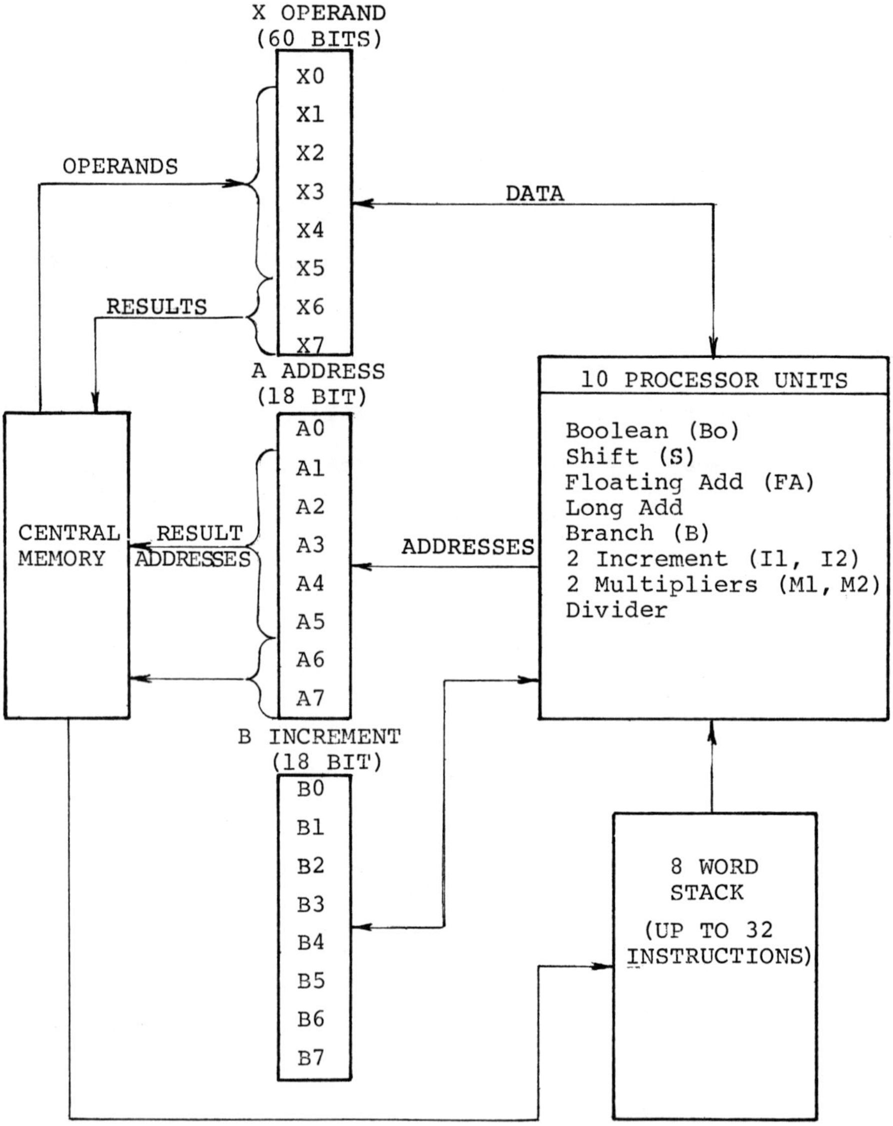

FIG. 2.5. CDC 6600 CPU Architecture

TABLE 2.3

CDC 6600 Program Breakdown

Function (% Cycles)	Multi-plies	Adds	Divi-sions	Reads	Writes	Exp.	Notes
Sensor Update (~0)	2m	m	0	3m	m	m	Not in stack (extn. refs.)
Interpolation (~16)	mn	2mn	0	4mn+3n	mn	0	Outer loop not in stack
Expansion (~0)	0	0	0	10m	10m	0	Instack
Convolution (~74)	5mn	10mn	0	16mn+15	6mn	0	Instack
Row Sums (~4)	0	mn-m	0	mn+m	m	0	Instack
Estimates (~0)	3m	3m-3	1	4m+11	3	0	Instack
Normalization (~6)	mn+m	0	0	mn+2m	mn	0	Instack
Approx. Totals	7mn	13mn	0	22mn	8mn	0	
Approx. No. of Minor Cycles	70mn	91mn	0	66mn*	24mn*	0	

Summary: Minor Cycles (Approx.)

Arithmetic:	161 mn	47.1%
Read/Write:	90 mn	26.3%
Waiting:	91 mn	26.6%
Measured:	342 mn	100.0%

(m = 32, n = 128 for test case) Measured time = 140 msec/est.

*The reads and writes are partially overlapped in time, however the total overhead of 52.9% is accurate.

nonlinear filters. In a later section of this paper we describe the fairly natural extension of our 2-state dimensional filter STAR 100 program to a 3-state dimensional problem, that of estimating phase, phase rate, and amplitude. One outgrowth of the software development was a determination of machine structure limitations for natural synthesis of higher dimensional filters. Basically, we found that Illiac IV was limited to the two-dimensional problem because of limited PE memory and the availability of only 64 PEs, and most importantly, because of the increase of program complexity and overhead with dimension. The STAR 100 program also becomes complex to program and overhead prone, but at a somewhat higher dimension with the basic limitation that the total number of grid points be less than 65K, the maximum length of vector for STAR vector operations, as we show later, a 3-state problem is feasible on STAR.

The implications for future machine design suitable for our problem are a pipeline machine which can compute with indexed vectors and with reduced operation cycle time.

§3. TIME FACTORS FOR 2-DIMENSIONAL PHASE DEMODULATION

In [4] a complete description of the two-state dimensional phase demodulation problem is given as well as numerous numerical results and a listing of the cyclic nonlinear filtering program. Because we wished to compare the Illiac, Star, and 6600 directly, the grid of 32 x 128 was chosen, i.e., 32 subdivisions in phase and 128 subdivisions in phase rate, to represent JN the conditional density of phase and phase rate given the observations. This grid was natural for the Illiac as it has 64 parallel channels so that the integral equation

$$J_n(x,y) = \frac{D_n(x)}{k} \int_{-\pi/\Delta}^{\pi/\Delta} S(y - \xi)\, J_{n-1}(x - \xi\Delta, \xi)\, d\xi \qquad (3.1)$$

$$y \in (-\pi/\Delta,\ \pi/\Delta), \qquad x \in (-\pi,\ \pi)$$

with the values of $J_n(x_i, \cdot)$ could be found in two passes; the first simultaneously giving $\{J_n(x_i, y_j)\}_{j=1...64}$ and the second giving simultaneously $\{J_n(x_i,y_j)\}_{j=65...128}$ and then repeated for each i. We knew before that this mesh was fine enough for accuracy. The serial cyclic point mass filter with interpolation required .700 seconds per estimate. The Illiac on the other hand produced estimates every .0308 second. The Illiac was not run at full capacity for our runs and we were unable to run it under the fastest possible conditions because other jobs had tied up the machine in August, so that our timing should be looked at as an upper bound to actual performance, clearly far away from the predicted theoretical performance. It is unclear whether in the production mode Illiac can achieve anywhere near theoretical performance, i.e., 64 times faster than CDC 6600 on a full parallel job. It may be that Illiac hardware cannot be made to operate at its theoretical speeds.

As described in a previous section for the Star 100 program, the density was carried as a 32 x 128 = 4096 component vector and the convolution developed as a sum of scalar times translates of this vector. This was done in order to take advantage of the pipeline speed of Star. At first with a direct Star Fortran program, a rate of .0118 second per estimate was obtained. By examination of the assembly language version of the Star program and timing each element, we replaced certain Star Fortran calls by more efficient assembly language routines, e.g., VGather and VSUM were replaced. After this was done, we achieved a rate of .0049 second per estimate. From this latter number it became clear that the serial 6600 program was not efficient. We were exceeding the factor of 25 theoretical speed ratio of Star to the 6600. This ratio is the ratio of Star multiplication time to 6600 multiply time. A new program, Starrun, was written which was the serial version of the Star program. Starrun achieved .178 second per estimate with the FTN, OPT = 2, Level 410 operating system. The FTN 4.5 compiler produces significantly different and faster assembly language programs than previous optimizing compilers. By elimination of some multiplications and arranging things so that everything is in the stack, the Starrun program was made to run at .14 second per estimate. The following table gives the timing of our programs on various machines. With considerable effort in each case devoted to generating fast code:

Machines	Time per Estimate (in milliseconds)	Compiler
CDC 6600	137.0	FTN 4.5 Level 410, OPT = 2
STAR 100	4.7	Assembly language Optimization
IBM 370-168	130.0	h-extended
ILLIAC IV	30.8	Glypnir Language Assembly listing optimization
PDP 11-70	870.0	FORTRAN 4 plus
CDC 7600	25	FTN 4.5 Level 410, OPT = 2

In order to get a feel for the relative speeds of the Star and the 6600, the ratio of multiply times, when the pipe on the Star is filled is 25 to 1, i.e., 6600 takes 10 minor cycles at 100 nanoseconds per cycle, while a multiply on Star takes 1 cycle at 40 nanoseconds per cycle. However, the 6600 has several look-ahead features as well as a stack which can speed up the processing. The FTN, OPT = 2 compiler seeks to put all loops in the stack and to fill all the registers so that the ratio we have found, 28.57 to 1, is very reasonable for a problem such as ours which is structured so that vector operations can be done most of the time on Star. Of course, using Star for a problem which does not have parallel structure would be wasteful as one would do as well or better on the 7600.

§4. BENCHMARK APPLICATION FOR DIFFERENT APPROXIMATIONS

It is very typical in engineering studies to settle on a compromise filter realization, which purports to approach the accuracy of the optimum nonlinear filter with tremendous amount of computation savings, without first exhibiting a truly accurate optimum scheme to serve as a benchmark against which any other approximation schemes can be judged. The development of such approximations is possible when the exact performance and amount of computations is known. Then and only then, the engineering compromise can be studied intelligently.

One of these compromise studies has recently been published--see [14]--as an approximation for the optimum two-dimensional phase demodulation. It can be generalized to three-dimensional phase-amplitude demodulation problems. This approximation is based on fitting the probability density function at each time step to a Gaussian distribution which produces the optimum first and second moments for the same error criterion. This means that the update is accurate and required only for the first and second moments. In order to be able to investigate the extension of this Gaussian approximation to the three-dimensional problem, it is essential to study very carefully the behavior of the optimum filter and establish a performance benchmark. This three-dimensional problem was neither analyzed in the optimum way nor simulated on any digital computer before because of its size and complexity. A first attempt was carried out to realize the optimum demodulator on the CYBER C175-T and CDC STAR-100 in order to establish boundary limits on the size of the program storage and the computation time.

The mathematical model, optimum filter and preliminary results for the three-dimensional problem will be discussed in the following sections.

§5. INTRODUCTION TO 3D-PROBLEM AND PRELIMINARY RESULTS

Achieving a good speed-up factor for the two-dimensional phase demodulation problem was the signal to move one step ahead and do the three-dimensional problem (phase, phase rate, and amplitude demodulation).

5.1 Mathematical Problem

The three-dimensional state vector consists of phase x, phase-rate y with additive white Gaussian noise u, and amplitude A corrupted with another independent white Gaussian noise v additive to the amplitude mean value.

$$x_n = x_{n-1} + y_{n-1}\,\Delta$$

$$y_n = y_{n-1} + u_{n-1}$$

$$A_n = \alpha(a - A_{n-1})\Delta + A_{n-1} + v_{n-1}$$

where

Δ is the time step

α is the damping constant

a is the amplitude mean value

$$E\, u_n = 0$$

$$E\, u_n^2 = q_1\Delta$$

$$E\, v_n = 0$$

$$E\, v_n^2 = q_2\Delta$$

$$E\, x_o = 0$$

$$E\, x_o^2 = c_1\sqrt{2}\, q_1^{1/4}\, r^{3/4}$$

$$E\, y_o = 0$$

$$E\, y_o^2 = \sqrt{2}\, q_1^{3/4}\, r^{1/4}$$

$$E\, A_o = a$$

$$E\, A_o^2 = \frac{c_2 q_2}{2\alpha}$$

The observation z is modelled in the following way:

$$z_n^1 = A_n \cos x_n + w_n^1$$

$$z_n^2 = A_n \sin x_n + w_n^2$$

$w_n^{1,2}$ is white Gaussian noise independent of u_n, v_n with $E\, w_n^{1,2} = 0$

$$E\left(w_n^{1,2}\right)^2 = \frac{r}{\Delta}$$

Δ is chosen to be the same as in the two-dimensional problem, i.e.,

$$\Delta = .1\left[\sqrt{2}\left(\frac{r}{q_1}\right)^{1/4}\right]$$

c_1 is chosen to be 2.19 (same as two-dimensional problem); c_2 is chosen to be 4; and $\alpha = 1$, $a = 1$, $q_2 = .01$.

5.2 Optimum Filter

The optimum filter realization is carried out by a recursive update of the density J_n using the discrete form of the presentation theorem (see [4]).

$$J_n(x,y,A) = K_n S_n(x,A) \int_{-\infty}^{\infty}\int_{-\pi/\Delta}^{\pi/\Delta} a_1(y-z)a_2(A-\beta\eta-a\alpha\Delta)$$

$$J_{n-1}(x-z\Delta,z,\eta)\,dz\,d\eta$$

where

$$S_n(x,A) = \mathrm{Exp}\left[\frac{\Delta}{r}\left[A\left(z_n^1 \cos x + z_n^2 \sin x\right) - \frac{A^2}{2}\right]\right]$$

$$a_1(y-z) = \sum_{k=-\infty}^{\infty} \mathrm{Exp}\left[-\frac{1}{2q_1\Delta}\left(y - z + \frac{2\pi k}{\Delta}\right)^2\right]$$

$$a_2(A-\beta\eta-a\alpha\Delta) = \mathrm{Exp}\left[-\frac{1}{2q_2\Delta}\left(A-\beta\eta-a\alpha\Delta\right)^2\right]$$

$$\beta = 1 - a\alpha\Delta$$

K_n is normalizing constant chosen so that

$$\int_{-\infty}^{\infty}\int_{-\pi/\Delta}^{\pi/\Delta}\int_{-\pi}^{\pi} J_n(x, Y, A)\,dx\,dY\,dA = 1$$

5.3 Optimum Estimates

The phase estimates are produced in the same manner as in two-dimensional problems. They are obtained by minimizing the error function $L_1(e) = 2(1 - \cos e)$

$$x_n^* = \arg \int_{-\infty}^{\infty}\int_{-\pi/\Delta}^{\pi/\Delta}\int_{-\pi}^{\pi} e^{ix} J_n(x,y,A)\,dx\,dy\,dA$$

The amplitude estimates are produced by simply minimizing the quadratic loss function $L_2(e) = e^2$, leading to

$$A_n^* = \int_{-\infty}^{\infty} \int_{-\pi/\Delta}^{\pi/\Delta} \int_{-\pi}^{\pi} A\, J_n(x,y,A)\,dx\,dy\,dA$$

5.4 Simulation

A Fortran program was written to simulate the model described in 5.1 and produce optimum density and estimates as described in Sections 5.2 and 5.3, respectively. The random number generator used is the same as in the two-dimensional problem. This program was debugged and executed on the next fastest computer available at NASA Langley--CYBER C 175-T, which is faster than the CDC 6600 by a factor of approximately 3. Then a final program was written for the Star which produced output which checked with the C 175-T results.

5.4.1 CYBER C 175-T Program

A point mass realization of the conditional density was simulated by choosing 16 x 96 x 16 masses located inside a box representing the three axes, the phase, phase rate, and amplitude, respectively (see Figure 5.1). The 16 x 96 masses for the phase and phase rate are located inside the rectangle bounded by $(-\pi, \pi]$ in phase and $(-\pi/\Delta, \pi/\Delta]$ in phase rate. The grid in the amplitude direction is floating according to the location of the prior estimate position. For every new amplitude estimate, the whole box will be centered on the amplitude estimate with eight masses equally distributed above the center and the same number below. The mesh size between the amplitude masses is chosen to be equal to half the square root of the variance of the estimate. The reason we chose the

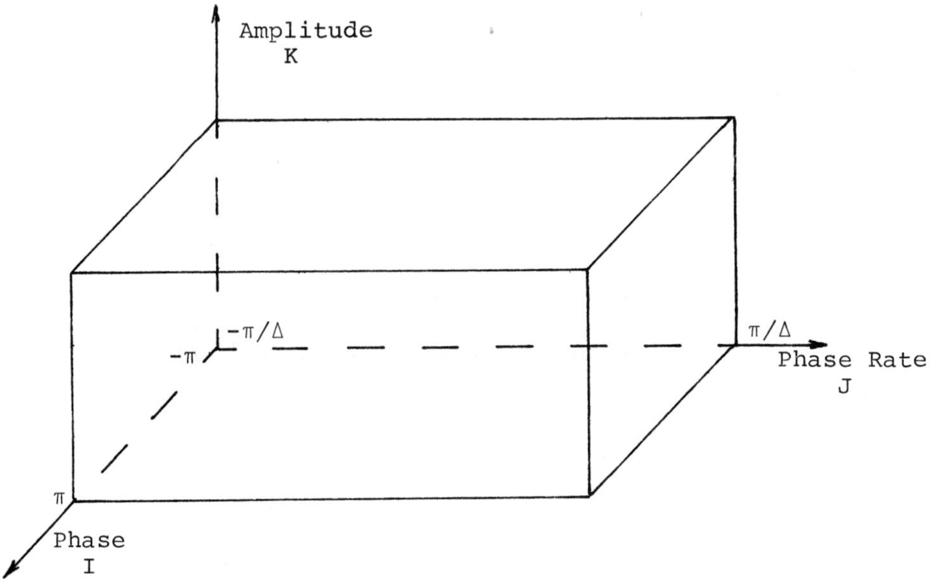

FIG. 5.1. Three-Dimensional State Space

number of discrete masses to be 16 x 96 x 16 is that the maximum vector length for Star operation is limited to 65K as well as corresponding also to storage limitations on the CYBER. Because we know from the two-dimensional problem that 16 x 96 masses was not accurate enough for density representation, the purpose of the 16 x 96 x 16 scheme is just to debug the program on the CYBER and try to produce the same results on the Star later. This way, we can have a valid comparison for the speed of the Star as compared to other computers for the same amount of computations.

The main structure of the program follows essentially the original program for the two-dimensional phase demodulator. The interpolative and scrambling matrices were precomputed once and for all, and called out inside the mass convolution loop for each iteration. Because of the choice of the dynamic model for the phase rate, $a_1(y - z)$ is symmetric and constant. The cut-off limit for the significant terms of $a_1(y - z)$ is determined by a radius of length = $\sqrt{50\ q_1 \Delta}$. This is not the case for the amplitude dynamics which does not have a symmetry because of the damping factor and all the sixteen terms have to be computed for every iteration. When this program was executed, it produced estimates every 12 seconds.

5.4.2 CDC Star-100 Program

The standard FORTRAN program in Section 5.4.1 was changed to adapt to the vector notations for the Star. The three-dimensional array was swept in the right order; I (phase), J (phase rate), K (amplitude). The main steps of the program follow the same structure of the two-dimensional Star program, except for the added third dimension as follows:

(i) In order to perform the scrambling according to the rule (I-J) inside the convolution in the most efficient way, a block mapping VXTO is used with length 17 for 96 x 16 blocks (see Figure 5.2). Of course, that required an initial step of copying the box on itself by using the 'MERGE' COMMAND and carrying a copy of the three axis array for the density masses.

(ii) The interpolation between two successive rows was carried on the box of size 17 x 96 x 16.

(iii) 'COMPRESS' COMMAND was finally used to get rid of the extra 17th element and to bring the box into its normal size, 16 x 96 x 16.

(iv) At this step the box was expanded in J direction far enough to cover the maximum number of terms of $a_1(y - z)$ as shown in Figure 5.3.

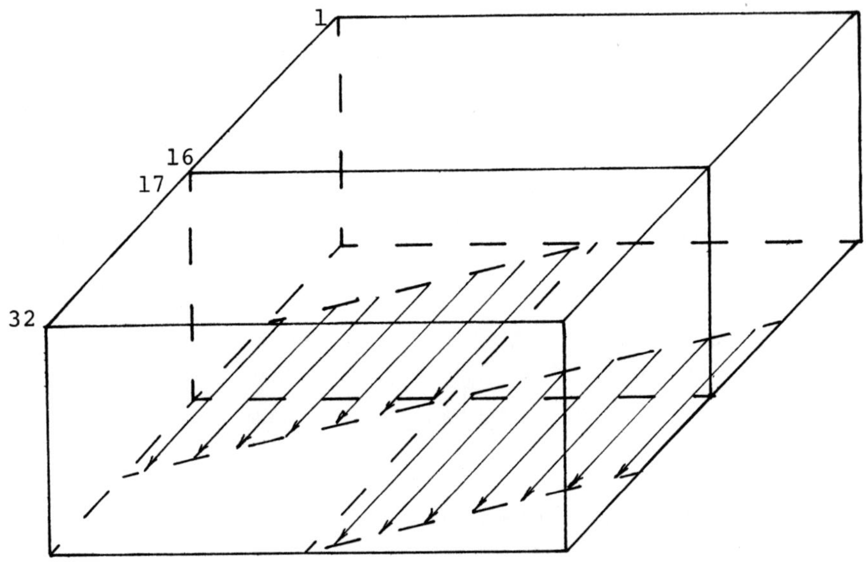

FIG. 5.2. Scrambling from Three Dimensions

(v) The convolution step was executed inside a triple loop for the number of terms of $a_1(y - z)$, the third dimension k and the dummy variable for integration over the third dimension. This means that the length of vector for each step was only 16 x 96.

(vi) Multiplication by the sensor terms was done in one vector operation of length 24,576.

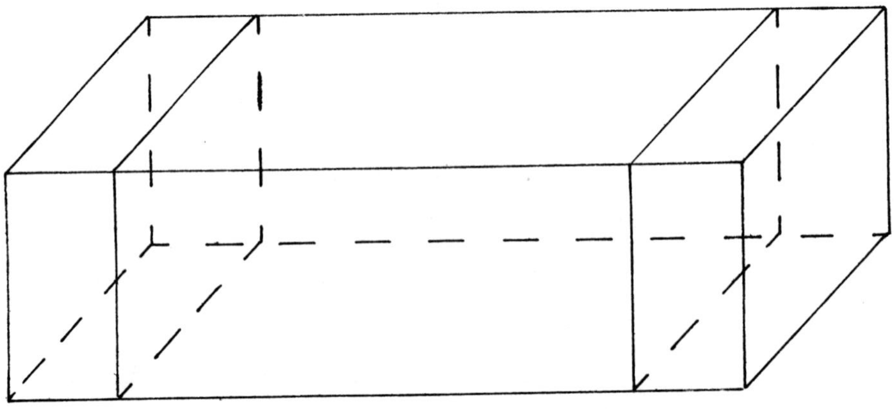

FIG. 5.3. Cyclic End Extention of Density

(vii) Normalization constant was obtained by folding the density successively on itself until it collapsed to a single element.

(viii) The density then was multiplied by the inverse of the constant and transferred to the old density for new iteration.

(ix) This normalized density is used for producing estimates for phase and amplitude. It took 80 milliseconds cpu time for every iteration to produce the same results as on the CYBER 175-T machine. This present speed factor between the Star and CYBER will be cut down drastically after converting the standard FORTRAN program for the CYBER into a serial version in an analogous way to the Star program where all the arrays are treated as single vectors.

§6. EXTENSIONS AND FUTURE RESEARCH

As is clear from equation (3.1), the computational problem of realization of nonlinear filters in general consists of two tasks; first, the passage from the filter density to the one-step predictor density, a convolution task; and secondly, using the new piece of data to construct from the one-step predictor density the next filter density. Independently of how the densities are finitely represented, these two partitions of the density update are always present. The convolution task is time consuming when done serially and calls for parallel or pipeline implementation. In general, strictly parallel devices are exorbitantly expensive as dimension and/or gridding increase.

Currently, we are investigating optical realization of the convolution task for two state dimensional problems with Drs. Steier and Sawchuck of the University of Southern California. The problem here is one of accuracy and loop closure. We hope to achieve with the optical methods the accuracy obtained with the hybrid realization. Another approach for the convolution task would be to map space on the line (see [9]) and achieve multi-dimension convolutions by using high frequency surface wave devices to realize single dimension convolutions; again, here the problem of loop closure is important and unsolved. However, this latter method is not confined to two dimensions as the optical approach is.

Probably, the most promising methods in terms of high-speed, low-cost systems are contemporary mini-computers used with special floating point array processors, with the convolution task essentially done in the array processor and the mini-computer used to accomplish the second task. An example of such an array processor with theoretical floating point arithmetic speeds only 2.36 slower than the Star 100 is the AP-120B of Floating Point Systems Inc. We plan to program and evaluate such a system in the course of the next year. Most probably, special purpose digital pipeline devices offer the most promise when accuracy is important, while optical and other analog devices would be useful for filtering problems when accuracy is not that important.

Another problem which merits attention is what are the interactions of other methods of representing the density besides the point mass used here, and the use of pipeline machines. It is not at all clear that either the Spline or

Fourier filters can be made to run 30 times faster on the Star than on the 6600; essentially because the vectors involved are a factor of 10 smaller and start-up times are no longer negligible.

Finally, there are two other contemporary machines which may offer interesting possibilities for producing fast nonlinear filters; they are the Cray machine and a vector machine produced by Texas Instruments. We hope to be able to report on these in the course of the year.

ACKNOWLEDGMENTS

This research was supported in part by United States Air Force, Air Force Systems Command under Grant AF-AFOSR-2141 and Contract F44620-76-C-0085; M.I.T. Lincoln Laboratory; Lockheed Aircraft; Institute for Computer Applications in Science and Engineering (ICASE), NASA Langley Research Center, Hampton, Virginia, under NASA Contract NAS1-14101; and Institute for Advanced Computation (IAC), Sunnyvale, California.

We received help from many people which made this comparison study possible. At IAC, Allan Birholtz was helpful in getting us on the Illiac system and Jerry Marin extended programming support. At ICASE Jim Ortega and Bob Voigt provided helpful discussions. John Knight, Rudine Smith, Everett Johnson, and J. Lambiotte of the Langley Research Center provided significant help in algorithm development. Humberto Torres helped us with getting turn-around and learning the system.

The month of June residence at ICASE was supported for Bucy by ICASE, for Senne by M.I.T. Lincoln Laboratory, and for Youssef by Lockheed. Help with the CDC 6600 program and systems questions was given us by Herb Spies of Eglin Air Force Base. Colonel W. J. Rabe of AFOSR was responsible for getting time for us on the Illiac and the 6600 at Eglin Air Force Base.

REFERENCES

1. Bucy, R. S. and K. D. Senne, "Digital Synthesis of Nonlinear filters," Automatica, 7 (1971), 287-298.

2. Bucy, R. S., M. J. Merritt, and D. S. Miller, "Hybrid Synthesis of the Optimal Discrete Filter," Stochastics, 1 (1974), 151-211.

3. Basañez, L., P. Brunet, R. S. Bucy, R. Huber, D. S. Miller, and J. Pagés, "A Hybrid Computer Optimal Filter," Proceedings of the Sixth Symposium on Nonlinear Estimation Theory and Its Applications (1975), San Diego, California.

4. Bucy, R. S., C. Hecht, and K. D. Senne, "An Engineer's Guide to Building Nonlinear Filters," Final Report SRL-TR-72-0004, Project 7904, Frank J. Seiler Research Laboratory, USAF Academy, Colorado (1972).

5. L. Basañez, P. Brunet, R. S. Bucy, R. Huber, D. S. Miller, and J. Pagés, "Simulation and Implementation of a Hybrid Computer Algorithm for an Optimal Nonlinear Filtering," Proceedings of the 9th Symposium on System Science, Honolulu, Hawaii (1976).

6. Bucy, R. S., A. J. Mallinckrodt, and H. Youssef, "High Speed Convolution of Periodic Functions," to appear, J. SIAM Applied Math.

7. Bucy, R. S. and H. Youssef, "Optimal Phase Demodulation," IEEE Trans. Autom. Contr., AC-21, 5 (1976), 732-736.

8. Youssef, H., "Interpolative Spline Filters," Ph.D. Thesis, Aerospace Eng. Dept., University of Southern California, 1975.

9. Bucy, R. S. and H. Youssef, "Fourier Realization of the Optimal Phase Demodulator," Proc. 4th Symp. on Nonlinear Estimation Theory and Its Applications, San Diego, Western

Periodicals, 1973, 34-38.

10. Bucy, R. S., "Linear and Nonlinear Filtering," Proc. IEEE, 58 (1970), 854-864.

11. Bucy, R. S. and J. Pagés, "A Priori Bounds for the Cubic Sensor Problem," to appear.

12. Bucy, R. S. and H. Youssef, "Dependence of the Optimal Phase Demodulator on Statistical Parameters," IEEE Trans. Autom. Contr., AC-20, 2 (1975), 259-260.

13. Bucy, R. S., C. Hecht, and K. D. Senne, "New Methods for Nonlinear Filtering," Rev. Français d'Automatique, J-1 (1973), 3-54.

14. Youssef, H. M., "Suboptimal Phase Demodulation," Proc. of 6th Symp. on Nonlinear Estimation and Its Applications, San Diego, 1975.

15. Casseres, D., "Illiac IV Machine Reference Manual for the Programmer," Inst. for Advanced Computation Doc. #PG-11700-0000-A, June, 1975.

16. STAR-100 Computer System Reference Manual, Control Data Corporation, Publication No. 60256000, 1975.

17. Bucy, R. S., K. D. Senne, and H. M. Youssef, "Pipeline Software for Filtering," to appear.

GENERALIZED STOCHASTIC APPROXIMATION AND ITS APPLICATION TO PARAMETER IDENTIFICATION OF DISCRETE STOCHASTIC PROCESSES

Pawel J. Szablowski

Institute of Mathematics
Warsaw Technical University
Warsaw, Poland

§I. INTRODUCTION

The paper is devoted mainly to the analysis of iterative methods of identification of parameters in discrete stochastic processes.

Before an abstract statement of the problem will be given, let us consider the following examples of processes for which parameters are to be identified. The last two of them have been "identified numerically," i.e., convergence of appropriate identification procedures has been confirmed by the convergence of a computer procedure (see [6]).

Sequence $\{y_n\}$, n = 0, 1, ..., of random variables satisfies, with probability 1, the following relationship:

$$y_{n+1} = \sum_{i=1}^{r} a_i y_{n+1-i} + \sum_{j=1}^{k} b_j \eta_{n+1-j} \tag{1.1}$$

where $\{\eta_n\}$, n = 0, 1, ..., is a sequence of independent zero-mean random variables.

One is interested in determining on the basis of observations, such sequences of random variables as $\{a_i^{(n)}\}$, $\{b_j^{(n)}\}$, n = 0, 1, ...; i = 1, ..., r; j = 1, ..., k (sequences of estimates of parameters $a_1, \cdots, a_r; b_1, \cdots, b_k$) as:

$$a_i^{(n)} \xrightarrow[n\to\infty]{} a_i \quad \text{a.s. or in probability; } i = 1, \cdots, r$$

$$b_j^{(n)} \xrightarrow[n\to\infty]{} b_j \quad \text{a.s. or in probability; } j = 1, \cdots, k$$

Sequences $\{a_i^{(n)}\}$ and $\{b_j^{(n)}\}$ must satisfy also the following practical conditions:

1) there exists a sequence $\{m_n\}$ of integers such that $a_i^{(n)}$ and $b_j^{(n)}$ are $\sigma(y_o, \cdots, y_{m_n})$ measurable (an on-line property)
2) successive estimates should be "generated" in the following recursive way

$$a_i^{(n+1)} = a_i^{(n)} + \qquad b_j^{(n+1)} = b_j^{(n)} + \boxed{\begin{array}{l}\text{function of nonincreasing}\\ \text{number of observations and}\\ a_i^{(n)},\ b_j^{(n)},\ i = 1, \ldots, r;\\ j = 1, \ldots, k\end{array}}$$

(iterative structure of an algorithm of calculating successive estimators)

The last property is very important when identifying parameters of process (1.1) on the basis of sequentially coming observations and solving on a computer with bounded size of memory.

As it turns out, the above problem of estimation is quite difficult and is not satisfactorily solved in general. Some interesting results are obtained in the case when $k = 1$, $b_1 = 1$ (autoregressive model; see [1,2]) and in the case when $r = 0$ (moving average model; see [3,5]).

Assume now that one is interested in an estimation of parameter p on the basis of observations $\{y_i\}$ which satisfy the relationship

$$y_{n+1} = y_n/(1 + py_n) + \eta_n, \quad n = 0, 1, 2, \ldots$$

where $y_o \geq 0$ a.s. and $\eta_n \geq 0$ a.s. Suppose that

$$\frac{1}{n} \sum_{i=0}^{n} \eta_i \xrightarrow[n\to\infty]{} c \quad \text{a.s.}$$

and that constant c is known, then it is possible to prove that sequence $\{\hat{p}_n\}$ (p_o given) generated by the relationship

$$\hat{p}_{n+1} = \hat{p}_n - \frac{1}{n} (y_{n+1} - y_n/(1 + \hat{p}_n y_n) - c)$$

converges with probability 1 to p.

Suppose that one is interested in estimation of parameter p on the basis of the observations $\{y_n\}$ satisfying the relationship

$$y_{n+1} = f_p(y_n) + \eta_n$$

where $\{\eta_n\}$ is the sequence of zero-mean random variables satisfying the law of large numbers, i.e.,

$$\frac{1}{n} \sum_{i=0}^{n} \eta_i \xrightarrow[n\to\infty]{} 0 \quad \text{a.s.}$$

and the function $f_p(x)$ has the following diagram:

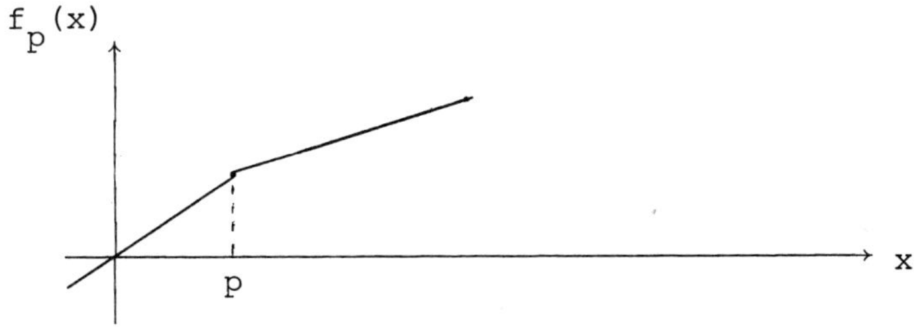

One is able to prove (see [6]) that procedure

$$\hat{p}_{n+1} = \hat{p}_n - \frac{1}{n}\left(f_{\hat{p}_n}(y_n) - y_{n+1}\right)$$

gives a sequence of random variables converging to p with probability 1.

§2. STATEMENT OF THE PROBLEM

Let (Ω, B, P) be fixed probability space on which sequence $\{y_n\}$ $n = 0, 1, \ldots$ of random vectors taking values in R^m is defined. Consider a family $\{Y_s\}$ $s = 1, 2, \ldots$ of finite subsets of the set of observations $\{y_n\}$ such that

$$\bigcup_{s=1}^{\infty} Y_s = \{y_n\}_{n=0}^{\infty}$$

Assume that for each s a joint distribution of random variables from the set $Y_s = \{y_{i_1^s}, \cdots, y_{i_{k_s}^s}\}$ depends on a constant number of parameters $\underline{p} = (p_1, \cdots, p_p) \in R^p$. Euclidean space R^p will be called parameter space. Hence $\underline{p}$ is a fixed point in R^p. Consider a sequence of Borel mappings:

$$F_s : \underbrace{R^m x \cdots x R^m}_{k_s \text{ times}} x R^p \longrightarrow R^p$$

Values of that mapping will depend, for each s, on random variables from the set $\{Y_s\}$ and $\hat{\underline{p}} \in R^p$, and will be denoted $F_s(Y_s, \hat{\underline{p}})$. Assume moreover that

$$EF_s(Y_s, \underline{p}) \xrightarrow[s\to\infty]{} 0$$

and $\underline{p}$ is the only point in R^p having the above property.

Properties of the following procedures are analyzed in this paper:

$$\hat{\underline{p}}_{s+1} = \hat{\underline{p}}_s - \mu_s F_{s+1}(Y_{s+1}, \hat{\underline{p}}_s) \qquad s = 0, 1, \ldots \tag{2.1}$$

where $\hat{\underline{p}}_o \in R^n$ and $\{\mu_s\}_{s=0}^{\infty}$ is a given sequence of positive reals. Conditions imposed on the functions F_s under which sequence $\{\hat{\underline{p}}_s\}$ converges to $\underline{p}$ either with probability 1 or in probability are given in [6].

In order to determine the class of problems covered by the question of convergence of the sequence $\{\hat{\underline{p}}_s\}$, we introduce the following notation:

$$R_s(\hat{\underline{p}}) = EF_s(Y_s, \hat{\underline{p}})$$

$$T_s(\omega, \hat{\underline{p}}) = F_s(Y_s(\omega), \hat{\underline{p}}) - R_s(\hat{\underline{p}})$$

Procedure (2.1) can be presented in the following form:

$$\hat{\underline{p}}_{s+1} = \hat{\underline{p}}_s - \mu_s(R_{s+1}(\hat{\underline{p}}_s) + T_{s+1}(\omega, \hat{\underline{p}}_s)) \tag{2.2}$$

As it was stated, if procedure (2.2) converges, it converges to such a point $\underline{p} \in R^p$ for which $R_s(\underline{p}) \xrightarrow[s\to\infty]{} 0$.

Assume now that $R_i(\hat{\underline{p}}) = R_j(\hat{\underline{p}})$ for $i \neq j$ that is a set of functions $\{R_s(\hat{\underline{p}})\}$ consist of one element. Hence, one can easily notice that procedure (2.2) is a well known procedure considered in stochastic approximation theory. However, unlike in the case of stochastic approximation, random vectors $T_s(\omega,\hat{\underline{p}})$ are neither assumed to be independent, nor to have the property of martingale differences. Thus, sequence $\hat{\underline{p}}_s$ neither has Markov property nor can margingale theory be applied to prove its convergence. Therefore, the above procedures are called generalized stochastic approximation procedures.

On the other hand, if it is assumed that $\forall\ \omega, \hat{\underline{p}}\ T_s(\omega,\hat{\underline{p}}) = 0$ and $R_i(\hat{\underline{p}}) = R_j(\hat{\underline{p}})$, $i,j = 1, 2, \ldots,$ then procedure (2.2) is reduced to the well known procedure which has to solve systems of equations $R(\underline{p}) = 0$. This trivial remark turns out to be very important in the interpretation of the convergence properties of the sequence $\{\hat{\underline{p}}_s\}$.

§3. REMARKS ON THE CONVERGENCE OF THE SEQUENCE $\{\hat{\underline{p}}_s\}$

Assume that a sequence of reals $\{\mu_s\}$ is chosen in such a way that $\mu_o = 1$; $\forall_s\ \mu_s \in (0,1)$. Then the following inequality can be proved:

$$|\hat{\underline{p}}_{s+1} - \underline{p} + c_{s+1}| \le \lambda_s \max(|\hat{p}_s - p + c_s|, \varepsilon_s)$$

where the sequences of random variables $\{\lambda_s\}$ and $\{\varepsilon_s\}$ depend on some quantities characterizing those functions F_s for which procedure (2.1) converges. Sequence $\{c_s\}$ consists of random vectors depending also on some quantities describing the class of functions F_s for which procedure (2.1) converges. One proves that if

$$\sum_{s=0}^{\infty} \mu_s = \infty$$

then

$$\prod_{s=0}^{n} \lambda_s \xrightarrow[n\to\infty]{} 0$$

with probability 1. However, what is more important, sequences $\{c_s\}$ and $\{\varepsilon_s\}$ depend also on the elements of the sequence

$$f_s = \sum_{i=0}^{s-1} a_i F_{i+1}(Y_{i+1}, \underline{p}) \Big/ \sum_{i=0}^{s-1} a_i$$

where sequence $\{a_i\}$ is such that

$$a_n \Big/ \sum_{i=0}^{n} a_i = \mu_n$$

Moreover, if $f_s \xrightarrow[s\to\infty]{} 0$ a.s. (in prob.), then $c_s \xrightarrow[s\to\infty]{} 0$ a.s. (in prob.) and $\varepsilon_s \xrightarrow[s\to\infty]{} 0$ a.s. (in prob.). One shows also that if $f_s = 0$ a.s., then $c_s = 0$ a.s. and $\varepsilon_s = 0$ a.s.

Note that the analyzed inequality indicates that if $f_s \xrightarrow[s\to\infty]{} 0$ a.s. (in prob.), then process $|\hat{\underline{p}}_s - \underline{p} + c_s|$ converges to zero as on the diagram below.

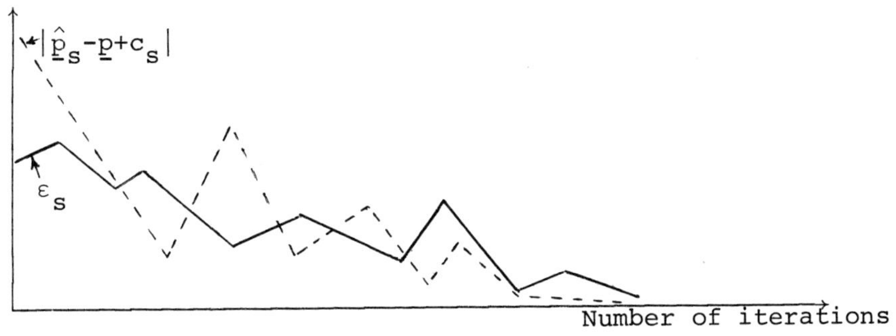

One can prove that if a sequence of reals $\{\mu_s\}$ is chosen in such a way that

$$\sum_{i=0}^{\infty} \mu_i = \infty; \qquad \mu_i \xrightarrow[i\to\infty]{} 0$$

then it occurs only finitely many times that $|\hat{\underline{p}}_s - \underline{p} + c_s| > \varepsilon_s$.

On the other hand, note that if one considers deterministic procedures, that is, if $T_s(\omega, \hat{\underline{p}}) = 0$ and all functions of $R_s(\hat{\underline{p}})$ do not depend on s, then $f_s = 0$ because at point $\underline{p}$ we have $R(\underline{p}) = 0$ by definition of point $\underline{p}$. Hence, $c_s = 0$ and $\varepsilon_s = 0$. Thus, in the deterministic case one obtains the well known estimation

$$|\hat{\underline{p}}_{s+1} - \underline{p}| \leq \lambda_s |\hat{\underline{p}}_s - \underline{p}|$$

The above considerations show that convergence of procedure

(2.1) has two aspects, one deterministic connected with the sequence $\{\lambda_s\}$ and another connected with the sequences $\{c_s\}$ and $\{\varepsilon_s\}$ or equivalently with sequence $\{f_s\}$.

The above interpretation allows proper choice of the sequence $\{\mu_s\}$. It is not difficult to note that the rate of convergence of the sequence $\{\hat{\underline{p}}_s\}$ to $\underline{p}$ depends heavily on the proper choice of the sequence $\{\mu_s\}$. This choice has to be the result of a compromise because sequence

$$\left\{\prod_{i=0}^{s} \lambda_i\right\}$$

converges to zero more "quickly" if the sequence $\{\mu_s\}$ tends to zero more "slowly." On the other hand, sequence $\{f_s\}$ tends to zero more "quickly," if sequence $\{\mu_s\}$ tends to zero more "quickly."

§4. THEOREM GUARANTEEING CONVERGENCE OF SEQUENCE $\{\underline{p}_s\}$

In order to make this paper more precise we shall quote one of several theorems guaranteeing convergence of the sequence $\{\hat{\underline{p}}_s\}$. The theorem given below is not the most general one. In particular, stochastic approximation in its most general formulation does not follow from this theorem. More general versions of the theorem are proved in [6].

Theorem 4.1 Let sequence $\{\mu_s\}$ be chosen in such a way that

$$\mu_o = 1;\ \forall_s\ \mu_s \in (0,1);\ \sum_{s=0}^{\infty} \mu_s = \infty;\ \mu_s \xrightarrow[s\to\infty]{} 0$$

Let $\{a_i\}$ be such a sequence of reals that $a_o = 1$; $a_s \Big/ \sum_{i=0}^{s} a_i = \mu_s$. Suppose, sequence $\{F_s(Y_s, \hat{\underline{p}}\}$ is such that

a) there exists only one point $\underline{p} \in R^p$ such that

$$\sum_{i=0}^{s-1} a_i F_{i+1}(Y_{i+1}, \underline{p}) \Big/ \sum_{i=0}^{s-1} a_i \xrightarrow[s\to\infty]{} 0 \quad \text{a.s.}$$

b) there exist sequences of random variables $\{\alpha_s\}$, $\{\zeta_s\}$, $\{\gamma_s\}$, $\{t_s\}$ such that

$$\lim_{s\to\infty} (\sup_{\hat{\underline{p}}\in R^p} \alpha_s) = 0 \quad \text{a.s.}$$

$$\limsup_{s\to\infty} (\sup_{\hat{\underline{p}}\in R^p} \gamma_s) < \infty \quad \text{a.s.}$$

$$\limsup_{s\to\infty} (\sup_{\hat{\underline{p}}\in R^p} t_s) < \infty \quad \text{a.s.}$$

$$\zeta_s = \zeta_s' + \zeta_s'' \quad \text{a.s.}$$

$$\liminf_{s\to\infty} (\inf_{\hat{\underline{p}}\in R^p} \zeta_s') > 0 \quad \text{a.s.}$$

$$\limsup_{\substack{m,n\to\infty \\ m>n}} \left(\sup_{\hat{\underline{p}}\in R^p} \left| \sum_{i=n}^{m} \mu_i \zeta_i'' \right| \right) < \infty$$

and random vectors β_s such that

$$\limsup_{s\to\infty} (\sup_{\hat{\underline{p}}\in R^p} |\beta_s|) < \infty \text{ a.s.}$$

$$\sum_{i=0}^{s-1} a_i \beta_{i+1} \Big/ \sum_{i=0}^{s-1} a_i \xrightarrow[s\to\infty]{} 0 \quad \text{a.s.}$$

and moreover the following estimations hold with probability 1:

$$\forall_{\hat{\underline{p}}\in R^p} \quad (\hat{\underline{p}} - \underline{p})'(F_s(Y_s,\hat{\underline{p}}) - F_s(Y_s,\underline{p}) - \beta_s) \geq \zeta_s |\hat{\underline{p}} - \underline{p}|^2 + \alpha_s$$

$$|F_s(Y_s,\hat{\underline{p}}) - F_s(Y_s,\underline{p})| \leq \gamma_s |\hat{\underline{p}} - \underline{p}| + t_s$$

<u>then</u> sequence $\{\hat{\underline{p}}_s\}$ given by formula (2.1) converges with probability 1 to $\underline{p}$.

Note that convergence of the procedures considered in the introduction can be proved with the help of the above theorem.

REFERENCES

1. D. W. Clarke, Generalized Least-Squares Estimation of Parameter Dynamic Model, IFAC Symposium on Identification in Aut. Contr. Systems, Prague, 1967.

2. James R. Hastings and M. W. Sage, Recursive Generalized Least-Squares Procedures for On-line Ident. of Process Param.," Proc. IEE 1970, 116N°12.

3. E. B. Lee, A. Manitius, and R. Triggiani, Final Report on Ident. and Control of Dynamic Processes, Rep. to 3M Comp., Minneapolis, Minnesota, 1973.

4. M. B. Nevelson and J. Hasminski, Stochasticheskaia Approximatsia i Riekurientnoie Otsenivanie, Izd Nauka, Moskwa 1972 (in Russian).

5. V. Peterka and A. Halousova, Tally Estimate of Astrom Model for Stoch. Processes, IFAC, Prague, 1972, pp. 2, 3.

6. P. J. Szablowski, Wykorzystanie projecia funkcji estymujacej do identyfikacji nieznanych parametrow w ukladach sterowania, Ph.D. thesis, Warsaw, 1975 (in Polish).

FILTERING AND CONTROL PROBLEMS FOR PARTIAL DIFFERENTIAL EQUATIONS

A. V. Balakrishnan

University of California at Los Angeles
Los Angeles, California

§1. INTRODUCTION

Many problems of filtering and control involving partial differential equations can be described abstractly in a manner similar to the ordinary differential equation case by invoking the theory of semigroups of operators [1]. We have thus a 'state equation' of the form

$$\dot{x}(t) = Ax(t) + Bu(t) + Fn(t)$$

where A is the infinitesimal generator of a strongly continuous semigroup, $u(\cdot)$ is the 'input,' $n(\cdot)$ is the noise, and B and F are bounded linear operators. The 'observation' $y(t)$ has the form

$$y(t) = Cx(t) + Gn(t)$$

where C and G are bounded linear operators. An optimal control and filtering theory for such a system has been developed in [1]. However, the extension to the case where the operators B, F, and C are <u>not</u> bounded turns out to be quite significant. In fact, they need to be not only unbounded; but also uncloseable. We shall treat two typical cases illustrating such extensions. The first problem (B unbounded) occurs when the control is on the boundary (of which there is no finite dimensional analog). The second problem, of C unbounded, uncloseable, occurs whenever we wish to consider "point-wise" observations (in the domain or on the boundary).

We begin with the boundary control problem.

§2. BOUNDARY CONTROL

In this section we shall consider the linear (deterministic)* quadratic regulator problem for a class of diffusion equations when the control is on the boundary. Lions [3] has treated such a problem but he requires that the control be in $H^{1/2}$ on the boundary (in particular requiring C_o^∞ boundaries). This restriction seems somewhat artificial and in any case here we require only that the controls be in L_2 on the boundary. Also we exploit semigroup theory. The main tool is the construction of a generalized solution for boundary inputs (see [1] for an elementary exposition).

Let $\mathcal{D}$ denote a bounded domain with boundary Γ in real Euclidean space R^n. Points in the space will be denoted by ξ, with components ξ_i. Let τ denote a second order strongly elliptic operator:

$$\tau f = \sum_1^m \sum_1^m a_{ij}(\xi) \frac{\partial^2}{\partial \xi_i \partial \xi_j} + \sum_1^m a_i(\xi) \frac{\partial f}{\partial \xi_i} \text{ in } \mathcal{D} \tag{2.1}$$

where the coefficients are continuous in the closure of $\mathcal{D}$.

We consider the control problem for the equation:

---

*For a treatment of stochastic boundary control see [2].

$$\frac{\partial f}{\partial f} = \tau f \qquad 0 < t, \qquad \xi \text{ in } \mathcal{D}$$

$$f(t, \cdot)\big|_{\Gamma} = u(t, \cdot) \qquad f(0, \xi) \text{ given} \tag{2.2}$$

where $u(t, \cdot)$ is in $L_2(\Gamma)$ for each t. Before we specify the cost function, we shall develop the semigroup theoretic approach to the solution of (2.2). For this purpose we first consider the special case of controls $u(t, \cdot)$ which are strongly continuously differentiable in $0 \le t < \infty$.

Let A denote the 'zero boundary value' restriction of τ, more specifically, let A denote the smallest closed extension in $L_2(\mathcal{D})$ of τ restricted to $C_o^\infty(\mathcal{D})$. Assume that τ is strongly elliptic so that the quadratic form:

$$\sum_1^m \sum_1^m a_{ij}(\xi) X_i X_j > \beta \sum_1^m X_i^2, \quad \beta > 0, \text{ for all } \xi \text{ in } \mathcal{D}$$

Then A generates a strongly continuous semigroup over $L_2(\mathcal{D})$, analytic in a sector in the right-half plane. Denote the semigroup S(t). We assume next that the boundary Γ is such that the Dirichlet problem

$$\tau f = 0 \text{ in } \mathcal{D} \qquad f\big|_{\Gamma} = g$$

where g is in $L_2(\Gamma)$ has a unique solution given by

$$f = Dg$$

where, furthermore, D is a linear bounded transformation mapping $L_2(\Gamma)$ into $L_2(\Omega)$.

Then following a technique due to Fattorini [4] we can express the solution of (2.2) in abstract form as (see [1]):

$$x(t) = S(t)(x(o) - Du(o)) + \int_o^t S(t-s)D\dot{u}(s)ds \tag{2.3}$$

where x(o) is the initial function (assumed to be in $L_2(\mathcal{D})$. For each t, x(t) describes the solution as an element of $L_2(D)$ and is strongly continuous in t. For x(o) in the domain of A, the solution satisfies (2.2) and moreover

$$\|x(t) - x(o)\| \to 0, \quad \text{as } t \to 0+$$

and of course x(t) 'assumes' the boundary value u(t). See [1] for the uniqueness of the solution (2.3).

While (2.3) is an acceptable solution, it has the disadvantage that the derivative of u(t) appears and we want to avoid this. It is here that we introduce our notion of a 'generalized' solution, sacrificing the "pointwise" interpretation of (2.2).

<u>The Generalized Solution of the Boundary Input Problem</u>

We exploit now the fact that the semigroup S(t) is analytic. From this it follows (see [5] for a proof) that for any u(·) in

$$W_b = L_2[[0, T]; L_2(\Gamma)]$$

we have that

$$\int_0^t S(t - \sigma)Du(\sigma)d\sigma$$

belongs to the domain of A a.e. in [0, T] and that a.e.:

$$A\int_0^t S(t-\sigma)Du(\sigma)d\sigma = \int_0^t AS(t-\sigma)Du(\sigma)d\sigma = g(t)$$

$$0 < t < T$$

where g(t) belongs to

$$W_s = L_2[0, T; L_2(\mathcal{D})]$$

Moreover

$$Lu = g; \; g(t) = -\int_0^t AS(t - \sigma)Du(\sigma)d\sigma \quad \text{a.e.} \tag{2.4}$$

defines a linear bounded transformation mapping W_b into W_s. Integrating by parts in (2.3) and exploiting the fact that L is bounded, we can obtain the 'generalized solution' which is now valid for any $u(\cdot)$ in W_b, as:

$$x(t) = S(t)x(0) - \int_0^t AS(t - s)Du(s)ds \text{ a.e. } 0 < t \tag{2.5}$$

and here the a.e. qualification is crucial. For the details, we refer once again to [1]. Note that A commutes with S(t) on the domain of A and hence it is *as if* we have

$$AS(t - s)Du(s) \approx S(t - s)Bu(s)$$

where B, roughly speaking, is "AD", although of course A is simply not defined on the range of D!

We can now specify the optimization problem. Given Q, linear bounded, mapping $L_2(\mathcal{D})$ into another Hilbert space, it is required to minimize:

$$q(u) = \int_0^T [Qx(t), Qx(t)]dt + \int_0^T [u(t), u(t)]dt \tag{2.6}$$

for $u(\cdot)$ in W_b, with $x(\cdot)$ given by (2.5). Note that we can recast (2.5) in the form:

$$x(t) = S(t)x(0) - Ay(t) \text{ a.e.}$$
$$\dot{y}(t) = Ay(t) + Du(t) \text{ a.e.}$$
$$y(0) = 0 \tag{2.7}$$

As the first step towards the characterization of the optimal control, let us calculate first L^*, the adjoint of L since it will play a key role. We have

$$L^*x = u; \; u(t) = \int_t^T - D^*(AS(s-t))^*x(s)ds \quad \text{a.e.} \qquad (2.8)$$

Now for $t > 0$

$$D^*(AS(t))^* = D^*A^*S(t)^*$$

Let us set

$$C = D^*A^*$$

Then C is clearly defined on the domain of A^* (and hence also in the range of $S(t)^*$ for $t > 0$) but C is not closeable. In fact, in the simple case where τ is actually the Laplacian, we can see (by Green's theorem) that C reduces to the normal derivative on the boundary Γ and hence is clearly not closeable. Indeed if C were closeable, its adjoint would be AD which is of course not defined except on the zero element.

Of course

$$C\; S(t)^*$$

is linear bounded for each $t > 0$ and

$$\int_t^T CS(s-t)^*Q^*Qx(s)ds \quad \text{a.e. } 0 < t < T$$

defines an element of W_b for $x(\cdot)$ in W_s.

By the same kind of theory as in the finite dimensional case, we can establish the existence of unique optimal control $u_o(\cdot)$ given by (see [1] for instance):

$$u_o = -\; (QL)^*(QLu_o + Qw) = -\; (QL)^*Qx_o(\cdot)$$

where w is the element in W_s defined by

$$S(t)x(0) \qquad 0 < t < T$$

We can express $u_o(\cdot)$ as:

$$u_o(t) = - Cz(t)$$

where z(t) is the solution of

$$\dot{z}(t) + A^*z(t) = - Q^*Qx_o(t); \quad z(T) = 0 \cdots \tag{2.9}$$

By analogy with the finite dimensional case, we expect that we must be able to express z(t) as

$$z(t) = P(t)x_o(t)$$

where P(t) is a linear bounded self-adjoint transformation mapping $L_2(\mathcal{D})$ into itself, satisfying, for x, y in domain of A, a.e. in t:

$$[\dot{P}(t)x,y] + [P(t)Ay,x] + [P(t)x,Ay]$$

$$- [CP(t)x,CP(t)y] + (Qx,Qy] = 0$$

$$P(T) = 0 \tag{2.10}$$

Note this implies in particular that for any x in $L_2(\mathcal{D})$

$$P(t)x \in \text{Domain of } C, \text{ a.e. } 0 < t < T$$

for which it is enough if for $x \in \mathcal{D}(A)$

$$P(t) \in \text{Domain of } A^*, \text{ a.e. } 0 < t < T$$

Assume now that we have found such a P(t). Then we can show that $P(t)x_o(t)$ satisfies (2.9). We have

$$\frac{d}{dt} [P(t)x_o(t),y] = \frac{d}{dt} [S(t)x(0) + Ay_o(t),P(t)y]$$

where

$$\dot{y}_o(t) = Ay_o(t) + Du_o(t); \quad y(0) = 0$$

The main calculation is that

$$[P(t)\dot{x}_o(t),y] = [P(t)AS(t)x(0),y]$$
$$- [Ay(t) + Du(t),A^*P(t)y]$$
$$= [x_o(t),A^*P(t)y] - [u(t),CP(t)y]$$
$$= [x_o(t),A^*P(t)y] + [CP(t)x_o(t),CP(t)y]$$

A little arithmetic shows that

$$P(t)x_o(t) \tag{2.11}$$

does satisfy (2.9), which of course has a unique solution. We also have that $q(u_o) = [P(0)\ x\ (0),\ x\ (0)]$.

The main problem is then to show existence of solution of (The Riccati equation) (2.10). The main difficulty arises from the fact that C is unbounded, uncloseable, and is not treated in current works on the operator Riccati equation [6, 7]. Here we shall exploit the dual filtering problem for this purpose, which is also of interest on its own. We shall however need to make a 'smoothness' assumption on Q:

$$Q^* \text{ maps into the domain of } A^* \ldots \tag{2.12}$$

which implies that A^*Q^* is bounded.

§3. A FILTERING PROBLEM

Let us consider the stochastic system:

$$\dot{x}(t,\omega) = A^*x(t,\omega) + F\omega(t),\ \text{a.e. } 0 < t < T$$
$$x(0) = 0 \tag{3.1}$$

where A is the same generator as in Section 2, F is linear bounded, mapping $H_s \times H_b$, such that $FF^* = Q^*Q$ and

$$\omega(t) = \begin{Bmatrix} \omega_1(t) \\ \omega_2(t) \end{Bmatrix}$$

where $\omega_1(\cdot)$, $\omega_2(\cdot)$ are independent standard white Gaussian noises in W_s and W_b, respectively, and hence $\omega(\cdot)$ defines similar white noise in $W_s \times W_b$. We have of course that

$$x(t,\omega) = \int_0^t S^*(t - \sigma)F\omega(\sigma)d\sigma$$

and for each ω, $x(t,\omega)$ is continuous in t.

The observation is

$$y(t,\omega) = C\,x\,(t,\omega) + \omega(t) \quad \text{a.e.} \quad 0 < t < T \tag{3.2}$$

where

$$FG^* = 0$$

$$GG^* = \text{Identity}$$

We have 'distributed' state noise, and the 'observation' noise is on the boundary. The filtering problem for such a system when C is bounded is treated in [1]. Here we consider a case where C is unbounded, uncloseable; more specifically, we take C as defined in Section 2. Then as we have seen, for each $\omega(\cdot)$,

$$x(t,\omega) \in \text{Domain of } C \quad \text{a.e.} \quad 0 < t < T$$

so that (3.2) is well defined for each ω. Moreover

$$Cx(t,\omega) = D^*A^*x(t,\omega) = D^* \int_0^t A^*S^*(t - \sigma)F\omega(\sigma)d\sigma$$

$$= \int_0^t CS^*(t - \sigma)F\omega(\sigma)d\sigma \quad \text{a.e.} \quad 0 < t < T$$

and hence $Cx(\cdot,\omega)$ is an element of W_b and thus (3.2) can be looked upon as a mapping into W_b for each $\omega(\cdot)$ in $W_s \otimes W_b$ (Cartesian product). Because of our assumption (2.12), we have that A*F is bounded, and hence $x(t,\omega) \in$ Domain of A* for every t; in fact:

$$Cx(t,\omega) = \int_0^t D^*S^*(t - \sigma)A^*F\omega(\sigma)d\sigma$$

Let us use the notation:

$$W_b(t) = L_2[[0, t]; L_2(\Gamma)]$$

$$W_s(t) = L_2[[0, t]; L_2(\mathcal{D})]$$

Then

$$y(s,\omega) \qquad 0 < s < t$$

is in $W_b(t)$ and is a second order random variable therein. In fact denoting it by $\eta(t,\omega)$, we have

$$E[\eta(t,\omega)\eta(t,\omega)^*] = I + (CL(t))(CL(t))^*$$

where

$$L(t)f = g; \; g(s) = \int_0^s S^*(s - \sigma)Ff(\sigma)d\sigma \qquad 0 < s < t$$

$$CL(t)f = g; \; g(s) = \int_0^s CS^*(s - \sigma)Ff(\sigma)d\sigma \quad \text{a.e. } 0 < s < t$$

and L(t) and CL(t) define linear bounded transformations on $W_s(t) \otimes W_b(t)$ into $W_s(t)$ and $W_b(t)$, respectively. Hence we know that

$$\hat{x}(t,\omega) = E[x(t,\omega)|\eta(t,\omega)] = M(t)(CL(t))^* \cdot [I + (CL(t)(CL(t))^*]^{-1}\eta(t,\omega) \qquad (3.3)$$

where

$$M(t)f = x; \; x = \int_0^t S^*(t - \sigma)Ff(\sigma)d\sigma$$

However, exploiting our assumption (2.12), we have:

Theorem 1

$$\hat{x}(t,\omega) \in \text{Domain of } C \qquad 0 < t < T$$

and

$C\ \hat{x}(t,\omega)$ for each t is a second-order random variable in $L_2(\Gamma)$, and further defines an element of W_b for each ω, and is a second order random variable in W_b.

Proof We note that for each $\Delta > 0$, $A^*S^*(\Delta)$ is linear bounded and hence

$$\begin{aligned} A^*S^*(\Delta)\ x(t,\omega) &= E[A^*S^*(\Delta)\ x\ (t,\omega)\,|\,\eta(t)\omega)] \\ &= A^*S^*(\Delta)\ M(t)\ (CL(t))^* \\ &\quad \cdot\ [I + (CL(t))\ (CL(t))]^{-1}\eta(t,\omega) \end{aligned}$$

where the operator $CL(t)$ is linear bounded, and further

$$\begin{aligned} A^*S(\Delta)\ M(t)f &= \int_0^t A^*S^*(\Delta)S^*(t-\sigma)Ff(\sigma)d\sigma \\ &= \int_0^t S^*(\Delta)S^*(t-\sigma)A^*Ff(\sigma)d\sigma \end{aligned}$$

and converges as Δ goes to zero, to:

$$\int_0^t S^*(t-\sigma)A^*Ff(\sigma)d\sigma$$

which defines a linear bounded transformation on $W_s(t) \otimes W_b(t)$ into H_s. This shows that

$$A^*S^*(\Delta)\ \hat{x}\ (t,\omega)$$

converges as $\Delta \to 0$ for each ω and t and hence that

$$\hat{x}(t,\omega) \in \text{Domain of } A^*$$

and hence also to the domain of C. Moreover,

$$A^*M(t)f = \int_0^t S^*(t-\sigma)A^*Ff(\sigma)d\sigma$$

so that the range of $M(t)$ is contained in the domain of A^*. Hence $CM(t)$ is linear bounded. Hence

$$C\,\hat{x}(t,\omega) = CM(t)\ (CL(t))^*[I + (CL(t))(CL(t))]^{-1}\eta(t,\omega)$$

$$= E[C\ x\ (t,\omega)\,|\,\eta(t,\omega)] \qquad (3.4)$$

The rest of the statements of the theorem follow from this.

Q.E.D.

Next let

$$P(t) = E((x(t,\omega) - \hat{x}(t,\omega))\ (x(t,\omega) - \hat{x}(t,\omega))^*)$$

We shall show that for each x in H_s,

$$P(t)x \in \mathcal{D}(A^*)$$

For,

$$P(t) = E[x(t,\omega)\ x\ (t,\omega)^*] - E(\hat{x}(t,\omega)\ \hat{x}\ (t,\omega)^*)$$

$$P(t)x = \int_0^t S(\sigma)^*FF^*S(\sigma)x\,d\sigma - E(\hat{x}(t,\omega)x(t,\omega)^*)x \qquad (3.5)$$

since

$$E[\hat{x}(t,\omega)\ x\ (t,\omega)^*] = E[\hat{x}(t,\omega)\hat{x}\ (t,\omega)^*]$$

We shall show that each term in (3.5) is in the domain of A*.

First

$$A^*\int_0^t S(\sigma)^*FF^*S(\sigma)x\,d\sigma = \int_0^t S(\sigma)^*A^*FF^*S(\sigma)x\,d\sigma$$

Also

$$E[\hat{x}(t,\omega)\ x\ (t,\omega)^*]x$$

$$= M(t)(CL(t))^*(I + (CL(t)(CL(t))^*)^{-1}E[\eta(t,\omega)\ x\ (t,\omega)^*]x$$

and belongs to the domain of A*, since the range of M(t) is contained in the domain of A*. Hence P(t)x belongs to the domain of A*, and hence also to the domain of C, and CP(t) is linear bounded. We also have hence that

$$E((Cx(t,\omega) - C\hat{x}(t,\omega))\ x\ (t,\omega)^*) = CP(t) \qquad (3.6)$$

Next from the fact that we can write

$$Cx(t,\omega) = D^* \int_0^t S^*(t - \sigma)A^*F\omega(\sigma)d\sigma$$

where A*F is linear bounded, we can apply the theory already developed in [1] to show that, defining R by:

$$E[y(\cdot,\omega),f]^2 = [(I + R)f,f]$$

we have the Krein factorization:

$$(I + R)^{-1} = (I - L)(I - L)$$

where L is volterra and quasinilpotent. Also

$$z(\cdot,\omega) = y(\cdot,\omega) - C\hat{x}(\cdot,\omega)$$

$$= (I - L)y(\cdot,\omega)$$

defines white noise in W_b; moreover,

$$y(\cdot,\omega) = (I + M)z(\cdot,\omega)$$

where M is also volterra and quasinilpotent. Hence it follows that

$$\hat{x}(t,\omega) = E(x(t,\omega)|\xi(t,\omega)) = K(t)\xi(t,\omega)$$

where

$$\xi(t,\omega) = z(s,\omega) \qquad 0 < s < t$$

as an element of $W_b(t)$, and

$$E[x(t,\omega)\xi(t,\omega)^*] = K(t)$$

But for $h(\cdot)$ in W_b,

$$E([x(t,\omega),x] \int_0^t [z(\sigma,\omega),h(\sigma)]d\sigma)$$

$$= E[(x(t,\omega),x] \int_0^t [Cx(\sigma,\omega) - C\hat{x}(\sigma,\omega),\ h(\sigma)]d\sigma$$

$$= \int_0^t [CP(\sigma)S(t - \sigma)x,h(\sigma)]d\sigma$$

Hence

$$\hat{x}(t,\omega) = \int_0^t (CP(\sigma)S(t - \sigma))^*Z(\sigma,\omega)d\sigma$$

Hence also

$$E[\hat{x}(t,\omega),x]^2 = \int_0^t \|CP(\sigma)S(t - \sigma)x\|^2 d\sigma$$

so that

$$[P(t)x,y] = \int_0^t [S(\sigma)^*FF^*S(\sigma)x,y]d\sigma - \int_0^t [CP(\sigma)S(t - \sigma)x, CP(\sigma)S(t - \sigma)y]d\sigma$$

and hence P(t) satisfies, for x, y in the domain of A:

$$[\dot{P}(t)x,y] = [P(t)Ax,y] + [P(t)x,Ay] + [Qx,Qy] - [CP(t)x, CP(t)y]$$

and hence P(T - t) satisfies (2.10) as required.

We leave open the question whether (2.10) is valid without assumption (2.12).

ACKNOWLEDGMENT

This research was supported in part under AFOSR, USAF Grant No. 73-2492-C.

REFERENCES

1. Balakrishnan, A. V., Applied Functional Analysis, Springer-Verlag, 1976.

2. Balakrishnan, A. V., Identification and Stochastic Control of a Class of Distributed Systems with Boundary Noise, IRIA Symposium, June 1974, published in Lecture Notes in Economics and Mathematical Systems, No. 107, Springer-Verlag.

3. Lions, J. L., Control of Systems Governed by Partial Differential Equations, Springer-Verlag, Dunod, 1969.

4. Fattorini, H. O., Boundary Control Systems, SIAM J. on Control 6 (1968), 349-385.

5. De Simon, Luciano, Un'applicazione della teoria degli integrali singolari allo studio delle equazioni differenziali lineari astratte del primo ordine, Rendiconti Del Seminario Matematico Della Universita di Padova (1964), vol. 34.

6. Lukes, D. L. and D. L. Russell, The Quadratic Criterion for Distributed Systems, SIAM J. on Control, 1969.

7. Mitter, S. K. and R. B. Vinter, Filtering for Linear Stochastic Hereditary Differential Systems, Lecture Notes in Economics and Mathematical Systems, No. 107, Springer-Verlag, 1975.